# Lecture Notes in Artificial Intelligence    8478

## Subseries of Lecture Notes in Computer Science

### LNAI Series Editors

Randy Goebel
*University of Alberta, Edmonton, Canada*
Yuzuru Tanaka
*Hokkaido University, Sapporo, Japan*
Wolfgang Wahlster
*DFKI and Saarland University, Saarbrücken,*

### LNAI Founding Series Editor

Joerg Siekmann
*DFKI and Saarland University, Saarbrücken, Germany*

Cynthia Vera Glodeanu    Mehdi Kaytoue
Christian Sacarea (Eds.)

# Formal
# Concept Analysis

12th International Conference, ICFCA 2014
Cluj-Napoca, Romania, June 10-13, 2014
Proceedings

 Springer

Volume Editors

Cynthia Vera Glodeanu
Technische Universität Dresden
01062 Dresden, Germany
E-mail: cynthia-vera.glodeanu@tu-dresden.de

Mehdi Kaytoue
INSA-Lyon, CNRS, LIRIS UMR 5205
69621 Lyon, France
E-mail: mehdi.kaytoue@insa-lyon.fr

Christian Sacarea
Babes-Bolyai University
400084 Cluj, Romania
E-mail: csacarea@math.ubbcluj.ro

ISSN 0302-9743                          e-ISSN 1611-3349
ISBN 978-3-319-07247-0                  e-ISBN 978-3-319-07248-7
DOI 10.1007/978-3-319-07248-7
Springer Cham Heidelberg New York Dordrecht London

Library of Congress Control Number: 2014938711

LNCS Sublibrary: SL 7 – Artificial Intelligence

*Typesetting:* Camera-ready by author, data conversion by Scientific Publishing Services, Chennai, India

Printed on acid-free paper

Springer is part of Springer Science+Business Media (www.springer.com)

# Preface

Formal Concept Analysis (FCA) is a multi-disciplinary field built on the solid foundation of lattice and order theory. Besides this, FCA is strongly rooted in philosophical aspects of the mathematical formalization of concept and concept hierarchy. Since its emergence in the 1980s the field has developed into a constantly growing research area in its own right, with a thriving theoretical community further applying and developing this powerful framework of qualitative analysis of data. One of the initial goals of FCA was to promote better communication between lattice theorists and potential users of lattice theory. The increasing number of applications in diverse areas such as data visualization, information retrieval, data mining, and knowledge discovery demonstrates how that goal is being met.

In order to offer researchers the opportunity to meet and discuss developments and applications of FCA annually, the International Conference on Formal Concept Analysis (ICFCA) was established and held for the first time in Darmstadt, Germany in 2003. Since then, the ICFCA has been held in different countries from Europe, America, Africa, and in Australia.

The 12$^{th}$ ICFCA took place during 10$^{th}$ to 13$^{th}$ June, 2014 at the Babeş-Bolyai University, Cluj-Napoca, Romania. There were 39 submissions by authors from 14 different countries. Each paper was reviewed by three members of the Program Committee (exceptionally four). 16 high-quality papers were chosen for publication in this volume, amounting to an acceptance rate of 41%. Six other works in progress were considered valuable for presentation during the conference and included in a side volume entitled *Contributions to ICFCA 2014*. For 4 of these papers chosen for publication, the authors accepted the opportunity to publish their work in the journal *Studia Universitatis Babeş-Bolyai, Series Informatica* of the hosting university.

The articles of the present volume cover a rich range of FCA aspects. The first group of papers tackles mathematical problems related to the number of concepts in a context (Albano), order theoretic aspects (Kerkhoff and Schneider, García-Pardo et al.) and lattice theoretic aspects (Chornomaz). The second group presents recent advances in enhanced FCA: Relational Concept Analysis, dealing with relational data (Codocedo and Napoli, Dolques et al.) and Formal Fuzzy Concept Analysis, processing uncertain data (Glodeanu and Konecny). Five works bridge FCA to other fields: data-mining (Soldano, Bouzmakov et. al), knowledge spaces in learning theory (Albrecht and Körndle, Ganter and Glodeanu) and knowledge discovery in algebraic structures (Revenko). Methodologies and applications to real world problems are explored in biology (Wollbold et al., Coste et al.), web media mining (Agrawal et al.), and image analysis (de Fréin).

In addition to the regular contributions, this volume also contains a historical paper entitled *Subdirect decomposition of concept lattices* from Rudolf Wille. It is our pleasure to make this pioneer work of FCA easily available to the community.

We were also delighted that three prestigious researchers accepted to give an invited talk, and we also included their corresponding papers:

- *Learning Spaces, and How to Build Them* by Prof. Jean-Paul Doignon, Université Libre de Bruxelles, Belgium;
- *On the Succinctness of Closure Operator Representations* by Prof. Sebastian Rudolph, Technische Universität Dresden, Germany;
- *MDL for Pattern Mining A Brief Introduction to Krimp* by Prof. Arno Siebes, Universiteit Utrecht, The Netherlands.

Our deepest gratitude goes to all the authors of submitted papers. Choosing ICFCA 2014 as a forum to publish their research was key to the success of the conference. Besides the submitted papers, the high quality of this published volume would not have been possible without the strong commitment of the authors, the Program Committee and Editorial Board members, and the external reviewers. Working with the efficient and capable team of local organizers was a constant pleasure. We are deeply indebted to all of them for making this conference a successful forum on FCA.

Last, but not least, we are most grateful to Springer for showing, for the $12^{th}$ consecutive year, their reliance on the International Conference on Formal Concept Analysis, as well to the organizations that sponsored this event: the Bitdefender company, the iQuest company, the Babeş-Bolyai University, and the City of Cluj-Napoca, Romania. Finally, we would like to emphasize the great help of EasyChair for making the technical duties easier.

June 2014
Cynthia Vera Glodeanu<br>Mehdi Kaytoue<br>Christian Sacarea

# Organization

## Executive Committee

### Conference Chair

Christian Sacarea      Babeş-Bolyai University, Cluj-Napoca, Romania

### Conference Organizing Committee

Brigitte Breckner      Babeş-Bolyai University, Cluj-Napoca, Romania

Sanda Dragos      Babeş-Bolyai University, Cluj-Napoca, Romania

Diana Halita      Babeş-Bolyai University, Cluj-Napoca, Romania

Diana Troanca      Babeş-Bolyai University, Cluj-Napoca, Romania

Viorica Varga      Babeş-Bolyai University, Cluj-Napoca, Romania

## Program and Conference Proceedings

### Program Chairs

Cynthia Vera Glodeanu      Technische Universität Dresden, Germany
Mehdi Kaytoue      Université de Lyon, France

### Editorial Board

Peggy Cellier      IRISA, INSA Rennes, France
Felix Distel      Technische Universität Dresden, Germany
Florent Domenach      University of Nicosia, Cyprus
Peter Eklund      University of Wollongong, Australia
Sebastien Ferré      Université de Rennes 1, France
Bernhard Ganter      Technische Universität Dresden, Germany
Robert Godin      Université du Québec à Montréal, Canada
Robert Jäschke      Leibniz Universität Hannover, Germany
Sergei O. Kuznetsov      Higher School of Economics, Russia
Leonard Kwuida      Bern University of Applied Sciences, Switzerland
Rokia Missaoui      Université du Québec en Outaouais, Canada
Sergei Obiedkov      Higher School of Economics, Russia

| Uta Priss | Ostfalia University of Applied Sciences, Germany |
| Sebastian Rudolph | Technische Universität Dresden, Germany |
| Stefan E. Schmidt | Technische Universität Dresden, Germany |
| Gerd Stumme | University of Kassel, Germany |
| Petko Valtchev | Université du Québec Montréal, Canada |
| Karl Erich Wolff | University of Applied Sciences, Germany |

## Honorary Member

| Rudolf Wille | Technische Universität Darmstadt, Germany |

## Program Committee

| Simon Andrews | University of Sheffield, UK |
| Mike Bain | University of New South Wales, Australia |
| Jaume Baixeries | Polytechnical University of Catalonia, Spain |
| Radim Bělohlávek | Palacký University, Czech Republic |
| Karell Bertet | L3I Université de La Rochelle, France |
| François Brucker | Centrale Marseille, France |
| Claudio Carpineto | Fondazione Ugo Bordoni, Italy |
| Stephan Doerfel | University of Kassel, Germany |
| Vincent Duquenne | ECP6-CNRS, Université Paris 6, France |
| Alain Gély | Université Paul Verlaine, France |
| Marianne Huchard | LIRMM, Université Montpellier, France |
| Dmitry Ignatov | Higher School of Economics, Russia |
| Tim Kaiser | SAP AG, Germany |
| Markus Krötzsch | Technische Universität Dresden, Germany |
| Michal Krupka | Palacký University, Czech Republic |
| Marzena Kryszkiewicz | Warsaw University of Technology, Poland |
| Wilfried Lex | Universität Clausthal, Germany |
| Engelbert Mephu Nguifo | LIMOS, Université de Clermont Ferrand 2, France |
| Amedeo Napoli | LORIA, France |
| Lhouari Nourine | Université Blaise Pascal, France |
| Jan Outrata | Palacký University, Czech Republic |
| Jean-Marc Petit | LIRIS, INSA de Lyon, France |
| Jonas Poelmans | Katholieke Universiteit Leuven, Belgium |
| Sandor Radeleczki | University of Miskolc, Hungary |
| Laszlo Szathmary | University of Debrecen, Hungary |
| Andreja Tepavčević | University of Novi Sad, Serbia |

## External Reviewers

| Gabriela Arevalo | Universidad Nacional de La Plata, Argentina |
| Philippe Fournier-Viger | Université du Québec à Montreal, Canada |
| Clément Guérin | L3I Université de La Rochelle, France |

Mohamed Nader Jelassi     Université de Clermont, France
Jan Konecny     Palacký University, Czech Republic
Michel Krebs     Bern University of Applied Sciences,
    Switzerland
Branimir Šešelja     University of Novi Sad, Serbia
Romuald Thion     Université de Lyon, France

## Sponsoring Institutions

The Babeş-Bolyai University Cluj-Napoca, Romania
The City of Cluj-Napoca, Romania
The Bitdefender Company, Romania
iQuest GmbH & Co KG, Germany

# Table of Contents

# Knowledge Discovery and Knowledge Spaces

# Methods and Applications

# History

# Learning Spaces, and How to Build Them

Jean-Paul Doignon

Université Libre de Bruxelles,
Department of Mathematics c.p. 216,
Bd du Triomphe, 1050 Brussels, Belgium
doignon@ulb.ac.be

**Abstract.** In Knowledge Space Theory (KST), a knowledge structure encodes a body of information as a domain, consisting of all the relevant pieces of information, together with the collection of all possible states of knowledge, identified with specific subsets of the domain. Knowledge spaces and learning spaces are defined through pedagogically natural requirements on the collection of all states. We explain here several ways of building in practice such structures on a given domain. In passing we point out some connections linking KST with Formal Concept Analysis (FCA).

**Keywords:** knowledge space, learning space, QUERY routine, antimatroid, convex geometry, closure space, formal concept lattice.

## 1 Introduction

In Knowledge Space Theory (KST) a 'knowledge structure' encodes a body of information as a 'domain' together with 'states of knowledge'. The domain is the set of all the relevant, elementary pieces of information. Each knowledge state is a subset of the domain, which contains all the items mastered at some time by some (hypothetical) individual. For example, the empty set and the domain itself represent respectively a completely ignorant and an omniscient students. We assume here that in any knowledge structure, the empty set is a state[1]. In general, there will be many more knowledge states; their collection captures the overall structure of the body of information. If $Q$ is the domain and $\mathcal{K}$ the collection of states, the knowledge structure is the pair $(Q, \mathcal{K})$. An example with domain $Q = \{a, b, c, d\}$ is displayed in Figure 1: the boxes show the nine states forming $\mathcal{K}$, while the ascending lines indicate the covering relation among states.

Without further restrictions on the collection of states, knowledge structures are too poorly organized for the development of a useful theory. Fortunately, pedagogical considerations lead in a natural way to impose restrictions on the state collection. We now explain two natural requirements by looking at the knowledge structure $(Q, \mathcal{K})$ from Figure 1. The subset $\{c, d\}$ is a knowledge state in $\mathcal{K}$, but there is no way for a student to acquire mastery of items $c$ and $d$

---

[1] In KST, it is often required that the domain be also a state; we leave out this assumption here in order to ease in Section 3 the comparison with the closure spaces of FCA.

C.V. Glodeanu, M. Kaytoue, and C. Sacarea (Eds.): ICFCA 2014, LNAI 8478, pp. 1–14, 2014.

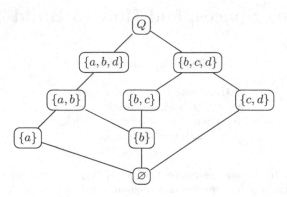

**Fig. 1.** An example of knowledge structure with domain $Q = \{a, b, c, d\}$ and nine knowledge states as shown

one after the other in any order (neither subset $\{c\}$ nor $\{d\}$ is a state in $\mathcal{K}$). This contradicts the (common) view that learning occurs progressively, that is one item at a time. For another singularity in the same knowledge structure $(Q, \mathcal{K})$, consider a student in state $\{b\}$. She may learn item $a$ to reach state $\{a, b\}$. On the other hand, while in state $\{b\}$ she may rather learn item $c$ first and reach state $\{b, c\}$; then, strangely enough, item $a$ is not learnable anymore to her (because the subset $\{a, b, c\}$ is not a state in $\mathcal{K}$). The definition of a 'learning space' as a particular type of knowledge structure rules out the two strange situations that we just illustrated on Figure 1. It imposes the following two conditions on the states of a knowledge structure $(Q, \mathcal{K})$.

[A] ACCESSIBILITY. Any state $K$ contains an item $q$ such that $K \setminus \{q\}$ is again a state.

[LC] LEARNING CONSISTENCY. For any state $K$ and items $q$, $r$, if $K \cup \{q\}$ and $K \cup \{r\}$ are states, then $K \cup \{q, r\}$ is also a state.

As we will explain in the next section, "learning space" happens to be just another name for "∪-stable antimatroids".

Knowledge Space Theory (KST) is at the basis of the computer-assisted teaching system **ALEKS**. Developed since around 1995 in a start-up company[2] with the same name in Irvine (California), **ALEKS** is now quite successful with $1,300,000$ single users in 2013. A special feature of **ALEKS** is its assessment module, whose foundation relies on the concept of a learning space. We will not expose the assessment principles, but rather consider the question of how to build a learning space for a specific body of information—this enterprise is a preliminary for the implementation of an efficient assessment module.

To give an example, suppose we have at hand 200 items which represent the topic of arithmetic at the ages 12–14. How can we build an adequate collection of knowledge states on these 200 items? The idea is to rely on

---

[2] The company was recently acquired by McGraw-Hill Education.

the advice of experts in the teaching of arithmetic, or as it is done today, on the huge database of past assessments of student mastery (in kind of a bootstrapping method, see details below in Section 4). Section 4 explains (at least the basic principles of) a general routine called QUERY. The QUERY routine emerges from work by Koppen and Doignon (1990) and Koppen (1993). Eppstein, Falmagne and Uzun (2009) were the first to apply it to build learning spaces. Then Falmagne and Doignon (2011) introduces another way of using the routine for the same goal. We sketch here a third way, maybe a more insightful one, of taking advantage of the QUERY routine.

In Section 3 we point out some links between KST and Formal Concept Analysis (FCA), thus complementing the works of Rusch and Wille (1996) and Spoto, Stefanutti and Vidotto (2010).

## 2  Learning Spaces and Knowledge Spaces

We first provide the formal definitions of concepts met in the Introduction.

**Definition 1.** A *knowledge structure* $(Q, \mathcal{K})$ consists of a finite, nonempty set $Q$ together with a collection $\mathcal{K}$ of subsets of $Q$. In the present text we make the only requirement $\varnothing \in \mathcal{K}$. The elements of the *domain* $Q$ are *items*, those of $\mathcal{K}$ *(knowledge) states*.

The restriction to finite domains $Q$ is made here because of our main goal—namely, the explanation of the QUERY routine.

**Definition 2.** A *learning space* $(Q, \mathcal{K})$ is a knowledge structure $(Q, \mathcal{K})$ in which the collection $\mathcal{K}$ of states satisfies[3] two conditions (as in the Introduction):

[A] ACCESSIBILITY. Any state $K$ contains an item $q$ such that $K \setminus \{q\}$ is again a state:
$$\forall K \in \mathcal{K}, \exists q \in K : K \setminus \{q\} \in \mathcal{K}; \tag{1}$$

[LC] LEARNING CONSISTENCY. For any state $K$ and items $q$, $r$, if $K \cup \{q\}$ and $K \cup \{r\}$ are states, then $K \cup \{q, r\}$ is also a state:

$$\forall K \in \mathcal{K}, \forall q, r \in Q : (K \cup \{q\}, K \cup \{r\} \in \mathcal{K}) \implies K \cup \{q, r\} \in \mathcal{K}. \tag{2}$$

A large collection of learning spaces derives from ordered sets. Let $(Q, \preceq)$ be a *partially ordered set* (in other words, $\preceq$ is a reflexive, transitive and antisymmetric relation, or a *partial order*, on $Q$). Define a *state of* $\preceq$ to be any subset $K$ of $Q$ such that

$$\forall q, r \in Q : (q \preceq r \text{ and } r \in K) \implies q \in K.$$

As it is easily checked, the collection $\mathcal{L}$ of states of $\preceq$ contains $\varnothing$ and $Q$, and it satisfies [A] and [LC] in Definition 2. So $(Q, \mathcal{L})$ is a learning space, that we call the *ordinal space (derived from $\preceq$)*. Notice that the collection of states of an

---

[3] In an unusual way, we do not require $Q \in \mathcal{K}$—see Footnote 1.

ordinal space is stable under both union and intersection, in the sense that any union and intersection of states are again states.

There are many other characterizations of learning spaces. To state one, we recourse to the notion of 'wellgradedness'. In rough terms, a collection of subsets of a finite domain $Q$ is well-graded when it is possible to move from any of its members to any other one by 'elementary' steps which, moreover, are in number equal to the 'distance' between the two members. Here, the distance means the 'symmetric-difference distance', and a step is elementary if it consists in either adding or deleting a single element.

**Definition 3.** The *(symmetric-difference) distance* between two subsets $K$ and $L$ of a finite domain $Q$ is equal to $d(K, L) = |K \Delta L|$ (this indeed defines a distance $d$ on the collection of subsets of $Q$). A collection $\mathcal{K}$ of subsets of a finite domain $Q$ is *well-graded* when for any two members $K$, $L$ of $\mathcal{K}$ with $d(K, L) = m$, there exist states $K_1, K_2, \ldots, K_{m-1}$ in $\mathcal{K}$ such that, with $K_0 = K$ and $K_m = L$, there holds $d(K_{i-1}, K_i) = 1$ for $i = 1, 2, \ldots, m$.

The notion of wellgradedness plays a role also outside KST. For instance, Doignon and Falmagne (1997) show that the collection of all partial orders (resp. "interval orders", "semiorders") on a finite domain is well-graded. Returning to our present topic, we notice that the states of a learning space form a well-graded collection, and even more:

**Proposition 1.** *Let $(Q, \mathcal{K})$ be a knowledge structure. Then the two following conditions are equivalent:*

*(i) $(Q, \mathcal{K})$ is a learning space;*
*(ii) the collection $\mathcal{K}$ is well-graded and stable under union.*

Stability under union is an important property in KST. For instance, knowledge structures whose collection of states is stable under union are closely related to "AND/OR graphs"[4]. They will be central in Section 4.

**Definition 4.** A *knowledge space* $(Q, \mathcal{K})$ is a knowledge structure whose collection $\mathcal{K}$ of states is stable under union.

In any knowledge space $(Q, \mathcal{K})$, some states can be written as unions of other ones, while some states cannot. We now characterize the latter.

**Definition 5.** In a knowledge structure $(Q, \mathcal{K})$, a *clause for an item* $q$ is any state which contains $q$ and is minimal for the latter property. A *clause* is a state which is a clause for some item. The set of all clauses is denoted as $\mathcal{B}$.

**Proposition 2.** *In a knowledge space $(Q, \mathcal{K})$, any state is a union of clauses, but no clause can be written as a union of other states. Moreover, any collection $\mathcal{A}$ of states having the property that any state in $\mathcal{K}$ is a union of members of $\mathcal{A}$ must contain all the clauses, that is $\mathcal{B} \subseteq \mathcal{A}$.*

---

[4] AND/OR graphs generalize partially ordered sets in that each item may have several set of predecessors; for details, see Doignon and Falmagne (1999), Chapter 3.

**Definition 6.** In a knowledge space $(Q, \mathcal{K})$, the *base* is the collection $\mathcal{B}$ of all clauses.

The following two other characterizations of learning spaces are due to Koppen (1998).

**Proposition 3.** *For a knowledge space $(Q, \mathcal{K})$, the three following assertions are equivalent:*

> (i) $(Q, \mathcal{K})$ *is a learning space;*
> (ii) *any clause is a clause for only one item;*
> (iii) *for any two distinct items $q$, $r$, the set of clauses for $q$ differ from the set of clauses for $r$.*

The names we introduced in Definitions 1, 2 and 4 reflect the motivation of KST. We now point out the links with more classical, mathematical structures. To do so, we associate to any knowledge structure $(Q, \mathcal{K})$ its *dual* structure $(Q, \overline{\mathcal{K}})$, where

$$\overline{\mathcal{K}} = \{ Q \setminus K \mid K \in \mathcal{K} \}.$$

Knowledge structures $(Q, \mathcal{K})$, if we remove the innocuous requirement $\varnothing \in \mathcal{K}$, are just "hypergraphs" (Berge, 1989). Knowledge spaces (with the requirement of stability under union) are exactly the duals of 'closure spaces' (see for instance Birkhoff, 1967; Buekenhout, 1967; van de Vel, 1993). Closure spaces play an important role in FCA as we will recall in Section 3 (Ganter and Wille, 1996).

**Definition 7.** A *closure space* $(Q, \mathcal{C})$ is a finite set $Q$ with a collection $\mathcal{C}$ of subsets of $Q$ which is closed under intersection and contains $Q$.

A closure space $(Q, \mathcal{C})$ corresponds to exactly one *closure operator on $Q$*, that is a map $2^Q \to 2^Q$ which is expansive, monotone and idempotent. To be precise, the closure operator of $(Q, \mathcal{C})$ is $2^Q \to 2^Q : A \to \overline{A} = \cap\{C \in \mathcal{C} \mid A \subseteq C\}$.

Learning spaces (characterized through well-gradedness and stability under union as in Proposition 1) are the $\cup$-stable antimatroids of Korte, Lovász and Schrader (1991) (except for the missing requirement $Q \in \mathcal{K}$— notice that by Proposition 1 any of our learning space has a maximum state containing all the other states, but this state may differ from $Q$). As a matter of fact, Proposition 1 can be found in Chapter III of the latter reference (see also Cosyn and Uzun, 2009). Notice that the duals of learning spaces are '$\cap$-stable antimatroids' in the sense of Edelman and Jamison (1985) (we give a definition which is different from, but equivalent to, the original one except that we admit the omission of $\varnothing$ from $\mathcal{C}$).

**Definition 8.** An *$\cap$-stable antimatroid* is a closure space $(Q, \mathcal{C})$ in which the collection $\mathcal{C}$ of closed sets satisfies

[E] EXTENDABILITY. For any closed set $C$ in $\mathcal{C}$, there exists an element $p$ of $Q \setminus C$ such that $C \cup \{q\}$ is again a closed set:

$$\forall C \in \mathcal{C}, \exists p \in Q \setminus C : \ C \cup \{q\} \in \mathcal{C}.$$

Note that in any $\cap$-stable antimatroid $(Q, \mathcal{C})$, there is a minimum closed set contained in all closed sets (which may be empty or not). Another characterization of $\cap$-stable antimatroids is as follows, in terms of the closure $A \to \overline{A}$ (this is the original definition in Jamison, 1980, 1982, however modified here to allow for the possible omission of $\varnothing$ from $\mathcal{K}$).

**Proposition 4.** *A closure space $(Q, \mathcal{C})$ is an $\cap$-stable antimatroid if and only if, for any closed set $C$ in $\mathcal{C}$ and distinct elements $p$, $q$ in $Q$:*

$$\left(p \in \overline{C \cup \{q\}} \text{ and } q \notin C\right) \quad \Longrightarrow \quad q \notin \overline{C \cup \{p\}}.$$

## 3   Knowledge Space Theory and Formal Concept Analysis

As is well known, closure spaces are useful in FCA. Given a context $(G, M, I)$ (thus $I$ is a relation from $G$ to $M$), define the two mappings

$$2^G \to 2^M : A \to A' = \{m \in M \mid \forall a \in A : a \, I \, m\}, \tag{3}$$

$$2^M \to 2^G : B \to B' = \{g \in G \mid \forall b \in B : g \, I \, b\}. \tag{4}$$

Then

$$2^G \to 2^G : A \to A'', \tag{5}$$

$$2^M \to 2^M : B \to B'' \tag{6}$$

are both closure operators. Their collections of closed sets,

$$\mathcal{C} = \{A \in 2^G \mid A = A''\}, \tag{7}$$

$$\mathcal{D} = \{B \in 2^G \mid B = B''\} \tag{8}$$

produce closure spaces, respectively $(G, \mathcal{C})$ and $(M, \mathcal{D})$. The two collections are put in one-to-one correspondence by the two mutually reciprocal, bijective mappings

$$\mathcal{C} \to \mathcal{D} : A \to A', \tag{9}$$

$$\mathcal{D} \to \mathcal{C} : B \to B'. \tag{10}$$

Moreover, these mappings are inclusion-reversing, so that the two partially ordered sets $(\mathcal{C}, \subseteq)$ and $(\mathcal{D}, \subseteq)$ are anti-isomorphic. In FCA, a pair $(A, B)$ with $A \in 2^G$, $B \in 2^M$, $A' = B$ and $B' = A$ is a *concept* with *extent* $A$ and *intent* $B$.

In line with Section 2, we may ask under which condition on the context $(G, M, I)$ the closure spaces $(G, \mathcal{C})$ and/or $(M, \mathcal{D})$ are $\cap$-stable antimatroids. Although Ganter and Wille (1996) do not mention the word "antimatroid", precise answers in lattice-theoretic terms appear in their Theorem 44. Here we directly derive from Proposition 3 similar answers in set-theoretic terms. Let us first look at an example.

*Example 1.* Let $G = \{a, b\}$, $M = \{m, n, p\}$, and $I$ be the relation described by the following boolean table:

|   | $m$ | $n$ | $p$ |
|---|---|---|---|
| $a$ | 1 | 1 | 0 |
| $b$ | 0 | 0 | 1 |

Figure 2 shows the resulting two closure spaces $(G, \mathcal{C})$ and $(M, \mathcal{D})$. Notice that

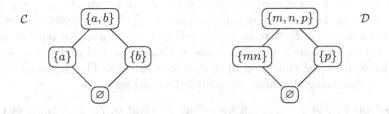

**Fig. 2.** The two closure spaces for the context in Example 1

$(G, \mathcal{C})$ is an $\cap$-stable antimatroid, while $(M, \mathcal{D})$ is not because $\mathcal{D}$ is not extendible in view of $\{p\}$.

**Definition 9.** In a context $(G, M, I)$, an attribute $m$ in $M$ is a *demarcator at an object* $g$ when $m'$ does not contain $g$ and is maximal among all the $b'$, for $b$ in $M$, which share this property. A *demarcator* is a demarcator at some object.

In the notation of Ganter and Wille (1996), the attribute $m$ is a demarcator at the object $g$ exactly when $g \nearrow m$; also, a demarcator is an attribute $m$ such that the concept $(m', m'')$ is $\wedge$-irreducible. Demarcators are dual to clauses (Definition 5). We now rephrase Proposition 2 and 3 in their dual versions.

**Proposition 5.** *For a context $(G, M, I)$, any closed set in $G$ is an intersection of demarcators. Moreover, the collection of demarcators is contained in any subset $N$ of $M$ having the property that any closed set $C$ in $G$ is the intersection of a subcollection of $\{n' \mid n \in N\}$.*

**Proposition 6.** *Given a context $(G, M, I)$, the following three assertions are equivalent:*

> *(i) the closure space $(G, \mathcal{C})$ is an $\cap$-stable antimatroid;*
> *(ii) any demarcator is a demarcator at only one object;*
> *(iii) for any two distinct objects $g$, $h$, the set of demarcators at $g$ differ from the set of demarcators at $h$.*

There is of course a similar criterion for $(M, \mathcal{D})$ to be an $\cap$-stable antimatroid. The demarcators $m'$, for $m \in M$, are crucial here because any closed set in $G$ is an intersection of such subsets (Proposition 5). They are also useful in relation with a quasi order that we now define on the collection of all relations from the finite set $G$ to the finite set $M$.

**Definition 10.** Let $\mathcal{I}$ be the collection of all relations from the finite set $G$ to the finite set $M$. Define a relation $\preccurlyeq$ on $\mathcal{I}$ by letting, for $I$, $J \in \mathcal{I}$:

$$I \preccurlyeq J \quad \text{when} \quad \forall m \in M, \exists m_1, m_2, \ldots, m_k \in M, \forall g \in G: \tag{11}$$

$$g\,I\,m \iff g\,J\,m_1, g\,J\,m_2, \ldots, g\,J\,m_k. \tag{12}$$

In other words, $I \preccurlyeq J$ exactly when each attribute extent $m'$ w.r.t $I$ (for each $m$ in $M$) is an intersection of some attribute extents $n'$ w.r.t. $J$ (where the $n$'s are in $M$); notice that an equivalent definition of $\preccurlyeq$ results when "$\forall m \in M$" is replaced with "for any demarcator $m$ w.r.t. $I$" and "$\exists m_1, m_2, \ldots, m_k \in M$" with "there exist demarcators $m_1, m_2, \ldots, m_k$ w.r.t. $J$". It is easily checked that $\preccurlyeq$ is a quasi order on $\mathcal{I}$ (that is, $\preccurlyeq$ is a reflexive and transitive relation on $\mathcal{I}$). Moreover, two relations $I$ and $J$ in $\mathcal{I}$ are equivalent ($I \preccurlyeq J$ and $J \preccurlyeq I$) exactly if they have the same collection of demarcator extents. The definition of $\preccurlyeq$ is tailored for delivering the following straightforward result.

**Proposition 7.** *For $I$ a relation from $G$ to $M$ (that is, $I \in \mathcal{I}$), denote by $\mathcal{C}_I$ the closed sets in $G$ of the context $(G, M, I)$. Then for $I$, $J$ in $\mathcal{I}$*

$$I \preccurlyeq J \quad \iff \quad \mathcal{C}_I \subseteq \mathcal{C}_J.$$

Thus $I$ and $J$ are equivalent in $\preccurlyeq$ exactly if they produce the same concept lattice.

## 4    The QUERY Routine to Build a Knowledge Space

Suppose we have all the items in an area of knowledge. How can we then build an adequate collection of (potential) knowledge states? In the first steps of the application of KST, relevant information came from experts in the area. A computer routine displays 'queries' on the screen, and collects experts' answers on the keyboard. A crucial feature of the QUERY routine is its ability to infer additional information from previous answers (it thus avoids setting forth too many queries). It is useful as well in current use of KST, where queries are addressed to a database of past assessment sessions rather than to human experts. There is a bootstrapping method at work here. The very first stage relies on a very crude collection $\mathcal{K}$ of potential knowledge states (for instance, if the domain $Q$ is not too large, all of its subsets are in the initial collection $\mathcal{K}$ of states). The database records the assessments based on $\mathcal{K}$. Then a call of the QUERY routine results (as we explain below) in the deletion of subsets from the collection $\mathcal{K}$. Next, the database records further assessments based on the new collection of states. The QUERY routine can then take advantage of the new assessment history, and again reduce the collection of states. There can be many repetitions of the assessment/QUERY sequence.

We focus here the exposition on the QUERY routine itself, however we leave many details aside. For instance, the two-stage process usually first works only with small subsets of the domain (say, with only 6 items). The information found about the states within the parallel subdomains delivers an initial list of (not too many) potential states on the whole domain. Then the two-stage process works on the full domain (for more about this, see Subsection 0.10.2 of Doignon and Falmagne, 2015).

In the present section, we explain how the original QUERY routine produces a knowledge space, that is a collection of states closed under union. In the next section, we explain how to adapt the routine in order that it produce a learning space (that is, a $\cup$-stable antimatroid): there, we want the collection of states to be not only $\cup$-closed, but also accessible (Definition 2).

A typical query to an expert or the database takes the following form, for some subset $A$ of $Q$ and some item $q$ in $Q$:

*Suppose that a student under examination has just provided wrong answers to all the items in $A$.*

*Is it practically certain that this student will also fail item $q$?*

*(Assume that the conditions are ideal in the sense that in the formulation of student answers, careless errors and lucky guesses are excluded.)*

We denote the above query by $(A, q)$. A positive answer to query $(A, q)$ rules out subsets from being potential knowledge states. Indeed, if there were a state $L$ with $A \cap L = \varnothing$ and $q \in L$, then the expert answer could not be positive. Thus, assuming that before query $(A, q)$ the available collection of states is $\mathcal{F}$, upon a positive answer to query $(A, q)$ we may delete from $\mathcal{F}$ all the elements of the collection

$$\mathcal{D}_\mathcal{F}(A, q) \quad = \quad \{L \in \mathcal{F} \mid A \cap L = \varnothing \text{ and } q \in L\}.$$

If the goal is to build a knowledge space, the next proposition shows that we may safely perform the deletion for as many queries as we want—if the initial collection is itself a knowledge space (it could be for instance $(Q, 2^Q)$).

**Proposition 8.** *For any knowledge space $(Q, \mathcal{K})$ and any query $(A, q)$, the collection $\mathcal{K} \setminus \mathcal{D}_\mathcal{K}(A, q)$ is stable under union, so that $(Q, \mathcal{K} \setminus \mathcal{D}_\mathcal{K}(A, q))$ is again a knowledge space.*

Now we briefly explain how inferences can be made from collected answers to past queries. Suppose first that we limit ourselves to queries $(A, q)$ having $|A| = 1$. In other words, we try to uncover a (binary) relation $R$ on $Q$: a positive answer to query $(\{q\}, r)$ acknowledges $q \, R \, r$, a negative answer entails (not $q \, R \, r$). The latent relation $R$, to be uncovered, captures prerequisites among the items. To be precise, for items $q$ and $r$, we have $q \, R \, r$ when $q$ is a prerequisite for $r$ (in the sense that any knowledge state containing $r$ also contains $q$). This is reminiscent of the definition of an ordinal space (see after Definition 2): because it is natural to assume that $R$ is a partial order on $Q$, all the states compatible with the answers to queries form the ordinal space derived from $R$. The transitivity of the relation $R$ makes it possible to infer additional information from answers to past queries. Indeed, if queries $(\{q\}, r)$ and $(\{r\}, s)$ were positively answered, then we may infer that query $(\{q\}, s$ will also be—and we may refrain from asking the latter query. Another case: if query $(\{q\}, s)$ receives a negative answer and query $(\{q\}, r)$ a positive one, then we may infer that query $(\{r\}, s)$ will receive a negative answer—and again we do not need asking the latter query. Similarly, query $(\{q\}, s)$ negatively answered, query $(\{r\}, s)$ positively answered entail that query $(\{q\}, r)$ will be negatively answered. There are similar inferences that can be made when no restriction is made on the size of $A$ in queries $(A, q)$; we refer the reader to Koppen (1998) for a detailed exposition.

In view of Proposition 8, the application of the QUERY routine, starting from any knowledge space, results after each query in the production of a knowledge space. But are we sure that the routine will uncover a latent knowledge space, if the expert answers are coherent with such a space? The answer is in the affirmative.

**Proposition 9.** *Suppose the* QUERY *routine starts with the initial knowledge space* $(Q, 2^Q)$, *and that the answers of the expert to queries are always compatible with a latent knowledge space* $(Q, \mathcal{L})$. *Then at any step, the knowledge space* $(Q, \mathcal{K})$ *built by the routine satisfies* $\mathcal{K} \subseteq \mathcal{L}$. *Moreover, if* $\mathcal{K} \subset \mathcal{L}$, *there is some query* $(A, q)$ *such that* $\mathcal{D}_{\mathcal{K}}(A, q)$ *is nonempty. Hence, after having collected or inferred the answers to all possible queries, the routine produces the latent knowledge space* $(Q, \mathcal{L})$.

## 5   Adapting the QUERY Routine to Build a Learning Space

In this section, we assume again that all the items forming the domain $Q$ are available. Our goal this time is to build a learning space, not just a knowledge space as in the previous section. A solution proposed by Eppstein, Falmagne and Uzun (2009) works in two steps: first use the QUERY routine to build a knowledge space $(Q, \mathcal{K})$, then add states until $\mathcal{K}$ becomes a learning space. Note that the second step involves arbitrariness in the choice of the additional states. One reason lies in the following observation: given a knowledge space $(Q, \mathcal{K})$, the collection of all learning spaces $(Q, \mathcal{L})$ such that $\mathcal{K} \subseteq \mathcal{L}$ contains in general several minimal elements (w.r.t. inclusion).

*Example 2.* For $Q = \{a, b\}$ and $\mathcal{K} = \{\varnothing\}$ (the smallest possible knowledge space on $\{a, b\}$), there are three learning spaces $(Q, \mathcal{L})$ such that $\mathcal{K} \subseteq \mathcal{L}$. Their collections $\mathcal{L}$ of states are $\{\varnothing, \{a\}, Q\}$, $\{\varnothing, \{b\}, Q\}$ and $\{\varnothing, \{a\}, \{b\}, Q\}$. Two of them are minimal.

Here is a fundamental difference between knowledge spaces and learning spaces: if $(Q, \mathcal{K}_1)$ and $(Q, \mathcal{K}_2)$ are knowledge spaces on the same domain, their intersection $(Q, \mathcal{K}_1 \cap \mathcal{K}_2)$ is again a knowledge space, while the similar assertion for learning spaces does not hold.

Falmagne and Doignon (2011) describes another solution to the problem of building a learning space, which adapts the QUERY routine according to the following general principle. Start from an initial learning space (for instance $(Q, 2^Q)$) and successively collect responses to queries; when a query receives a positive answer, delete subsets which cannot be states only if the resulting space is again a learning space—if it is not, keep the information provided by the positive answer for possible, later use. The resulting 'adapted QUERY' routine performs well, in particular it uncovers the latent learning space governing the expert answers—if there is any such latent space (see Proposition 12 below). To provide more details on the adapted QUERY routine, we need two more notions from KST.

**Definition 11.** Let $(Q, \mathcal{K})$ be a knowledge structure, and $K$ be a state in $\mathcal{K}$. The *inner fringe* of $K$ is

$$K^{\mathcal{I}} = \{q \in K \mid K \setminus \{q\} \in \mathcal{K}\}.$$

The *outer fringe* of $K$ is

$$K^{\mathcal{O}} = \{q \in Q \setminus K \mid K \cup \{q\} \in \mathcal{K}\}.$$

The outer fringe of the state $K$ contains the items which a student in state $K$ is ready to learn. In a general knowledge structure, both the inner and outer fringes can be empty. Accessibility (Definition 2) entails that the inner fringe of any nonempty state is nonempty, while extendability (Definition 8) entails that the outer fringe of any state different from the domain is nonempty.

**Proposition 10.** *In a learning space $(Q, \mathcal{L})$, no two states in $\mathcal{L}$ have the same pair of inner and outer fringes.*

In other words, a state in a learning space $(Q, \mathcal{L})$ is determined by its two fringes (and the availability of $\mathcal{L}$).

The design of the *adapted* QUERY routine relies on the following property. Remember that $\mathcal{D}_{\mathcal{K}}(A, q)$ is the collection of subsets ruled out by a positively answered query $(A, q)$:

$$\mathcal{D}_{\mathcal{F}}(A, q) \quad = \quad \{L \in \mathcal{F} \mid A \cap L = \emptyset \text{ and } q \in L\}.$$

**Proposition 11.** *For any learning space $(Q, \mathcal{L})$ and any query $(A, q)$, the collection $\mathcal{L} \setminus \mathcal{D}_{\mathcal{L}}(A, q)$ gives a learning space $(Q, \mathcal{L})$ if and only if there is no state $K$ in $\mathcal{L}$ such that $|K^{\mathcal{I}}| = 1$, $A \cap K = K^{\mathcal{I}}$ and $q \in K$.*

The adapted QUERY routine relies on Proposition 11 to test whether a positive answer to the query $(A, q)$ may be safely applied—that is, to test whether the knowledge space resulting from the deletion of the sets forming $\mathcal{D}_{\mathcal{F}}(A, q)$ is again a learning space. We refer the reader to Falmagne and Doignon (2011) for a full description of the adapted QUERY routine, but state here one of its fundamental properties.

**Proposition 12.** *If $\mathcal{L}$ is a latent learning space and the query answers are truthful with respect to $\mathcal{L}$, then the adapted QUERY routine will ultimately uncover $\mathcal{L}$.*

The proof of Proposition 12 relies on results of Edelman and Jamison (1985) and Caspard and Monjardet (2004) about the collection of all antimatroids on a given set.

We now briefly sketch a third way of using the QUERY routine for building a learning space. A fundamental property of the collection of all learning spaces on a domain $Q$ is that it forms a $\vee$-semilattice w.r.t. inclusion (Caspard and Monjardet, 2004). Let us denote by $\mathbf{L}$ the family of all collections $\mathcal{L}$ of subsets of $Q$ such that $(Q, \mathcal{L})$ is a learning space. If $\mathcal{L}_1$ and $\mathcal{L}_2$ are in $\mathbf{L}$, then among all the collections $\mathcal{L}$ in $\mathbf{L}$ such that $\mathcal{L}_1 \subseteq \mathcal{L}$ and $\mathcal{L}_2 \subseteq \mathcal{L}$, there is smallest one for inclusion, namely their *least upper bound*

$$\mathcal{L}_1 \vee \mathcal{L}_2 = \{L_1 \cup L_2 \,|\, L_1 \in \mathcal{L}_1, L_2 \in \mathcal{L}_2\}. \tag{13}$$

Notice that the partially ordered set $(\mathbf{L}, \subseteq)$ has no minimum element; its minimal elements are all the *full chains*, that is the collections $\{L_0, L_1, \ldots, L_m\}$ of subsets of $Q$ such that $L_0 \subset L_1 \subset \cdots \subset L_m$ and $m = |Q|$. We may turn $(\mathbf{L}, \subseteq)$ into a lattice[5] by adding to $\mathbf{L}$ a new element $\perp$ which becomes the minimum (we assume $\perp \subseteq \mathcal{L}$ for any $\mathcal{L}$ in $\mathbf{L} \cup \{\perp\}$). Then any two elements $\mathcal{L}_1$ and $\mathcal{L}_2$ in $\mathbf{L} \cup \{\perp\}$ have a *greatest lower bound*

$$\mathcal{L}_1 \wedge \mathcal{L}_2 = \bigvee \{\mathcal{L} \in \mathbf{L} \cup \{\perp\} \,|\, \mathcal{L} \subseteq \mathcal{L}_1, \mathcal{L} \subseteq \mathcal{L}_2\}. \tag{14}$$

Now when the QUERY routine has $\mathcal{L}$ as its actual collection of states, with $\mathcal{L} \in \mathbf{L}$, and it receives a positive answer to the query $(A, q)$, we know that only the subsets in $\mathcal{L}^* = \mathcal{L} \setminus \mathcal{D}_{\mathcal{F}}(A, q)$ should remain in the new collection of possible states. The latter collection $\mathcal{L}^*$ always provides a knowledge space on $Q$ (Proposition 8), but in general not a learning space (Proposition 11). There are two cases which our *adjusted* QUERY *routine* must handle:

(i)  $\mathcal{L}^*$ contains some element $\mathcal{F}$ of $\mathbf{L}$; then, in view of the definition of $\vee$ above, it contains for sure the learning space

$$\bigvee \{\mathcal{F} \in \mathbf{L} \,|\, \mathcal{F} \subseteq \mathcal{L}^*\}.$$

We then instruct the adjusted QUERY routine to replace $\mathcal{L}$ with the latter learning space.

(ii)  $\mathcal{L}^*$ does not contain any element from $\mathbf{L}$. We then keep $\mathcal{L}$ as the actual collection of states.

Although the best way to implement Step (i) in the adjusted QUERY routine is still under investigation, it is clear that the outcome of the adjusted QUERY routine is always a learning space. We also have a result similar to Proposition 12 (which was about the adapted QUERY routine).

---

[5] This is the usual way of turning a $\vee$-semilattice into a lattice.

**Proposition 13.** *If $\mathcal{L}$ is a latent learning space and the query answers are truthful with respect to $\mathcal{L}$, then the adjusted* QUERY *routine will ultimately uncover $\mathcal{L}$.*

Notice that in the setting of Proposition 13, Case (ii) never occurs in the execution of the routine. We leave for further work a comparison of the performances of the three ways of using the QUERY routine to uncover a latent learning space.

# References

Berge, C.: Hypergraphs. Combinatorics of finite sets, Transl. from the French. North-Holland, Amsterdam (1989)

Birkhoff, G.: Lattice Theory. American Mathematical Society, Providence (1967)

Buekenhout, F.: Espaces à fermeture. Bull. Soc. Math. Belg. 19, 147–178 (1967)

Caspard, N., Monjardet, B.: Some lattices of closure systems on a finite set. Discrete Math. Theor. Comput. Sci. 6(2), 163–190 (electronic) (2004)

Cosyn, E., Uzun, H.B.: Note on two necessary and sufficient axioms for a well-graded knowledge space. J. Math. Psych. 53, 40–42 (2009)

Doignon, J.-P., Falmagne, J.-C.: Well-graded families of relations. Discrete Math. 173, 35–44 (1997)

Doignon, J.-P., Falmagne, J.-C.: Knowledge Spaces. Springer, Berlin (1999)

Doignon, J.-P., Falmagne, J.-C.: Knowledge Spaces and Learning Spaces. To appear in: Batchelder, W.H., Colonius, H., Dzhafarov, E.N., Myung, J. (eds.) New Handbook of Mathematical Psychology (in press)

Edelman, P.H., Jamison, R.E.: The theory of convex geometries. Geom. Dedicata 19, 247–270 (1985)

Eppstein, D., Falmagne, J.-C., Uzun, H.B.: On verifying and engineering the wellgradedness of a union-closed family. J. Math. Psych. 53, 34–39 (2009)

Falmagne, J.-C., Doignon, J.-P.: Learning Spaces. Springer, Berlin (2011)

Ganter, B., Wille, R.: Formale Begriffsanalyse: Mathematische Grundlagen. Springer, Heidelberg (1996); English translation by Franske, C.: Formal Concept Analysis: Mathematical Foundations

Jamison, R.E.: Copoints in antimatroids. In: Proceedings of the Eleventh Southeastern Conference on Combinatorics, Graph Theory and Computing (Florida Atlantic Univ., Boca Raton, Fla., 1980), vol. II. Congr. Numer., vol. 29, pp. 535–544 (1980)

Jamison, R.E.: A perspective on abstract convexity: classifying alignments by varieties. In: Convexity and Related Combinatorial Geometry (Norman, Okla., 1980). Lecture Notes in Pure and Appl. Math., vol. 76, pp. 113–150. Dekker, New York (1982)

Koppen, M.: Extracting human expertise for constructing knowledge spaces: An algorithm. J. Math. Psych. 37, 1–20 (1993)

Koppen, M.: On alternative representations for knowledge spaces. Math. Social Sci. 36, 127–143 (1998)

Koppen, M., Doignon, J.-P.: How to build a knowledge space by querying an expert. J. Math. Psych. 34, 311–331 (1990)

Korte, B., Lovász, L., Schrader, R.: Greedoids. Algorithms and Combinatorics, vol. 4.
  Springer, Berlin (1991)
Rusch, A., Wille, R.: Knowledge spaces and formal concept analysis. In: Bock, H.-H.,
  Polasek, W. (eds.) Data Analysis and Information Systems. Studies in Classification,
  Data Analysis, and Knowledge Organization. Springer, Heidelberg (1996)
Spoto, A., Stefanutti, L., Vidotto, G.: Knowledge space theory, formal concept analysis,
  and computerized psychological assessment. Behavior Res. Meth. 42, 342–350 (2010)
van de Vel, M.L.J.: Theory of convex structures. North-Holland, Amsterdam (1993)

# On the Succinctness
# of Closure Operator Representations

Sebastian Rudolph

Technische Universität Dresden, Germany
sebastian.rudolph@tu-dresden.de

**Abstract.** It is widely known that closure operators on finite sets can be represented by sets of implications (also known as inclusion dependencies) as well as by formal contexts. In this paper, we consider these two representation types, as well as generalizations of them: extended implications and context families. We discuss the mutual succinctness of these four representations and the tractability of certain operations used to modify closure operators.

## 1 Introduction

Closure operators and closure systems are a basic notion in algebra and occur in various computer science scenarios such as logic programming or databases. One central task when dealing with closure operators is to represent them in a succinct way while still allowing for their efficient computational usage. Formal concept analysis (FCA) naturally provides two complementary ways of representing closure operators: by means of *formal contexts* on one side and *implication sets* on the other. Although being complementary, these two representations share the property that they allow for tractable closure computation. In fact, this property is also exhibited by further representation types, which properly generalize the ones mentioned above: *context families* consist of several contexts and the closure is specified as the "simultaneous fixpoint" of all the separate contexts' closures; *extended implications* are implications where auxiliary elements are allowed.

For a given closure operator, the space needed to represent it in one or the other way may differ significantly: it is well known that there are closure operators whose minimal implicational representation is exponentially larger than their minimal contextual one and vice versa (see Section 3).

Thus, when designing algorithms which store and manipulate closure operators (as many FCA algorithms do) it is important to know which of the possible representation types allow for efficient storage and still guarantee fast (that is: PTIME) execution of typical computations.

This paper investigates the four representation types in this respect. To this end, we will consolidate known results from diverse areas into one framework and provide some findings which are – to the best of our knowledge – novel and original to fill the remaining gaps. Our main results can be generalized as follows:

C.V. Glodeanu, M. Kaytoue, and C. Sacarea (Eds.): ICFCA 2014, LNAI 8478, pp. 15–36, 2014.

- We show that context families allow for succinct representation of both contexts and implications, and that extended implication sets can succinctly represent all the three other representation types. We also show that a succinct translation (i.e., one where the size of the result is polynomially bounded by the input) in all other directions is not possible.
- We clarify the complexities for comparing closure operators in different representations in terms of whether one is a refinement of the other. Interestingly, some of the investigated comparison tasks are tractable (i.e., time-polynomial), others are not (assuming P $\neq$ NP). We provide algorithms for the tractable cases and CONP-hardness arguments for the others.
- We go through standard manipulation tasks for closure operators (refinement by adding a closed set, coarsening through an implication, projection, meet and join in the lattice of closure operators) and clarify which are tractable and which are not.

Parts of this paper are based on an earlier publication [20].

## 2  Preliminaries

We start providing a condensed overview of the notions used in this paper.

### 2.1  Closure Operators

**Definition 1.** *Let $M$ be an arbitrary set. A function $\varphi : 2^M \to 2^M$ is called a closure operator* on $M$ *if it is*

1. *extensive, i.e., $A \subseteq \varphi(A)$ for all $A \subseteq M$,*
2. *monotone, i.e., $A \subseteq B$ implies $\varphi(A) \subseteq \varphi(B)$ for all $A, B \subseteq M$, and*
3. *idempotent, i.e., $\varphi(\varphi(A)) = \varphi(A)$ for all $A \subseteq M$.*

*A set $A \subseteq M$ is called* closed *(or $\varphi$-closed in case of ambiguity), if $\varphi(A) = A$. The set of all closed sets $\{A \mid A = \varphi(A) \subseteq M\}$ is called* closure system.

It is easy to show that for an arbitrary closure system $\mathcal{S}$, the corresponding closure operator $\varphi$ can be reconstructed by

$$\varphi(A) = \bigcap_{B \in \mathcal{S},\ A \subseteq B} B.$$

Hence, there is a one-to-one correspondence between a closure operator and the according closure system.

**Definition 2.** *Given two closure operators $\varphi$ and $\psi$ on $M$, $\varphi$ is called* finer *than $\psi$ (written $\varphi \preceq \psi$, alternatively we also say $\psi$ is* coarser *than $\varphi$) if every $\psi$-closed set is also $\varphi$-closed. We call $\varphi$ and $\psi$* equivalent *(written $\varphi \equiv \psi$), if $\varphi(A) = \psi(A)$ for all $A \subseteq M$.*

It is well-known that the set of all closure operators together with the "finer than" relation constitutes a complete lattice. The lattice operations can be defined as follows: $\varphi \wedge \psi$ is the closure operator mapping any $X \subseteq M$ to $\varphi(X) \cap \psi(X)$ whereas $\varphi \vee \psi$ is the closure operator that maps $X \subseteq M$ to the smallest set closed under $\varphi$ and $\psi$ (which, for finite sets, can be obtained by alternatingly applying $\varphi$ and $\psi$ to $X$ until a fixpoint is reached). The finest closure operator is the identity function mapping every set to itself. The coarsest closure operator maps every input set to $M$.

The precise numbers of closure operators on a finite sets are known up to $|M| = 7$:

| $|M|$ | number of closure operators on $M$ | reference |
|---|---|---|
| 1 | 2 | |
| 2 | 7 | |
| 3 | 61 | |
| 4 | 2,480 | |
| 5 | 1,385,552 | [12] |
| 6 | 75,973,751,474 | [11] |
| 7 | 14,087,648,235,707,352,472 | [3] |

Moreover, general lower and upper bounds have been determined [2], according to which the number of closure operators an an $n$-element set is between $2^{\left(\lfloor n/2 \rfloor \atop \lfloor n/2 \rfloor\right)}$ and $2^{2\sqrt{2}\left({n \atop \lfloor n/2 \rfloor}\right)(1+o(1))}$.

The lower bound can be exploited to obtain a first negative result regarding succinct representability of closure operators in general.

**Proposition 1.** *There is no uniform representation of closure operators that requires at most polynomial space w.r.t. $|M|$.*

*Proof.* Suppose the contrary, i.e., that there exists some fixed $k$ such that every closure operator on $M$ can be expressed by a string of length $|M|^k$ over some alphabet $\Sigma$ of bounded size, say $|\Sigma| = \ell$. Obviously, there are $\ell^{(|M|^k)}$ such strings in total. Thus we obtain

$$\ell^{(|M|^k)} = 2^{\log_2(\ell)(|M|^k)} < 2^{(2^{\lfloor |M|/2 \rfloor})} < 2^{\left(|M| \atop \lfloor |M|/2 \rfloor\right)}$$

for sufficiently large $M$. Therefore, there are less strings of the required length than there are distinct closure operators.     □

Finally, we introduce the notion of a projection of a closure operator.

**Definition 3.** *Given a closure operator $\varphi$ on a set $M$ and some set $N \subseteq M$, the projection of $\varphi$ to $N$, written $\varphi|_N$ is a closure operator on $N$ with $\varphi|_N(X) = \varphi(X) \cap N$ for all $X \subseteq N$.*

Next we introduce four ways of representing closure operators. Thereby and in what follows, we will restrict our considerations to closure operators over finite sets, which is a reasonable assumption when investigating succinctness and complexity properties.

## 2.2  Contexts and Context Families

Following the normal line of argumentation of FCA [8], we use formal contexts as data structure to encode closure operators.

**Definition 4.** *A formal context $\mathbb{K}$ is a triple $(G, M, I)$ with an arbitrary set $G$ called* objects, *an arbitrary set $M$ called* attributes, *and a relation $I \subseteq G \times M$ called* incidence relation. *The* size *of $\mathbb{K}$ (written: $\#\mathbb{K}$) is defined as $|G| \cdot |M|$, i.e., as the number of bits required to store $I$.*

This basic data structure can then be used to define operations on sets of objects or attributes, respectively.

**Definition 5.** *Let $\mathbb{K} = (G, M, I)$ be a formal context. We define a function $(\cdot)^I : 2^G \to 2^M$ with $A^I := \{m \mid gIm \ \text{for all } g \in A\}$ for $A \subseteq G$. Furthermore, we use the same notation to define the function $(\cdot)^I : 2^M \to 2^G$ where $B^I := \{g \mid gIm \ \text{for all } m \in B\}$ for $B \subseteq M$. For convenience, we sometimes write $g^I$ instead of $\{g\}^I$ and $m^I$ instead of $\{m\}^I$.*

Applied to an object set, this function yields all attributes common to these objects; by applying it to an attribute set we get the set of all objects having those attributes. The following facts are consequences of the above definitions:

- $(\cdot)^{II}$ is a closure operator on $G$ as well as on $M$.
- For $A \subseteq G$, $A^I$ is a $(\cdot)^{II}$-closed set and dually
- for $B \subseteq M$, $B^I$ is a $(\cdot)^{II}$-closed set.

In the following, we will focus only on the closure operator on attribute sets and exploit the fact that this closure operator is independent from the concrete object set $G$; it suffices to know the set of the context's object intents. Thus, we will directly use intent sets, that is: families $\mathcal{F}$ of subsets of $M$ to represent formal contexts.

**Definition 6.** *Given a family $\mathcal{F} \subseteq 2^M$, we let $\mathbb{K}(\mathcal{F})$ denote the formal context $(G, M, I)$ with $G = \mathcal{F}$ and, for an $A \in \mathcal{F}$, we let $AIm$ exactly if $m \in A$. Given $B \subseteq M$, we use the notation $B^{\mathcal{F}}$ to denote the attribute closure $B^{II}$ in $\mathbb{K}(\mathcal{F})$ and let $\#\mathcal{F} = \#\mathbb{K}(\mathcal{F}) = |\mathcal{F}| \cdot |M|$.*

For the sake of simplicity we will from now on to refer to $\mathcal{F}$ as contexts (on $M$). We recall the first basic complexity result:

**Proposition 2.** *For any context $\mathcal{F}$ on a set $M$ and any set $A \subseteq M$, the closure $A^{\mathcal{F}}$ can be computed in $O(\#\mathcal{F}) = O(|\mathcal{F}| \cdot |M|)$ time.*

Given an arbitrary context $\mathcal{F}$ representing some closure operator $\varphi$ on some set $M$, the question whether there exists another $\mathcal{F}'$ representing $\varphi$ and satisfying $\#\mathcal{F}' < \#\mathcal{F}$ – and if so, how to compute it – is straightforwardly solved by noting that this coincides with the question if $\mathbb{K}(\mathcal{F})$ is row-reduced [8] and how to row-reduce it. Hence we obtain:

**Proposition 3.** *Given a context $\mathcal{F}$ on $M$, a size-minimal context $\mathcal{F}'$ with $(\cdot)^{\mathcal{F}} \equiv (\cdot)^{\mathcal{F}'}$ can be computed in $O(|\mathcal{F}| \cdot \#\mathcal{F}) = O(|\mathcal{F}|^2 \cdot |M|)$ time.*

Algorithm 1 displays the according method cast in our representation via set families.

We note that for a given closure operator $\varphi$, the minimal $\mathcal{F}$ with $\varphi \equiv (\cdot)^{\mathcal{F}}$ is uniquely determined. We will denote it by $\mathcal{F}(\varphi)$.

The notion of contexts can be extended to that of context families.

**Definition 7.** *A context family on a set $M$ is a finite set $\mathfrak{F} = \{\mathcal{F}_1, \dots \mathcal{F}_n\}$ of formal contexts on $M$. The size of $\mathfrak{F}$ (written: $\#\mathfrak{F}$) is defined as $\sum_{i=1}^{n} \#\mathcal{F}_i$.*

*The closure operator $(\cdot)^{\mathfrak{F}}$ associated with $\mathfrak{F}$ is defined via its closed sets: $X$ is $(\cdot)^{\mathfrak{F}}$-closed if it is $(\cdot)^{\mathcal{F}}$-closed for every $\mathcal{F} \in \{\mathcal{F}_1, \dots \mathcal{F}_n\}$.*

---

**Algorithm 1.** `minimizeContext`

---

**Input:** context $\mathcal{F}$ on $M$
**Output:** size-minimal context $\mathcal{F}'$
    such that $(\cdot)^{\mathcal{F}} \equiv (\cdot)^{\mathcal{F}'}$

1. $\mathcal{F}' := \mathcal{F}$
2. **for each** $A \in \mathcal{F}'$ **do**
3.     **if** $A = A^{\mathcal{F}' \setminus \{A\}}$ **then**
4.         $\mathcal{F}' := \mathcal{F}' \setminus \{A\}$
5.     **end if**
6. **end for**
7. output $\mathcal{F}'$

---

Note that the provided definition of $(\cdot)^{\mathfrak{F}}$ can be equivalently expressed by $(\cdot)^{\mathfrak{F}} := \bigvee_{\mathcal{F} \in \mathfrak{F}} (\cdot)^{\mathcal{F}}$ using the join operation $\bigvee$ in the lattice of closure operators.

This kind of data structure has been investigated in another area of computer science called model-based reasoning [6] and as we will see, it is more succinct than plain contexts. On the other hand, this seems to comes at a prize: the obvious upper bound for closure computation is higher than for plain contexts:

**Proposition 4.** *For any context family $\mathfrak{F}$ on a set $M$ and any set $A \subseteq M$, the closure $A^{\mathfrak{F}}$ can be computed in $O(\#\mathcal{F} \cdot |M|) = O(|\mathcal{F}| \cdot |M|^2)$ time.*

*Proof.* Following from the definition, $A^{\mathfrak{F}}$ must be the smallest simultaneous fixpoint of $(\cdot)^{\mathcal{F}_1}, \dots, (\cdot)^{\mathcal{F}_n}$ that contains $A$. Thanks to monotonicity and finiteness of $M$, such a fixpoint can be obtained by $|M|$-fold application of $(\cdot)^{\mathcal{F}_1 \cdots \mathcal{F}_n}$ to $A$. Exploiting Proposition 2, a one-fold application requires $\sum_{i=1}^{n} O(\#\mathcal{F}) = O(\sum_{i=1}^{n} \#\mathcal{F}) = O(\#\mathfrak{F})$ time, which leads to the above result for $|M|$-fold application. $\square$

This finding can be seen as a special case of a more general result [6], according to which checking if a propositional formula in CNF with $m$ conjuncts is entailed by the Horn theory represented by a context family is feasible in $O(m \cdot \#\mathfrak{F})$ time.

Unlike for contexts, no canonical minimal representation for context families is known. We usually assume that each context $\mathcal{F} \in \mathfrak{F}$ is minimized, but there are still several such representations for one closure operator in the general case.

## 2.3   Implications and Extended Implications

Given a set of attributes, *implications* on that set are logical expressions that can be used to describe certain attribute correspondences which are valid for all objects in a formal context.

**Definition 8.** *Let $M$ be an arbitrary set. An* implication *on $M$ is a pair $(A, B)$ with $A, B \subseteq M$. To support intuition we write $A \to B$ instead of $(A, B)$. We say an implication $A \to B$ holds *for an attribute set $C$ (also: $C$ respects $A \to B$), if $A \not\subseteq C$ or $B \subseteq C$. Moreover, an implication i holds (or: is valid) in a formal context $\mathbb{K} = (G, M, I)$ if it holds for all sets $\{g\}^I$ with $g \in G$. We then write $\mathbb{K} \models i$. The size of an implication set $\mathfrak{I}$ (written: $\#\mathfrak{I}$) is defined as $|\mathfrak{I}| \cdot |M|$. Given a set $A \subseteq M$ and a set $\mathfrak{I}$ of implications on $M$, we write $A^{\mathfrak{I}}$ for the smallest set that contains $A$ and respects all implications from $\mathfrak{I}$. (Since those two requirements are preserved under intersection, the existence of a smallest such set is assured).*

It is obvious that for any set $\mathfrak{I}$ of implications on $M$, the operation $(\cdot)^{\mathfrak{I}}$ is a closure operator on $M$. Furthermore, it can be easily shown that an implication $A \to B$ is valid in a formal context $\mathbb{K} = (G, M, I)$ exactly if $B \subseteq A^{II}$.

The following result is an often noted and straightforward consequence from [17].

**Proposition 5 (Maier 1983).** *For any attribute set $B \subseteq M$ and set $\mathfrak{I}$ of implications, $B^{\mathfrak{I}}$ can be computed in $O(\#\mathfrak{I}) = O(|\mathfrak{I}| \cdot |M|)$ time.*

Like in the case of the contextual encoding, also here it is natural to ask for a size-minimal set of implications that corresponds to a certain closure operator.

Although there is in general no unique minimal implication set for a given closure operator $\varphi$, the so-called Duquenne-Guigues base or stem base [10] is often used as a (minimal) canonical representation. We follow this practice and denote it by $\mathfrak{I}(\varphi)$.

Algorithm 2 (cf. [4,23,19]) provides a well-known way to turn an arbitrary implication set into an equivalent Duquenne-Guigues base. Thus we can note the following complexity result.

**Proposition 6 (Day 1992).** *Given a set $\mathfrak{I}$ of implications on $M$, a size-minimal $\mathfrak{I}'$ with $(\cdot)^{\mathfrak{I}} \equiv (\cdot)^{\mathfrak{I}'}$ can be computed in $O(|\mathfrak{I}| \cdot \#\mathfrak{I}) = O(|\mathfrak{I}|^2 \cdot |M|)$ time.*

A closer look at the algorithm reveals that the $O(|\mathfrak{I}| \cdot |M|)$ space bound comes about by the necessity of a 2-pass processing of the implication set. Note that both passes can be performed *in situ* (i.e., by overwriting the input with the output) which would require only $O(|M|)$ additional memory.

---

**Algorithm 2.** `minimizeImpSet`

---

**Input:** implication set $\mathfrak{I}$ on $M$
**Output:** size-minimal implication set $\mathfrak{I}'$
   such that $(\cdot)^{\mathfrak{I}} \equiv (\cdot)^{\mathfrak{I}'}$

1. $\widetilde{\mathfrak{I}} := \emptyset$
2. **for each** $A \to B \in \mathfrak{I}$ **do**
3.    $\widetilde{\mathfrak{I}} := \widetilde{\mathfrak{I}} \cup \{A \to (A \cup B)^{\mathfrak{I}}\}$
4. **end for**
5. $\mathfrak{I}' := \emptyset$
6. **for each** $A \to B \in \widetilde{\mathfrak{I}}$ **do**
7.    delete $A \to B$ from $\widetilde{\mathfrak{I}}$
8.    $C := A^{\widetilde{\mathfrak{I}} \cup \mathfrak{I}'}$
9.    **if** $C \neq B$ **then**
10.       $\mathfrak{I}' := \mathfrak{I}' \cup \{C \to B\}$
11.    **end if**
12. **end for**
13. output $\mathfrak{I}'$

---

We will now slightly generalize the notion of implications by allowing for "auxiliary elements" that do not belong to $M$.

**Definition 9.** *An* extended implication set *on $M$ is an implication set over some set $N \supseteq M$ where the elements of $N \setminus M$ are called* auxiliary attributes. *The size of an extended implication set $\mathfrak{I}$ (written: $\#\mathfrak{I}$) is defined as $|\mathfrak{I}| \cdot |N|$. Given an extended implication set $\mathfrak{I}$ over $M$, we associate with it the closure operator $(\cdot)^{\mathfrak{I}}|_M$.*

We will see later that allowing for auxiliary attributes enables a more succinct representation of closure operators. The complexities for closure computation follow directly from those for plain implication sets.

## 3   Mutual Succinctness

Given the four encodings of closure operators, a question which arises naturally is whether one encoding is superior to the other in terms of memory required to store it. First of all, note that for a given $M$, we will find a representation of any of the four types whose size is bounded is bounded by $2^{|M|} \cdot |M|$, i.e., at most exponential in the size of $M$.

The following proposition shows that for some $\varphi$, $\#\mathcal{F}(\varphi)$ is exponentially larger than $\#\mathfrak{I}(\varphi)$.

**Proposition 7.** *There exists a sequence $(\varphi_n)_{n \in \mathbb{N}}$ of closure operators such that $\#\mathcal{F}(\varphi_n) \in \Theta(2^n)$ whereas $\#\mathfrak{I}(\varphi_n) \in \Theta(n^2)$.*

*Proof.* We define $\varphi_n$ as the closure operator on the set $M_n = \{1, \dots, 2n\}$ that corresponds to the implication set $\mathfrak{I}_b$ containing the implication $\{2i - 1, 2i\} \to$

|       | 1 | 2 | ... | 2n-3 | 2n-2 | 2n-1 | 2n |
|-------|---|---|-----|------|------|------|-----|
| $g_1$    | × |   | ... | ×    |      | ×    |     |
| $g_2$    | × |   | ... | ×    |      |      | ×   |
| $g_3$    | × |   | ... |      | ×    | ×    |     |
| $g_4$    | × |   | ... |      | ×    |      | ×   |
| $\vdots$ | $\vdots$ | $\vdots$ | $\vdots$ | $\vdots$ | $\vdots$ | $\vdots$ | $\vdots$ |
| $g_{2^n-1}$ |   | × | ... |      | ×    | ×    |     |
| $g_{2^n}$   |   | × | ... |      | ×    |      | ×   |

**Fig. 1.** Example for a context that is exponential in the size of its stem base

$M_n$ for every $i \in \{1,\ldots,n\}$. Then, we obtain $\#\mathfrak{I}(\varphi_n) = 2n^2$. On the other hand, $\mathcal{F}(\varphi_n) = \{\{2k - a_k \mid 1 \leq k \leq n\} \mid \langle a_1,\ldots,a_n\rangle \in \{0,1\}^n\}$ (as schematically displayed in Fig. 1) whence we obtain $\#\mathcal{F}(\varphi_n) = 2^n \cdot 2n$.    □

This shows that plain contexts cannot succinctly (that is: with only polynomial increase in size) represent closure operators defined via implication sets. However, we will next show that this can be achieved by context families. To this end we first define the notion of one-implication-context.

**Definition 10.** *For an implication* $i = A \rightarrow B$ *on some set* $M$, *the one-implication-context* $\mathcal{F}_i$ *is defined by* $\mathcal{F}_i = \{M \setminus \{m\} \mid m \in (M \setminus B) \cup A\} \cup \{M \setminus \{m, m'\} \mid m \in B \setminus A, m' \in A\}$.

It is not hard to verify that $\mathcal{F}_i$ is the unique context which is reduced, in which $i$ holds and that satisfies that every other implication holding therein is a logical consequence of $i$. In other words, whenever $B \setminus A$ is nonempty, the stem base of $\mathcal{F}_{A \rightarrow B}$ will contain exactly the implication $A \rightarrow A \cup B$. We omit a proof here as this is a special case of Proposition 22 presented later. Furthermore, we obtain $\#\mathcal{F}_i < |M|^3$.

**Definition 11.** *For an implication set* $\mathfrak{I} = \{i_1,\ldots,i_n\}$ *on some set* $M$, *the associated context family* $\mathfrak{F}(\mathfrak{I})$ *is defined by* $\mathfrak{F}(\mathfrak{I}) = \{\mathcal{F}_{i_1},\ldots,\mathcal{F}_{i_n}\}$.

**Proposition 8.** *For any implication set* $\mathfrak{I}$ *on some set* $M$ *holds* $(\cdot)^{\mathfrak{I}} \equiv (\cdot)^{\mathfrak{F}(\mathfrak{I})}$. *Moreover,* $\#\mathfrak{F}(\mathfrak{I}) < \#\mathfrak{I} \cdot |M|^2$.

This shows that for every implication set, there exists a context family that is only polynomially larger and represents the same closure operator.

We now turn our attention to the other direction, asking if implications allow for a succinct representation of contextually specified closure operators. It is known that this is not the case: as a consequence of a result on the number of pseudo-intents [14,18], we know that for some $\varphi$, $\#\mathfrak{I}(\varphi)$ is exponentially larger than $\#\mathcal{F}(\varphi)$.

**Proposition 9 (Kuznetsov 2004, Mannila & Räihä 1992).** *There exists a sequence $(\varphi_n)_{n \in \mathbb{N}}$ of closure operators such that $\#\mathcal{F}(\varphi_n) \in \Theta(n^2)$ but $\#\mathfrak{I}(\varphi_n) \in \Theta(2^n)$.*

This result implies that in general, one cannot avoid the exponential blowup if a contextually represented closure operator is to be represented by means of implications on the set $M$.

However, as the following definition and theorem show, this does not hold for extended implication sets, i.e., if auxiliary attributes are allowed. In fact we show that the exponential blowup can then be avoided.

**Definition 12.** *Given a context $\mathcal{F}$ on a set $M$, let $M^+$ denote the set $M$ extended by a one new attribute $m_F$ for each $F \in \mathcal{F}$. Then we define $\mathfrak{I}_{\mathcal{F}}$ as the extended implication set containing for every $m \in M$ the two implications $\{m\} \to \{m_F \mid F \in \mathcal{F}, m \notin F\}$ and $\{m_F \mid F \in \mathcal{F}, m \notin F\} \to \{m\}$.*

**Theorem 10.** *Given a context $\mathcal{F}$ on a set $M$, the following hold*

1. $\#\mathfrak{I}_{\mathcal{F}} = 2 \cdot |M| \cdot |M^+| = 2 \cdot |M| \cdot (|M| + |\mathcal{F}|) \leq 4 \cdot (\#\mathcal{F})^2.$
2. $(\cdot)^{\mathcal{F}} \equiv (\cdot)^{\mathfrak{I}_{\mathcal{F}}}|_M$, *that is,* $A^{\mathcal{F}} = A^{\mathfrak{I}_{\mathcal{F}}} \cap M$ *for all* $A \subseteq M$.

*Proof.* The first claim is obvious.

For the second claim, we first show that for an arbitrary set $A \subseteq M$ holds $A^{\mathfrak{I}_{\mathcal{F}}} = B \cup C$ with $B = \{m_F \mid F \in \mathcal{F}, A \not\subseteq F\}$ and $C = \{m \mid \{m_F \mid F \in \mathcal{F}, m \notin F\} \subseteq B\}$. To show $A^{\mathfrak{I}_{\mathcal{F}}} \subseteq B \cup C$ we note that $A \subseteq B \cup C$ and that $B \cup C$ is $\mathfrak{I}_{\mathcal{F}}$-closed: $B \cup C$ satisfies all implications of the type $\{m_F \mid F \in \mathcal{F}, m \notin F\} \to \{m\}$ by definition of $C$. To check implications of the second type, $\{m\} \to \{m_F \mid F \in \mathcal{F}, m \notin F\}$, we note that

$$
\begin{aligned}
C &= \{m \mid \{m_F \mid F \in \mathcal{F}, m \notin F\} \subseteq B\} \\
&= \{m \mid \{m_F \mid F \in \mathcal{F}, m \notin F\} \subseteq \{m_F \mid F \in \mathcal{F}, A \not\subseteq F\}\} \\
&= \{m \mid \forall F \in \mathcal{F} : m \notin F \to A \not\subseteq F\}
\end{aligned}
$$

Now, picking an $m \in C$, we find that every $m_F$ for which $m \notin F$ must also satisfy $A \not\subseteq F$ and therefore $m_F \in B$ so we find all implications of the second type satisfied.

Further, we show $B \cup C \subseteq A^{\mathfrak{I}_{\mathcal{F}}}$, by proving $B \subseteq A^{\mathfrak{I}_{\mathcal{F}}}$ and $C \subseteq A^{\mathfrak{I}_{\mathcal{F}}}$ separately. We obtain $B = \{m_F \mid F \in \mathcal{F}, A \not\subseteq F\} \subseteq A^{\mathfrak{I}_{\mathcal{F}}}$ due to the following: given an $F \in \mathcal{F}$ with $A \not\subseteq F$, we find an $m \in A$ with $m \notin F$ and thus an implication $m \to \{m_F, \dots\}$ contained in $\mathfrak{I}_{\mathcal{F}}$, therefore $A^{\mathfrak{I}_{\mathcal{F}}}$ must contain $m_F$.

We then also obtain $C := \{m \mid \{m_F \mid F \in \mathcal{F}, m \notin F\} \subseteq B\} \subseteq A^{\mathfrak{I}_{\mathcal{F}}}$ by the following argument: picking an $m \in C$, we find the implication $\{m_F \mid F \in \mathcal{F}, m \notin F\} \to \{m\}$ contained in $\mathfrak{I}_{\mathcal{F}}$. On the other hand, we already know $B \subseteq A^{\mathfrak{I}_{\mathcal{F}}}$ and $B \supseteq \{m_F \mid F \in \mathcal{F}, m \notin F\}$, hence $m \in A^{\mathfrak{I}_{\mathcal{F}}}$.

Finally, we obtain $A^{\mathfrak{I}_{\mathcal{F}}}|_M = A^{\mathfrak{I}_{\mathcal{F}}} \cap M = C = \{m \mid \forall F \in \mathcal{F} : m \notin F \to A \not\subseteq F\} = \{m \mid \forall F \in \mathcal{F} : A \subseteq F \to m \in F\} = \bigcap_{F \in \mathcal{F}, A \subseteq F} F = A^{\mathcal{F}}$ for any $A \subseteq M$. $\square$

Thus, we obtain a polynomially size-bounded implicational representation of a context. In our view this is a remarkable – although not too intricate – insight as it seems to challenge the practical relevance of computationally hard problems w.r.t. pseudo-intents (recognizing, enumerating, counting), on which theoretical FCA research has been focusing lately [15,19,16,22,21,5].

What remains to be clarified is the mutual succinctness of context families vs. extended implication sets. Can they be polynomially transformed into each other, is one strictly more succinct than the other or are they incomparable in that respect?

We will first show that extended implication sets can indeed polynomially express closure operators which are defined via context families.

**Definition 13.** *Given a context family* $\mathfrak{F} = \{\mathcal{F}_1, \ldots, \mathcal{F}_n\}$, *we obtain the corresponding extended implication set* $\mathfrak{I}_{\mathfrak{F}}$ *as the union* $\bigcup_{i=1}^{n} \mathrm{rename}(\mathfrak{I}_{\mathcal{F}}, i)$ *where* rename *is a function replacing all auxiliary attributes* $m \notin M$ *occurring in* $\mathfrak{I}_{\mathcal{F}}$ *by a fresh attribute denoted by* $(m, i)$, *thus forcing the auxiliary attribute sets of* $\mathfrak{I}_{\mathcal{F}_1}, \ldots, \mathfrak{I}_{\mathcal{F}_n}$ *to be mutually disjoint.*

**Proposition 11.** *Given a context family* $\mathfrak{F}$, *we obtain* $(\cdot)^{\mathfrak{I}_{\mathfrak{F}}} \equiv (\cdot)^{\mathfrak{F}}$. *Moreover,* $\#\mathfrak{I}_{\mathfrak{F}} = 2(|M|^2 + \#\mathfrak{F})$.

The final question, if every extended implication set has a polynomial-sized context family counterpart, is the last missing piece to the big picture about succinctness of representation types. The question must be answered negatively and we do so by providing a sequence of closure operators having a size-polynomial representation as extended implication set but not as context family.

**Proposition 12.** *There exists a sequence* $(\varphi_n)_{n \in \mathbb{N}}$ *of closure operators that can be represented by a sequence* $(\mathfrak{I}_n)_{n \in \mathbb{N}}$ *of extended implication sets with* $\#\mathfrak{I}_n \in \Theta(4n^2)$ *but not by a sequence of context families whose size is bounded by a polynomial in* $n$.

*Proof.* Let $M_n = \{even\} \cup \{zero_i, one_i \mid 1 \leq i \leq n\}$. Next, for any $S \subseteq \{1, \ldots, n\}$ we define $Y_S := \{one_i \mid i \in S\} \cup \{zero_i \mid i \notin S\}$. Now, let $\varphi_n(X) = X \cup \{even\}$ whenever there is some $S \subseteq \{1, \ldots, n\}$ of even cardinality for which $Y_S \subseteq X$. Otherwise, let $\varphi_n(X) = X$. It can be easily verified that $\varphi$ is indeed a closure operator.

We next note that $\varphi$ can be represented by the extended implication set $\mathfrak{I}_n$ with auxiliary attributes $\{even_i, odd_i \mid 1 \leq i \leq n\}$ containing the implications

$$one_1 \rightarrow odd_1$$
$$zero_1 \rightarrow even_1$$
$$odd_i, one_{i+1} \rightarrow even_{i+1} \quad \text{for all } 1 \leq i < n$$
$$even_i, one_{i+1} \rightarrow odd_{i+1} \quad \text{for all } 1 \leq i < n$$
$$odd_i, zero_{i+1} \rightarrow odd_{i+1} \quad \text{for all } 1 \leq i < n$$
$$even_i, zero_{i+1} \rightarrow even_{i+1} \quad \text{for all } 1 \leq i < n$$
$$even_n \rightarrow even$$

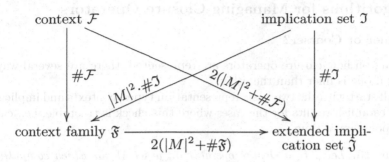

**Fig. 2.** Overview about polynomial translatability between the four representation types. Arrows indicate the existence of a polynomial translation and the arrow labels indicate the upper bounds for the size of the resulting data structure. If no arrow exists between two representation types, no polynomial translation exists.

It is rather easy to see that $\mathfrak{J}_n$ has the given size and implements the wanted behavior.

For the second part, toward the contrary, assume there were a context family $\mathfrak{F}$ of polynomial size with the desired behavior. Then, by definition, for any set $S \subseteq \{1, \ldots, n\}$ holds $even \in Y_S^{\mathfrak{F}}$ iff[1]

$$\bigvee_{\mathcal{F} \in \mathfrak{F}} \bigwedge_{\substack{A \in \mathcal{F} \\ even \notin A}} \bigvee_{m' \in M_n \setminus A} m' \in Y_S.$$

Consequently, $S$ contains an even number of elements, iff $\{p_i \mapsto true \mid i \in S\} \cup \{p_i \mapsto false \mid i \notin S\}$ is a truth assignment for the propositional formula

$$\bigvee_{\mathcal{F} \in \mathfrak{F}} \bigwedge_{\substack{A \in \mathcal{F} \\ even \notin A}} \bigvee_{m_k \in M_n \setminus A} \begin{cases} p_k & \text{if } m_k = one_k \\ \neg p_k & \text{otherwise} \end{cases}.$$

Note that this propositional formula has linear size compared to $\mathfrak{F}$ and, by definition, it encodes a parity function over $p_1, \ldots, p_n$. However, this yields a contradiction with the known result that parity cannot be computed by constant-depth, polynomial-size Boolean circuits [7]. □

Figure 3 provides a summary of this section. Note that the non-existence of polynomial translations from context families to contexts and from extended implication sets to implication sets follows from the existing translations and the known non-existence of polynomial translations between contexts and implication sets.

---

[1] Note that in this proof, the symbols $\bigvee$ and $\bigwedge$ stand for logical connectives, whereas in the rest of the paper, they denote lattice operations.

# 4   Algorithms for Managing Closure Operators

## 4.1   Finer or Coarser?

Depending on how closure operators are represented, there are several ways of checking if one is finer than the other.

We will start with the two basic representation types, contexts and implication sets and establish results for the cases where this check is tractable, i.e., can be done in polynomial time.

**Theorem 13.** *Let $\varphi$ be a closure operator on a set $M$ for which computing of closures can be performed in $t_\varphi$ time. Then, the following hold:*

- *For a context $\mathcal{F}$ on $M$, the problem $\varphi \preceq (\cdot)^\mathcal{F}$ can be decided in $|\mathcal{F}| \cdot t_\varphi$ time.*
- *For an implication set $\mathfrak{I}$ on $M$, the problem $(\cdot)^\mathfrak{I} \preceq \varphi$ can be decided in $|\mathfrak{I}| \cdot t_\varphi$ time.*

*Proof.* Algorithm 3 provides a solution for the first case. It verifies that every element (in other words: every object intent) of $\mathcal{F}$ is $\varphi$-closed, this suffices to guarantee that all $\mathcal{F}$-closed sets are $\varphi$-closed since every $\mathcal{F}$-closed set is an intersection of elements of $\mathcal{F}$ and $\varphi$-closed sets are closed under intersections (since this holds for every closure operator).

Algorithm 4 provides a solution for the second case. To ensure that every $\varphi$-closed set is also $(\cdot)^\mathfrak{I}$-closed, it suffices to show that every $\varphi$-closed set respects all implications from $\mathfrak{I}$. If every $\varphi$-closed set respects an implication $A \to B \in \mathfrak{I}$ can in turn be verified by checking if $B \subseteq \varphi(A)$.                     $\square$

The results established in the above theorem give rise to precise polynomial complexity bounds for seven of the 16 possible comparisons between the different representation types of closure operators.

**Corollary 14.** *Given contexts $\mathcal{F}, \mathcal{F}'$, a context family $\mathfrak{F}$, implication sets $\mathfrak{I}, \mathfrak{I}'$ and an extended implication set $\mathfrak{J}$ on some set $M$, it is possible to check*

| **Algorithm 3.** finerThanContext | **Algorithm 4.** coarserThanImpSet |
|---|---|
| **Input:** closure operator $\varphi$ on set $M$, context $\mathcal{F}$ | **Input:** closure operator $\varphi$ on set $M$, implication set $\mathfrak{I}$ |
| **Output:** YES if $\varphi \preceq (\cdot)^\mathcal{F}$, NO otherwise | **Output:** YES if $(\cdot)^\mathfrak{I} \preceq \varphi$, NO otherwise |
| 1. **for each** $A \in \mathcal{F}$ **do** | 1. **for each** $A \to B \in \mathfrak{I}$ **do** |
| 2.    **if** $A \neq \varphi(A)$ **then** | 2.    **if** $B \not\subseteq \varphi(A)$ **then** |
| 3.       output NO | 3.       output NO |
| 4.       exit | 4.       exit |
| 5.    **end if** | 5.    **end if** |
| 6. **end for** | 6. **end for** |
| 7. output YES | 7. output YES |

- $(\cdot)^{\mathcal{F}} \preceq (\cdot)^{\mathcal{F}'}$ in time $O(|\mathcal{F}| \cdot |\mathcal{F}'| \cdot |M|) = O(\#\mathcal{F} \cdot \#\mathcal{F}'/|M|)$,
- $(\cdot)^{\mathfrak{I}} \preceq (\cdot)^{\mathfrak{I}'}$ in time $O(|\mathfrak{I}| \cdot |\mathfrak{I}'| \cdot |M|) = O(\#\mathfrak{I} \cdot \#\mathfrak{I}'/|M|)$,
- $(\cdot)^{\mathfrak{I}} \preceq (\cdot)^{\mathcal{F}}$ in time $O(|\mathcal{F}| \cdot |\mathfrak{I}| \cdot |M|) = O(\#\mathcal{F} \cdot \#\mathfrak{I}/|M|)$,
- $(\cdot)^{\mathfrak{F}} \preceq (\cdot)^{\mathcal{F}}$ in time $O(\sum_{\mathcal{F}' \in \mathfrak{F}} |\mathcal{F}'| \cdot |\mathcal{F}| \cdot |M|^2) = O(\#\mathfrak{F} \cdot \#\mathcal{F})$,
- $(\cdot)^{\mathfrak{I}} \preceq (\cdot)^{\mathfrak{I}}$ in time $O(|\mathfrak{I}| \cdot |\mathfrak{I}'| \cdot |N|) = O(\#\mathfrak{I} \cdot \#\mathfrak{I}'/|M|)$,
- $(\cdot)^{\mathfrak{I}} \preceq (\cdot)^{\mathfrak{F}}$ in time $O(\sum_{\mathcal{F}' \in \mathfrak{F}} |\mathcal{F}'| \cdot |\mathfrak{I}| \cdot |M|^2) = O(\#\mathfrak{F} \cdot \#\mathfrak{I})$, and
- $(\cdot)^{\mathfrak{I}} \preceq (\cdot)^{\mathcal{F}}$ in time $O(|\mathcal{F}| \cdot |\mathfrak{I}| \cdot |N|) = O(\#\mathcal{F} \cdot \#\mathfrak{I}/|M|)$.

Surprisingly, the ensuing question – whether it is possible to establish a polynomial time complexity bound for the missing comparison cases – has to be denied assuming P $\neq$ NP. The corresponding findings are based on the following theorem. This result in a slightly different formulation is already known in other communities [9], but we give a direct proof for the sake of self-containedness.

**Theorem 15.** *The problem of deciding if $(\cdot)^{\mathcal{F}} \preceq (\cdot)^{\mathfrak{I}}$ for some context $\mathcal{F}$ and an implication set $\mathfrak{I}$ on some set $M$ is* CONP*-complete.*

*Proof.* To show coNP membership, we note that $(\cdot)^{\mathcal{F}} \not\preceq (\cdot)^{\mathfrak{I}}$ if and only if there is a set $A$ and which is $(\cdot)^{\mathfrak{I}}$-closed but not $(\cdot)^{\mathcal{F}}$-closed. Clearly, we can guess such a set and check the above properties in polynomial time.

We show coNP hardness by a polynomial reduction of the problem to 3SAT [13]. Given a set $\mathcal{C} = \{C_1, \ldots, C_k\}$ of 3-clauses (i.e. $|C_i| = 3$) over a set of literals $L = \{p_1, \neg p_1, \ldots p_\ell, \neg p_\ell\}$, we let $M = L$ and define

$$\mathfrak{I} := \{\{p_i, \neg p_i\} \to M \mid p_i \in L\}$$

as well as

$$\mathcal{F} := \{M \setminus (C_i \cup \{m\}) \mid C_i \in \mathcal{C}, m \in M\}.$$

We now show that there is a set $A$ with $A^{\mathfrak{I}} = A$ but $A^{\mathcal{F}} \neq A$ if and only if there is a valuation on $\{p_1, \ldots, p_\ell\}$ for which $\mathcal{C}$ is satisfied.

For the "if" direction assume $val : \{p_1, \ldots, p_\ell\} \to \{true, false\}$ to be that valuation and define $A := \{p_i \mid val(p_i) = true\} \cup \{\neg p_i \mid val(p_i) = false\}$. Obviously, $A$ is $(\cdot)^{\mathfrak{I}}$-closed. On the other hand, since by definition $A$ must contain one element from each $C_i \in \mathcal{C}$, we have that $F \not\subseteq A$ for all $F \in \mathcal{F}$ and hence $A^{\mathcal{F}} = M \neq A$.

For the "only if" direction, assume $A^{\mathfrak{I}} = A$ but $A^{\mathcal{F}} \neq A$. By construction of $\mathcal{F}$, the latter can only be the case if $A$ contains one element of each $C_i \in \mathcal{C}$. Thus, the valuation $val : \{p_1, \ldots, p_\ell\} \to \{true, false\}$ with

$$val(p_i) = \begin{cases} true & \text{if } p_i \in A \\ false & \text{otherwise} \end{cases}$$

witnesses the satisfiability of $\mathcal{C}$. $\qquad\qquad\square$

In fact, this negative result allows us to infer equally negative results for all remaining eight open cases.

**Table 1.** Upper bounds for time complexities for checking the $\preceq$ relation depending on the representation types

| $\preceq$ | context | implications | context family | extended implications |
|---|---|---|---|---|
| context | $\#\mathcal{F} \cdot \#\mathcal{F}'/\lvert M\rvert$ | coNP-hard | coNP-hard | coNP-hard |
| implications | $\#\mathcal{F} \cdot \#\mathfrak{I}/\lvert M\rvert$ | $\#\mathfrak{I} \cdot \#\mathfrak{I}'/\lvert M\rvert$ | $\#\mathfrak{F} \cdot \#\mathfrak{I}$ | $\#\mathfrak{I} \cdot \#\mathfrak{I}/\lvert M\rvert$ |
| context family | $\#\mathfrak{F} \cdot \#\mathcal{F}$ | coNP-hard | coNP-hard | coNP-hard |
| extended implications | $\#\mathcal{F} \cdot \#\mathfrak{I}/\lvert M\rvert$ | coNP-hard | coNP-hard | coNP-hard |

**Proposition 16.** *For arbitrary context $\mathcal{F}$, context families $\mathfrak{F}, \mathfrak{F}'$, implication set $\mathfrak{I}$, and extended implication sets $\mathfrak{I}, \mathfrak{I}'$ on some set $M$, each of the following checks is* coNP-*hard:* $\mathcal{F} \preceq \mathfrak{F}$, $\mathcal{F} \preceq \mathfrak{I}$, $\mathfrak{F} \preceq \mathfrak{I}$, $\mathfrak{F} \preceq \mathfrak{I}$, $\mathfrak{F} \preceq \mathfrak{F}'$, $\mathfrak{F} \preceq \mathfrak{I}$, $\mathfrak{I} \preceq \mathfrak{I}$, $\mathfrak{I} \preceq \mathfrak{F}$, *and* $\mathfrak{I} \preceq \mathfrak{I}'$.

*Proof.* The coNP-hard problem $\mathcal{F} \preceq \mathfrak{I}$ can be polynomially translated in any of the above problems employing the translations given in Section 3.    □

Table 1 summarizes the situation providing the time complexities for the tractable cases.

## 4.2    Adding a Closed Set

We now consider the task of making a closure operator $\varphi$ minimally "finer" by requiring that a given set $A$ be a closed set.

**Definition 14.** *Given a closure operator $\varphi$ on $M$ and some $A \subseteq M$, the $A$-refinement of $\varphi$ (written $\varphi{\downarrow}A$) is defined as the coarsest closure operator $\psi$ with $\psi \preceq \varphi$ and $\psi(A) = A$.*

It is straightforward to show that $B$ is a $\varphi{\downarrow}A$-closed set exactly if it is $\varphi$-closed or the intersection of $A$ and a $\varphi$-closed set. Clearly, if a closure operator is represented as formal context, refinements can be computed by simply adding a row, i.e., for any context $\mathcal{F}$ on $M$ and set $A \subseteq M$ we have for $\mathcal{F}' := \mathcal{F} \cup \{A\}$ that $(\cdot)^{\mathcal{F}}{\downarrow}A \equiv (\cdot)^{\mathcal{F}'}$. Of course, $\mathcal{F}'$ will in general not be size-minimal even if $\mathcal{F}$ is.

**Fact 17.** *Given a context $\mathcal{F}$ on $M$ and some $A \in M$, there is an $\mathcal{F}'$ with $(\cdot)^{\mathcal{F}'} \equiv (\cdot)^{\mathcal{F}}{\downarrow}A$ and $\#\mathcal{F}' \leq \#\mathcal{F} + \lvert M\rvert$.*

If the closure operator is represented as a context family, $A$ has to be added to each of its contexts. This ensures that $A$ is indeed a closed set of the whole context family.

**Fact 18.** *Given a context family $\mathfrak{F}$ on $M$ and some $A \in M$, there is a context family $\mathfrak{F}'$ with $(\cdot)^{\mathfrak{F}'} \equiv (\cdot)^{\mathfrak{F}}{\downarrow}A$ and $\#\mathfrak{F}' \leq \#\mathfrak{F} + \lvert\mathfrak{F}\rvert \cdot \lvert M\rvert$*

Surprisingly, if the closure operator is represented in terms of implications, adding a closed set may incur exponential blow up as shown in some recent work on belief revision in propositional Horn logic [1].[2]

**Proposition 19 (Adaricheva et al. 2012).** *For some natural number $n$, let $M_n = \{w\} \cup \{t_i, u_i, v_i \mid 1 \le i \le n\}$, let $\mathfrak{I}_n = \{t_i \to v_i;\ u_i \to v_i \mid 1 \le i \le n\} \cup \{v_1, \ldots, v_n \to w\}$ and let $A_n = \{w\} \cup \{t_i, u_i \mid 1 \le i \le n\}$. Then representing $(\cdot)^{\mathfrak{I}_n}\!\downarrow A$ requires exponentially many implications.*

As it turns out, the situation again changes when auxiliary attributes can be used. In this case a polynomial size implicational representation can be found. The intuition behind the encoding presented in the following definition is to introduce copies of all attributes outside $A$ and to use an implication set in which all those attributes are renamed into their copies. Moreover, a specific "trigger attribute" $tr$ is implied by any of the original attributes from $M \setminus A$. Whenever $tr$ is activated, all the introduced copies imply their original counterpart.

**Definition 15.** *Let $\mathfrak{I}$ be an extended implication set on $M$ with the total attribute set $N$. Let $A \subseteq M$. Then we define an extended implication set $\mathfrak{I}\!\downarrow A$ on $M$ with total attribute set $N' := N \cup \{m' \mid m \in N \setminus A\} \cup \{tr\}$ as follows:*

$$\mathfrak{I}\!\downarrow A = \{m \to m',\ m \to tr,\ m', tr \to m \mid m \in N \setminus A\} \cup$$
$$\{(B \cap A) \cup \{m' \mid m \in B \setminus A\} \to (C \cap A) \cup \{m' \mid m \in C \setminus A\} \mid B \to C \in \mathfrak{I}\}$$

**Proposition 20.** *Given an extended implication set $\mathfrak{I}$ on $M$ with $N$ the total attribute set and some $A \in M$, we have $(\cdot)^{\mathfrak{I}\!\downarrow A}|_M \equiv (\cdot)^{\mathfrak{I}}|_M\!\downarrow A$ and $\#(\mathfrak{I}\!\downarrow A) \le 2 \cdot \#\mathfrak{I} + 6|N|^2$ and for the total attribute set $N'$ of $\mathfrak{I}\!\downarrow A$ holds $|N'| \le 2|N|$.*

*Proof.* Checking the provided size bounds is straightforward.

We now show the first claim by proving that a subset $S \subseteq M$ is $(\cdot)^{\mathfrak{I}'}|_M$-closed iff it is $(\cdot)^{\mathfrak{I}}|_M$-closed or the intersection of $A$ and some $(\cdot)^{\mathfrak{I}}|_M$-closed set.

We start with the "only if" direction, distinguishing two cases. Given some set $S \in M$ with $S \subseteq A$, we obtain $S^{\mathfrak{I}'} = S^{\mathfrak{I}} \cap A \cup \{m' \mid m \in S^{\mathfrak{I}} \cap (N \setminus A)\}$ and, in particular, $S^{\mathfrak{I}'}$ does not contain $tr$ and hence also no $m \in N \setminus A$. Therefore $S^{\mathfrak{I}'}|M = S^{\mathfrak{I}}|M \cap A$. Next assume $S \not\subseteq A$, i.e., there is some $m \in M$ with $m \in S \setminus A$. Then $tr \in S^{\mathfrak{I}'}$ and therefore $S^{\mathfrak{I}'} = S^{\mathfrak{I}} \cap A \cup \{m, m' \mid m \in S^{\mathfrak{I}} \cap (N \setminus A)\} \cup \{tr\} = S^{\mathfrak{I}} \cup \{m' \mid m \in S^{\mathfrak{I}} \cap (N \setminus A)\} \cup \{tr\}$. Hence we get $S^{\mathfrak{I}'}|M = S^{\mathfrak{I}}|M$.

For the "if" direction, we distinguish the two cases. First, assume $S$ is $(\cdot)^{\mathfrak{I}}|_M$-closed, i.e., $S^{\mathfrak{I}} \cap M = S$. Toward a contradiction, suppose that $S$ is not $(\cdot)^{\mathfrak{I}'}|_M$-closed, hence there is some $m \in M$ with $m \in S^{\mathfrak{I}'} \setminus S$. If $m$ is brought about by an implication of type $tr, m' \to m$, we also have $tr \in S^{\mathfrak{I}'}$ and therefore $S^{\mathfrak{I}'}|M = S^{\mathfrak{I}}|M = S$, a contradiction. Otherwise $m \in A$ but then we obtain $m \in S^{\mathfrak{I}} = S$ another contradiction. Second, assume $S$ is the intersection of $A$ and some $(\cdot)^{\mathfrak{I}}|_M$-closed set $S'$. Then we obtain $S^{\mathfrak{I}'} = (A \cap S')^{\mathfrak{I}'} \subseteq A^{\mathfrak{I}'} \cap S'^{\mathfrak{I}'} = A \cap S'^{\mathfrak{I}'} = A \cap S'^{\mathfrak{I}} = A \cap S' = S$, which, together with the trivial $S \subseteq S^{\mathfrak{I}'}$, shows $S^{\mathfrak{I}'} = S$. $\qquad\square$

---

[2] The author is indebted to Kira V. Adaricheva pointing him to a severe flaw in his earlier publication on the subject [20], where he erroneously claimed that a polynomial solution exists.

## 4.3   Adding an Implication

The task dual to the one from the preceding section is to make a given closure operator coarser by requiring that all closed sets of the coarsened version respect a given implication. In other words, all closed sets not respecting the implication are removed.

**Definition 16.** *Given a closure operator $\varphi$ on $M$ and some implication* $i = A \to B$ *with $A, B \subseteq M$, the i-coarsening of $\varphi$ (written $\varphi{\uparrow}i$) is defined as the finest closure operator $\psi$ with $\varphi \preceq \psi$ and $B \subseteq \psi(A)$.*

Clearly, if a closure operator is represented as implication set (extended or not), coarsenings can be computed by simply adding the implication to the set. Note that $\mathfrak{J}' := \mathfrak{J} \cup \{i\}$ will in general not be size-minimal.

**Fact 21.** *Given a (possibly extended) implication set $\mathfrak{J}$ on $M$ and some implication i on $M$, there is an $\mathfrak{J}'$ with $(\cdot)^{\mathfrak{J}'} \equiv (\cdot)^{\mathfrak{J}}{\uparrow}i$ and $|\mathfrak{J}'| \leq |\mathfrak{J}| + 1$.*

If the closure operator is represented by a context, a little more work is needed for this task. The idea behind the following definition is as follows: a set is closed w.r.t. the updated context $\mathcal{F}'$ iff it is closed w.r.t. the original context $\mathcal{F}$ and respects the new implication i. Thus, all $C \in \mathcal{F}$ respecting i will be in $\mathcal{F}'$. For the other $C$, we have to add their i-respecting intersections with other sets, which essentially can only be intersections with sets $D$ that do not contain the premise of i.

**Definition 17.** *Given a context $\mathcal{F}$ on $M$ and some implication $i = A \to B$ on $M$, we define a new context $\mathcal{F}{\uparrow}i$ as follows*

$$\mathcal{F}{\uparrow}i := \{C \mid C \in \mathcal{F} \text{ and } C \text{ respects } A \to B\} \cup$$
$$\{C \cap D \mid C, D \in \mathcal{F}, \ A \not\subseteq D \text{ and } C \text{ does not respect } A \to B\}$$

**Proposition 22.** *Given a context $\mathcal{F}$ on $M$ and some implication i on $M$, we have $(\cdot)^{\mathcal{F}{\uparrow}i} \equiv (\cdot)^{\mathcal{F}}{\uparrow}i$. Moreover, we have $|\mathcal{F}{\uparrow}i| \leq |\mathcal{F}|^2$ and hence $\#\mathcal{F}{\uparrow}i \leq (\#\mathcal{F})^2/|M|$.*

*Proof.* It is easy to check that $\mathcal{F}{\uparrow}i$ satisfies the given size bounds. We show its correctness by verifying that a set is $(\cdot)^{\mathcal{F}{\uparrow}i}$-closed if and only if it is $(\cdot)^{\mathcal{F}}$-closed and respects $A \to B$.

For the "if" direction, let $S$ be an $(\cdot)^{\mathcal{F}}$-closed set that respects $A \to B$. This means that either $B \subseteq S$ or $A \not\subseteq S$. In the first case, note that every $C \in \mathcal{F}$ with $S \subseteq C$ respects $A \to B$ and thus each such $C$ is contained in $\mathcal{F}{\uparrow}i$ as well. Since $S$ is the intersection of all these $C$, it must itself be $(\cdot)^{\mathcal{F}{\uparrow}i}$-closed. In the second case, there must be some $C \in \mathcal{F}$ with $S \subseteq C$ and $A \not\subseteq C$. Thus we obtain

$$S = \bigcap\nolimits_{S \subseteq D \in \mathcal{F}} F'$$
$$= \left( \bigcap\nolimits_{S \subseteq D \in \mathcal{F}, \ D \text{ respects } A \to B} D \right) \cap \left( \bigcap\nolimits_{S \subseteq D \in \mathcal{F}, \ D \text{ violates } A \to B} D \right) \cap F$$
$$= \left( \bigcap\nolimits_{S \subseteq D \in \mathcal{F}, \ D \text{ respects } A \to B} D \right) \cap \left( \bigcap\nolimits_{S \subseteq D \in \mathcal{F}, \ D \text{ violates } A \to B} D \cap C \right)$$

and see that $S$ is an intersection of $(\cdot)^{\mathcal{F}\uparrow i}$-closed sets and hence itself $(\cdot)^{\mathcal{F}\uparrow i}$-closed.

For the "only if" direction, consider an arbitrary $(\cdot)^{\mathcal{F}\uparrow i}$-closed set $S$. It can be easily checked that all $C \in \mathcal{F}\uparrow i$ respect $A \to B$, hence also $S$ does. Moreover, by definition, every $C \in \mathcal{F}\uparrow i$ is an intersection of elements of $\mathcal{F}$ and thus $(\cdot)^{\mathcal{F}}$-closed.    □

For a context family $\mathfrak{F}$, there are two options of computing an implication-coarsening. One option is to exchange one context $\mathcal{F} \in \mathfrak{F}$ by $\mathcal{F}\uparrow i$. We will present the second option which will lead to a smaller blowup under reasonable assumptions.

**Definition 18.** *Given a context family $\mathfrak{F}$ on $M$ and some implication i on $M$, we define a new context $\mathfrak{F}\uparrow i$ as $\mathfrak{F} \cup \{\mathcal{F}_i\}$.*

**Proposition 23.** *Given a context family $\mathfrak{F}$ on $M$ and some implication i on $M$, we have $(\cdot)^{\mathfrak{F}\uparrow i} \equiv (\cdot)^{\mathcal{F}}\uparrow i$. Moreover, we have $\#\mathfrak{F}\uparrow i = \#\mathfrak{F} + |M|^3$.*

## 4.4  Projection

Next, we investigate for all four representation types, if a succinct presentation of the projection of a closure operator to a subset $N \subseteq M$ of the attributes exists. The findings are mostly trivial or simple consequences of earlier results. We start with contexts, where it is straightforward that the result can be obtained by element-wise projection.

**Proposition 24.** *Given a formal context $\mathcal{F}$ on a set $M$, its projection to some set $N$ can be expressed by $\mathcal{F}_N = \{A \cap N \mid A \in \mathcal{F}\}$. Moreover, $\#\mathcal{F}_N = |\mathcal{F}| \cdot |N| = \#\mathcal{F} \cdot |N|/|M| \leq \#\mathcal{F}$.*

Turning to implications, we note that if a polynomial-size representation of the projection existed, this would imply the existence of a polynomial translation from extended implication sets into implication sets (by projecting away all the auxiliary attributes), contradicting our finding in Section 3.

**Proposition 25.** *There is no polynomial-size representation of projections of closure operators when representing them via implications.*

For extended implications, the case is trivial: the attributes which are to be projected away are simply redefined to be auxiliary attributes.

**Fact 26.** *Given an extended implication set $\mathfrak{J}$ on a set $M$, its projection to some set $N$ can be expressed by itself, i.e., $\mathfrak{J}_N = \mathfrak{J}$. We obtain $\#\mathfrak{J}_N = \#\mathfrak{J}$.*

Last, we consider the context family representation type. Again we can show indirectly that no polynomial representation of projections can exist: assuming its existence, we could polynomially translate extended implications on $M$ with total attribute set $M'$ into context families by first using the polynomial implication-to-context family translation detailed in Section 3 to arrive at a context family on $M'$ and then polynomially project away the auxiliary attributes in $M'$. However we know from Section 3 that such a translation cannot exist.

**Proposition 27.** *There is no polynomial-size representation of projections of closure operators when representing them via context families.*

## 4.5 Lattice Operations

Last but not least, we will examine succinctness of the diverse representation types when applying the lattice operations $\vee$ and $\wedge$ in the lattice of closure operators described in Section 2. We will distinguish between binary and $n$-ary application.

For contexts, $\wedge$ with arbitrary arity is very easy to compute and incurs no blowup whatsoever: one simply needs to concatenate all input contexts.

**Proposition 28.** *Given $n$ contexts $\mathcal{F}_1, \ldots, \mathcal{F}_n$, we let $\mathcal{F} = \mathcal{F}_1 \cup \ldots \cup \mathcal{F}_n$. Then, $(\cdot)^{\mathcal{F}_1} \wedge \ldots \wedge (\cdot)^{\mathcal{F}_n} \equiv (\cdot)^{\mathcal{F}}$ and $\#\mathcal{F} = \#\mathcal{F}_1 + \ldots + \#\mathcal{F}_n$.*

On the other hand already the binary application of $\vee$ may result in exponential blowup, a result shown in the context of model-based reasoning [6].

**Proposition 29 (Eiter et al. 1998).** *There exist sequences $(\varphi_n)_{n\in\mathbb{N}}$ and $(\psi_n)_{n\in\mathbb{N}}$ of closure operators such that $\#\mathcal{F}(\varphi_n \vee \psi_n) \in \Theta(2^n)$ whereas $\#\mathcal{F}(\varphi_n) = \#\mathcal{F}(\psi_n) \in \Theta(n^2)$.*

We provide the construction used by [6], but omit the proof. They let $M = \{1, \ldots, 4n\}$ and define $\varphi_n$ via the context

$$\mathcal{F}_n = \{M \setminus (\{2n{+}1, \ldots, 3n\} \cup \{i, (i \bmod n) + 3n\}) \mid i \in \{1, \ldots, 2n\}\}$$

and $\psi_n$ via the context

$$\mathcal{F}'_n = \{M \setminus (\{3n{+}1, \ldots, 4n\} \cup \{i, (i \bmod n) + 2n\}) \mid i \in \{1, \ldots, 2n\}\}.$$

For implications, conversely, $\vee$ is very easily computable by just taking the union of the implication sets.

**Proposition 30.** *Given $n$ implication sets $\mathfrak{I}_1, \ldots, \mathfrak{I}_n$, we let $\mathfrak{I} = \mathfrak{I}_1 \cup \ldots \cup \mathfrak{I}_n$. Then, $(\cdot)^{\mathfrak{I}_1} \vee \ldots \vee (\cdot)^{\mathfrak{I}_n} \equiv (\cdot)^{\mathfrak{I}}$ and $\#\mathfrak{I} = \#\mathfrak{I}_1 + \ldots + \#\mathfrak{I}_n$.*

On the other hand, $\wedge$ may result in exponential blowup even if applied only binarily:

**Proposition 31.** *There exist sequences $(\varphi_n)_{n\in\mathbb{N}}$ and $(\psi_n)_{n\in\mathbb{N}}$ of closure operators such that $\#\mathfrak{I}(\varphi_n \wedge \psi_n) \in \Theta(2^n)$ whereas $\#\mathfrak{I}(\varphi_n) \in \Theta(n^2)$ and $\#\mathfrak{I}(\psi_n) \in \Theta(n)$.*

*Proof.* Let $M_n = \{a_i, b_i \mid 1 \leq i \leq n\} \cup \{c, d\}$, let $\varphi_n$ be represented by $\mathfrak{I}_1$ containing the implications

$$a_i \to b_i \quad 1 \leq i \leq n,$$
$$b_1, \ldots, b_n \to d,$$

and let $\psi_n$ be represented by $\mathfrak{J}_2 = \{c \to d\}$. We now show that $\mathfrak{J}(\varphi_n \vee \psi_n)$ contains $2^n$ implications by showing that there are $2^n$ pseudo-closed sets. For every set $S \subseteq \{1, \ldots, n\}$ let $A_S := \{a_i \mid i \in S\} \cup \{b_i \mid i \notin S\} \cup \{c\}$. It can be easily verified that $A_S$ is pseudo-closed, since it is not closed (as the closure must contain $d$) and it cannot not properly contain pseudo-closed sets since each of its subsets is closed. Clearly, there are $2^n$ distinct subsets of $\{1, \ldots, n\}$. On the other hand, every minimal implicational representation of $\varphi_n \vee \psi_n$ must contain at least as many implications as there are pseudo-closed sets [10,8].     □

Switching to extended implications improves the situation. Computing $\vee$ remains easy and can be done by taking the union of the implication sets, one just has to take care (possibly via a renaming) that the auxiliary attributes of the separate sets are disjoint. The quadratic blowup comes from the fact that both the number of implication and the auxiliary attribute sets add up.

**Proposition 32.** *Given $n$ extended implication sets $\mathfrak{J}_1, \ldots, \mathfrak{J}_n$ with total attribute sets $M_1, \ldots, M_n$, we let $\mathfrak{J} = \bigcup_{i=1}^{n} \mathrm{rename}(\mathfrak{J}_i, i)$. Then, $(\cdot)^{\mathfrak{J}_1} \vee \ldots \vee (\cdot)^{\mathfrak{J}_n} \equiv (\cdot)^{\mathfrak{J}}$ and $\#\mathfrak{J} = (\sum_{i=1}^{n} |\mathfrak{J}_i|) \cdot (|M| + \sum_{i=1}^{n} |M_1 \setminus M|)$.*

Computing $\wedge$ for extended implication sets is remarkably easier than for implication sets. The idea here is to introduce disjoint "copies" of all implication sets such that closure computation is done independently. Finally one has to add some "confluence rules" which make sure that a proper attribute is added to the closure if it is contained in each of the separate independently computed closures.

**Definition 19.** *Let renameall be the function that takes an extended implication set $\mathfrak{J}$ and a natural number $i$ as input and returns the implication set with every (proper or auxiliary) attribute $m$ in $\mathfrak{J}$ replaced by a new attribute denoted $(m, i)$.*
*Given $n$ extended implication sets $\mathfrak{J}_1, \ldots, \mathfrak{J}_n$ on $M$, let*

$$\bigwedge \{\mathfrak{J}_1, \ldots, \mathfrak{J}_n\} := \mathfrak{J}_{\mathrm{in}} \cup \mathfrak{J}_{\mathrm{out}} \cup \bigcup_{1 \leq i \leq n} \mathrm{renameall}(\mathfrak{J}_i, i),$$

*define a new extended implication set on $M$ where $\mathfrak{J}_{\mathrm{in}} = \{m \to (m, i) \mid m \in M, 1 \leq i \leq n\}$ and $\mathfrak{J}_{\mathrm{out}} = \{(m, 1), \ldots, (m, n) \to m \mid m \in M\}$.*

**Proposition 33.** *Given $n$ extended implication sets $\mathfrak{J}_1, \ldots, \mathfrak{J}_n$ on $M$, we let $\mathfrak{J} = \bigwedge \{\mathfrak{J}_1, \ldots, \mathfrak{J}_n\}$. Then, $(\cdot)^{\mathfrak{J}_1} \wedge \ldots \wedge (\cdot)^{\mathfrak{J}_n} \equiv (\cdot)^{\mathfrak{J}}$ and $\#\mathfrak{J} = (|M| + \sum_{i=1}^{n} (|\mathfrak{J}_i| + |M|)) \cdot (|M| + \sum_{i=1}^{n} |M_1|)$*

Finally, we turn to context families. Like for implications, $\vee$ is very easily computable by just taking the union of the separate context families.

**Proposition 34.** *Given $n$ context families $\mathfrak{F}_1, \ldots, \mathfrak{F}_n$, we let $\mathfrak{F} = \mathfrak{F}_1 \cup \ldots \cup \mathfrak{F}_n$. Then, $(\cdot)^{\mathfrak{F}_1} \vee \ldots \vee (\cdot)^{\mathfrak{F}_n} \equiv (\cdot)^{\mathfrak{F}}$ and $\#\mathfrak{F} = \#\mathfrak{F}_1 + \ldots + \#\mathfrak{F}_n$.*

On the other hand, computing $\wedge$ is a bit more intricate. We can ensure polynomial size for the binary version, but not for $n$-ary application.

**Definition 20.** *Given context families* $\mathfrak{F}_1, \ldots, \mathfrak{F}_n$ *on* $M$, *we let*

$$\bigwedge\{\mathfrak{F}_1, \ldots, \mathfrak{F}_n\} := \{\mathcal{F}_1 \cup \ldots \cup \mathcal{F}_n \mid (\mathcal{F}_1, \ldots, \mathcal{F}_n) \in \mathfrak{F}_1 \times \ldots \times \mathfrak{F}_n\}$$

**Proposition 35.** *Given* $n$ *context families* $\mathfrak{F}_1, \ldots, \mathfrak{F}_n$ *on* $M$, *let* $\mathfrak{F} = \bigwedge\{\mathfrak{F}_1, \ldots, \mathfrak{F}_n\}$. *Then, we have* $(\cdot)^{\mathfrak{F}_1} \wedge \ldots \wedge (\cdot)^{\mathfrak{F}_n} \equiv (\cdot)^{\mathfrak{F}}$ *and* $\#\mathfrak{F} = \prod_{i=1}^{n} |\mathfrak{F}_i| \cdot \sum_{i=1}^{n} \sum_{\mathcal{F} \in \mathfrak{F}_i} \#\mathcal{F}/|\mathfrak{F}_i|$

*Proof.* By definition of $\bigwedge\{\mathfrak{F}_1, \ldots, \mathfrak{F}_n\}$, exploiting Proposition 34 and Proposition 28 and using distributivity of the lattice operations, we obtain

$$(\cdot)^{\mathfrak{F}} \equiv \bigvee_{(\mathcal{F}_1, \ldots, \mathcal{F}_n) \in \mathfrak{F}_1 \times \ldots \times \mathfrak{F}_n} (\cdot)^{\mathcal{F}_1} \wedge \ldots \wedge (\cdot)^{\mathcal{F}_n} \equiv \bigwedge_{i=1}^{n} \bigvee_{\mathcal{F} \in \mathfrak{F}_i} (\cdot)^{\mathcal{F}} \equiv \bigwedge_{i=1}^{n} (\cdot)^{\mathfrak{F}_i}$$

$\square$

# 5   Conclusion

In this paper we have investigated two archetypic and two more exotic representations of closure operators with respect to their mutual succinctness and their suitability for performing certain operations in terms of computation time and output size. The results are summarized in Table 2. Therein, for closure computation and comparison via $\preceq$, upper bounds for the computation time are given in case poly-time algorithms exist, whereas "intractable" indicates cUNP-hardness. For the other computations, the expressions give an upper bound on the output size in case a polynomial such bound exists (for all those cases, the computation time is linearly bounded by the output size), "exponential" denotes that exponential blow-up can be demonstrated, whereas "superpolynomial" merely means that it is known that a polynomial bound cannot exist. Note that for computation of $n$-ary $\bigwedge$ of context families, $n$ must be considered fixed to ensure polynomiality.

**Table 2.** Upper bounds for computations with the four representation types

|  | context $\mathcal{F}$ | implication set $\mathfrak{I}$ | context family $\mathfrak{F}$ | extended implication set $\mathfrak{J}$ |
|---|---|---|---|---|
| closure | $\#\mathcal{F}$ | $\#\mathfrak{I}$ | $\#\mathfrak{F} \cdot |M|$ | $\#\mathfrak{J}$ |
| check $\preceq$ | $\#\mathcal{F} \cdot \#\mathcal{F}'/|M|$ | $\#\mathfrak{I} \cdot \#\mathfrak{I}'/|M|$ | intractable | intractable |
| add implication | $(\#\mathcal{F})^2/|M|$ | $\#\mathfrak{I} + |M|$ | $\#\mathfrak{F} + |M|^3$ | $\#\mathfrak{J} + |M|$ |
| add closed set | $\#\mathcal{F} + |M|$ | exponential | $\#\mathfrak{F} + |\mathfrak{F}| \cdot |M|$ | $2 \cdot \#\mathfrak{J} + 6|N|$ |
| project | $\#\mathcal{F}$ | exponential | superpolynomial | $\#\mathfrak{J}$ |
| $n$-ary $\bigwedge$ | $\sum_i \#\mathcal{F}_i$ | exponential $n{=}2$ | $\prod_i |\mathfrak{F}_i| \cdot \sum_i \#\mathfrak{F}_i$ | $(\sum_i \#\mathfrak{J}_i)^2$ |
| $n$-ary $\bigvee$ | exponential $n{=}2$ | $\sum_i \#\mathfrak{I}_i$ | $\sum_i \#\mathfrak{F}_i$ | $(\sum_i \#\mathfrak{J}_i)^2$ |

There are many open questions left. On the theoretical side, central open questions are if – in the cases where an exponential blowup may occur – there are algorithms transforming one representation into another in *output polynomial time*, that is, if the time required for the computation is polynomially bounded by the size of the output. Note that a negative answer to this question would also disprove the existence of polynomial-delay algorithms.

On the practical side, coming back to our initial motivation, it should be experimentally investigated if variants of standard FCA algorithms can be improved by adding the option of working with alternative closure operator representations.

**Acknowledgements.** The author is thankful to Kira Adaricheva and Mikhail A. Babin, who gave very valuable hints to existing related work, and to Markus Krötzsch for his thorough proof-reading.

# References

1. Adaricheva, K.V., Sloan, R.H., Szörényi, B., Turán, G.: Horn belief contraction: Remainders, envelopes and complexity. In: Proceedings of the 13th International Conference on Principles of Knowledge Representation and Reasoning, KR 2012 (2012)
2. Burosch, G., Demetrovics, J., Katona, G.O.H., Kleitman, D.J., Sapozhenko, A.A.: On the number of databases and closure operations. Theor. Comput. Sci. 78(2), 377–381 (1991)
3. Colomb, P., Irlande, A., Raynaud, O.: Counting of Moore Families for n=7. In: Kwuida, L., Sertkaya, B. (eds.) ICFCA 2010. LNCS, vol. 5986, pp. 72–87. Springer, Heidelberg (2010)
4. Day, A.: The lattice theory of functional dependencies and normal decompositions. International Journal of Algebra and Computation 2(4), 409–431 (1992)
5. Distel, F.: Hardness of enumerating pseudo-intents in the lectic order. In: Kwuida, L., Sertkaya, B. (eds.) ICFCA 2010. LNCS, vol. 5986, pp. 124–137. Springer, Heidelberg (2010)
6. Eiter, T., Ibaraki, T., Makino, K.: Computing intersections of Horn theories for reasoning with models. Tech. Rep. IFIG research report 9803, Universität Gießen (1998), http://bibd.uni-giessen.de/ghtm/1998/uni/r980014.htm
7. Furst, M.L., Saxe, J.B., Sipser, M.: Parity, circuits, and the polynomial-time hierarchy. Mathematical Systems Theory 17(1), 13–27 (1984)
8. Ganter, B., Wille, R.: Formal Concept Analysis: Mathematical Foundations. Springer (1997)
9. Gottlob, G., Libkin, L.: Investigations on Armstrong relations, dependency inference, and excluded functional dependencies. Acta Cybernetica 9(4), 385–402 (1990)
10. Guigues, J.L., Duquenne, V.: Familles minimales d'implications informatives resultant d'un tableau de données binaires. Math. Sci. Humaines 95, 5–18 (1986)
11. Habib, M., Nourine, L.: The number of Moore families on n=6. Discrete Mathematics 294(3), 291–296 (2005)
12. Higuchi, A.: Lattices of closure operators. Discrete Mathematics 179(1-3), 267–272 (1998)

13. Karp, R.M.: Reducibility Among Combinatorial Problems. In: Miller, R.E., Thatcher, J.W. (eds.) Complexity of Computer Computations, pp. 85–103. Plenum Press (1972)
14. Kuznetsov, S.O.: On the intractability of computing the Duquenne-Guigues base. Journal of Universal Computer Science 10(8), 927–933 (2004)
15. Kuznetsov, S.O., Obiedkov, S.: Counting pseudo-intents and #P-completeness. In: Missaoui, R., Schmidt, J. (eds.) ICFCA 2006. LNCS (LNAI), vol. 3874, pp. 306–308. Springer, Heidelberg (2006)
16. Kuznetsov, S.O., Obiedkov, S.A.: Some decision and counting problems of the Duquenne-Guigues basis of implications. Discrete Applied Mathematics 156(11), 1994–2003 (2008)
17. Maier, D.: The Theory of Relational Databases. Computer Science Press (1983)
18. Mannila, H., Räihä, K.J.: Design of Relational Databases. Addison-Wesley (1992)
19. Rudolph, S.: Some notes on pseudo-closed sets. In: Kuznetsov, S.O., Schmidt, S. (eds.) ICFCA 2007. LNCS (LNAI), vol. 4390, pp. 151–165. Springer, Heidelberg (2007)
20. Rudolph, S.: Some notes on managing closure operators. In: Domenach, F., Ignatov, D.I., Poelmans, J. (eds.) ICFCA 2012. LNCS, vol. 7278, pp. 278–291. Springer, Heidelberg (2012)
21. Sertkaya, B.: Some computational problems related to pseudo-intents. In: Ferré, S., Rudolph, S. (eds.) ICFCA 2009. LNCS, vol. 5548, pp. 130–145. Springer, Heidelberg (2009)
22. Sertkaya, B.: Towards the complexity of recognizing pseudo-intents. In: Rudolph, S., Dau, F., Kuznetsov, S.O. (eds.) ICCS 2009. LNCS, vol. 5662, pp. 284–292. Springer, Heidelberg (2009)
23. Wild, M.: Implicational bases for finite closure systems. In: Lex, W. (ed.) Arbeitstagung Begriffsanalyse und Künstliche Intelligenz, pp. 147–169. Springer (1991)

# MDL in Pattern Mining
## A Brief Introduction to KRIMP

Arno Siebes

Algorithmic Data Analysis Group
Universiteit Utrecht, The Netherlands
arno@cs.uu.nl

**Abstract.** In this short paper we sketch a brief introduction to our KRIMP algorithm. Moreover, we briefly discuss some of the large body of follow up research. Pointers to the relevant papers are provided in the bibliography.

## 1 Patterns

Arguably *patterns* are the most important contribution of the data mining community to data analysis. On way to define patterns is as *partial* models, i.e., they do not necessarily concern all tuples (a.k.a. objects, individuals, ...) nor do they necessarily comprise all attributes (a.k.a. variables, features, ...). Another way is to identify patterns with subsets of the data that are for some reason deemed interesting.

The prototypical pattern mining problem is undoubtedly frequent item set mining [1], which is usual formulated in the context of transaction data. Each such transaction is a set of items, e.g., the items a customer buys at a supermarket and a transaction database is simply a bag of transactions.

The patterns are also sets of items called item sets. An item set $I$ occurs in a transaction $t$ if $I \subseteq t$. The support of an item set $I$ in a database $db$, denoted by $supp_{db}(I)$ is the number of transactions in $db$ in which $I$ occurs. The frequent item set problem is then to find *all* item sets whose support exceeds some user defined threshold. Because of the apriori property

$$I \subseteq J \Rightarrow supp_{db}(I) \geq supp_{db}(J)$$

all frequent item sets can be found relatively efficiently using level-wise search; relatively because there may be exponentially many frequent item sets.

In [10] the authors generalize frequent item set mining into a much wider class of pattern mining problems known as *theory mining*. To keep our discussion simple, we stay in the realm of item set mining, but anything we do can be generalized to this wider context.

Frequent item set mining also illustrates rather well why pattern mining is important. The set of customers of a supermarket is hardly homogeneous. While the famous[1] frequent item set example {beer, diapers} may be applicable to

---

[1] Famous but highly likely an urban legend.

C.V. Glodeanu, M. Kaytoue, and C. Sacarea (Eds.): ICFCA 2014, LNAI 8478, pp. 37–43, 2014.

young couples it is probably not relevant to fifty-something couples. While the discerning palate implicated by the item set {Condrieu, Saint Marcellin} carves up the customer space in a completely different – and independent – way.

## 2    The Pattern Explosion

As our example item sets above show patterns can provide useful insight in the data, but, there is an problem. If the support threshold is set high, a few patterns will be discovered, but mostly patterns that are already well known to domain experts. If the support threshold is set low, however, the number of patterns discovered explodes. It is not uncommon to discover more patterns than one has transactions in the database!

Given that one of the main goals of pattern mining is to provide insight in the data, this explosion of the number of patterns is rather embarrassing. So it is not surprising that this problem received much attention and that a wide variety of more or less successful solutions have been put forward.

One of the earliest and best-known is actually closely related to Formal Concept Analysis, viz., *closed item set* mining [11]. Closed item sets are those item sets for which each superset has a strictly smaller support. In other words, they are the closure of the obvious Galois connection between transactions and item sets.

While the number of closed frequent item sets is clearly smaller that the number of frequent item sets, there is no guarantee that their number is far smaller – and often it isn't. The collection of closed frequent item sets has the property that all frequent item sets can be derived from it. That is, the collection of closed frequent item sets is a *condensed representation* of the set of all frequent item sets.

This observation gave rise to other condensed representations, such as *free* item sets [4] and non-derivable item sets [5]; the latter being the smallest condensed representation. However, all of these suffer from the fact that they may still yield very large result sets.

The popular alternative approach is through *constraints* [8]. Clearly, using filters a user can – in principle – easily search through the large set of frequent patterns to find the truly interesting ones. One of the main goals of constraint based pattern mining is to generate only interesting patterns, e.g., by pushing the constraints into the discovery process. While it is relatively easy to remove definitely uninteresting patterns using filters and/or constraints it turns out that it is hard to define – and thus mine for – the small group of interesting ones only.

## 3    What Is the Problem?

The reason why it is so hard to delineate the interesting patterns is that with low(er) support, many patterns describe essentially the same subset of the database. In more detail,

- a database has many small subsets,
- many of these subsets can be described by patterns
- many of the small subsets described by patterns differ in at most a few objects.

Let both $A$ and $A \cup \{a\}$ be closed item sets. If their difference in support is only 1, it is hard to imagine how both can be interesting to a user.

In other words, to decide whether or not a pattern is interesting we also have to look at other patterns as well as the subset of the data that is described by these patterns. That is, we should not be looking for interesting patterns, but for interesting *sets of patterns* and to make such a set of patterns interesting, its members should not describe the same subset of the data over and over again. To formalize this, we use MDL.

## 4   MDL for Pattern Sets

The MDL principle [7] can be paraphrased as: *Induction by Compression.* Slightly more formal, it can be described as follows. Given a set of models $\mathcal{H}$, the best model $H \in \mathcal{H}$ for data set $D$ is the one that minimises

$$L(H) + L(D|H)$$

in which

- $L(H)$ is the length, in bits, of the description of $H$
- $L(D|H)$ is the length, in bits, of the description of the data when encoded with $H$.

In the remainder of this section we briefly describe how we employ MDL to find small characteristic sets of patterns; see, e.g., [14,22] for more detail.

The key idea of our compression based approach is the code table. A code table has item sets on the left-hand side and a code for each item set on its right-hand side. The item sets in the code table are ordered descending on 1) item set length and 2) support. The actual codes on the right-hand side are of no importance: their lengths are. To explain how these lengths are computed we first have to introduce the coding algorithm. A transaction $t$ is encoded by KRIMP by searching for the first item set $c$ in the code table for which $c \subseteq t$. The code for $c$ becomes part of the encoding of $t$. If $t \setminus c \neq \emptyset$, the algorithm continues to encode $t \setminus c$. Since we insist that each code table contains at least all singleton item sets, this algorithm gives a unique encoding to each (possible) transaction. The set of item sets used to encode a transaction is called its cover. Note that the coding algorithm implies that a cover consists of non-overlapping item sets. The length of the code of an item in a code table $CT$ depends on the database we want to compress; the more often a code is used, the shorter it should be. To compute this code length, we encode each transaction in the database $db$. The frequency of an item set $c \in CT$ is the number of transactions $t \in db$ which have $c$ in their cover. The relative frequency of $c \in CT$ is the probability that $c$ is used to encode an arbitrary $t \in db$. For optimal compression of $db$, the higher $P(c)$, the shorter its code should be. In fact, from information theory [6], we have the Shannon code length for $c$, which is optimal, as:

$$l_{CT}(c) = -\log(P(c|db)) = -\log\left(\frac{freq(c)}{\sum_{d \in CT} freq(d)}\right)$$

The length of the encoding of a transaction is now simply the sum of the code lengths of the item sets in its cover. Therefore the encoded size of a transaction $t \in db$ compressed using a specified code table $CT$ is calculated as follows:

$$L_{CT}(t) = \sum_{c \in cover(t,CT)} l_{CT}(c)$$

The size of the encoded database is the sum of the sizes of the encoded transactions, but can also be computed from the frequencies of each of the elements in the code table:

$$L_{CT}(db) = \sum_{t \in db} L_{CT}(t) = -\sum_{c \in CT} freq(c) \log\left(\frac{freq(c)}{\sum_{d \in CT} freq(d)}\right)$$

To find the optimal code table using MDL, we need to take into account both the compressed database size as described above as well as the size of the code table. For the size of the code table, we only count those item sets that have a non-zero frequency. The size of the right-hand side column is obvious; it is simply the sum of all the different code lengths. For the size of the left-hand side column, note that the simplest valid code table consists only of the singleton item sets. This is the *standard encoding (st)* which we use to compute the size of the item sets in the left-hand side column. Hence, the size of the code table is given by:

$$L(CT) = \sum_{c \in CT: freq(c) \neq 0} l_{st}(c) + l_{CT}(c)$$

In [14] we defined the optimal set of (frequent) item sets as that one whose associated code table minimises the total compressed size:

$$L(CT) + L_{CT}(db)$$

KRIMP starts with a valid code table (only the collection of singletons) and a sorted list of candidates. These candidates are assumed to be sorted descending on 1) support and 2) item set length. Each candidate item set is considered by inserting it at the right position in $CT$ and calculating the new total compressed size. A candidate is only kept in the code table iff the resulting total size is smaller than it was before adding the candidate. If it is kept, we reconsider all other elements of $CT$ to see if they still contribute to compression. If not, they are permanently removed. The whole process is illustrated in Figure 1.

In [14,22] it is shown that this simple heuristic algorithm reduces the number of frequent item sets dramatically – e.g. by retaining only 1 in a million frequent patterns.

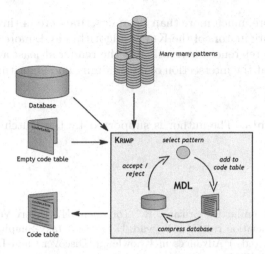

**Fig. 1.** KRIMP in action

Clearly, there are far simpler approaches that can reduce the number of frequent item sets, e.g., by randomly deleting almost all item sets. Hence, in [22,18] we show that KRIMP returns a *characteristic* set of patterns. This is illustrated by using code tables for classification. The idea is simply that we learn a code table for each class separately and assign a new transaction to the class whose code table compresses the transaction best.

## 5   There Is More

KRIMP is just one algorithm to compute code tables, it is rather wasteful as it first generates all frequent item sets and only then starts looking for the code table. SLIM [16] is much faster because it generates candidate item sets on the fly; by taking the current code table into account large parts of the search space can be ignored.

KRIMP (and SLIM) searches for one optimal model. GROEI [12] aims to find a collection of good models, each optimal for a given level of complexity.

We specified KRIMP in the context of transaction data. Because categorical data sets are easily transformed to transaction data, it is obvious that KRIMP also works for that type of data. The basic ideas underlying KRIMP have also been applied to different types of data, e.g., relational data [9], streaming data [17], data from evolutionary biology [2], and recently to seismographic data [3]

Next to classification, code tables can also be used for other kinds of data mining tasks, such as clustering [19], change detection [17], and outlier detection [15]. Moreover, they are also useful in traditionally more statistical tasks such as data generation [21] – which allows for a strong form of privacy protection – data imputation [20], and data smoothing [13].

And there is more, much more than I can describe here in this brief introduction; both by the originators of the KRIMP algorithm and, more importantly, by others. The papers referenced here will give the reader at least a start in the fascinating new area at the intersection of data mining and algorithmic information theory.

**Acknowledgements.** The author is supported by the Dutch national COMMIT project.

# References

1. Agrawal, R., Mannila, H., Srikant, R., Toivonen, H., Inkeri Verkamo, A.: Fast discovery of association rules. In: Fayyad, U.M., Piatetsky-Shapiro, G., Smyth, P., Uthurusamy, R. (eds.) Advances in Knowledge Discovery and Data Mining, pp. 307–328. AAAI/MIT Press (1996)
2. Bathoorn, R., Siebes, A.: Constructing (almost) phylogenetic trees from developmental sequences data. In: Boulicaut, J.-F., Esposito, F., Giannotti, F., Pedreschi, D. (eds.) PKDD 2004. LNCS (LNAI), vol. 3202, pp. 500–502. Springer, Heidelberg (2004)
3. Bertens, R., Siebes, A.: Characterising seismic data. In: ICDM 2014 Proceedings. IEEE (2014)
4. Boulicaut, J.-F., Bykowski, A., Rigotti, C.: Free-sets: a condensed representation of boolean data for the approximation of frequency queries. Data Mining and Knowledge Discovery 7(1), 5–22 (2003)
5. Calders, T., Goethals, B.: Non-derivable itemset mining. Data Mining and Knowledge Discovery 14(1), 171–206 (2007)
6. Cover, T.M., Thomas, J.A.: Elements of Information Theory. Wiley- Interscience, New York (2006)
7. Grünwald, P.: The Minimum Description Length Principle. MIT Press (2007)
8. Boulicaut, J.-F., De Raedt, L., Mannila, H. (eds.): Constraint-Based Mining and Inductive Databases. LNCS (LNAI), vol. 3848. Springer, Heidelberg (2005)
9. Koopman, A., Siebes, A.: Characteristic relational patterns. In: KDD 2009 Proceedings, pp. 437–446 (2009)
10. Mannila, H., Toivonen, H., Inkeri Verkamo, A.: Levelwise search and borders of theories in knowledge discovery. Data Mining and Knowledge Discovery 1(3), 241–258 (1997)
11. Pasquier, N., Bastide, Y., Taouil, R., Lakhal, L.: Discovering frequent closed itemsets for association rules. In: Beeri, C., Bruneman, P. (eds.) ICDT 1999. LNCS, vol. 1540, pp. 398–416. Springer, Heidelberg (1999)
12. Siebes, A., Kersten, R.: A structure function for transaction data. In: SDM 2011 Proceedings, pp. 558–569. SIAM (2011)
13. Siebes, A., Kersten, R.: Smoothing categorical data. In: Flach, P.A., De Bie, T., Cristianini, N. (eds.) ECML PKDD 2012, Part I. LNCS, vol. 7523, pp. 42–57. Springer, Heidelberg (2012)
14. Siebes, A., Vreeken, J., van Leeuwen, M.: Item sets that compress. In: SDM 2006 Proceedings, pp. 393–404. SIAM (2006)
15. Smets, K., Vreeken, J.: The odd one out: Identifying and characterising anomalies. In: SDM 2011Proceedings, pp. 804–815 (2011)

16. Smets, K., Vreeken, J.: SLIM: Directly mining descriptive patterns. In: SDM 2012 Proceedings, pp. 236–247 (2012)
17. van Leeuwen, M., Siebes, A.: Streamkrimp: Detecting change in data streams. In: Daelemans, W., Goethals, B., Morik, K. (eds.) ECML PKDD 2008, Part I. LNCS (LNAI), vol. 5211, pp. 672–687. Springer, Heidelberg (2008)
18. van Leeuwen, M., Vreeken, J., Siebes, A.: Compression picks item sets that matter. In: Fürnkranz, J., Scheffer, T., Spiliopoulou, M. (eds.) PKDD 2006. LNCS (LNAI), vol. 4213, pp. 585–592. Springer, Heidelberg (2006)
19. van Leeuwen, M., Vreeken, J., Siebes, A.: Identifying the components. Data Mining and Knowledge Discovery 19(2), 173–292 (2009)
20. Vreeken, J., Siebes, A.: Filling in the blanks: Krimp minimisation for missing data. In: ICDM 2008 Proceedings, pp. 1067–1072. IEEE (2008)
21. Vreeken, J., van Leeuwen, M., Siebes, A.: Preserving privacy through data generation. In: ICDM 2007 Proceedings, pp. 685–690. IEEE (2007)
22. Vreeken, J., van Leeuwen, M., Siebes, A.: KRIMP: Mining itemsets that compress. Data Mining and Knowledge Discovery 23(1), 169–214 (2011)

# Upper Bound for the Number of Concepts of Contranominal-Scale Free Contexts

Alexandre Albano

Institut für Algebra
Technische Universität Dresden

**Abstract.** We show an improvement of Prisner's upper bound for the number of concepts of a formal context. The improvement factor is of the order $(\max\{|G|, |M|\})^c$, where $c$ is the size of the biggest contranominal scale that can be found as a subcontext. We also prove that the $c \in O(1)$ condition is necessary to establish that an arbitrary sequence of contexts has a polynomial number of concepts, by constructing a lower bound. Complexity aspects of calculating $c$ are discussed.

**Keywords:** number of concepts, upper bound, contranominal scale.

## 1 Introduction

An important question in Formal Concept Analysis is to predict the size of a concept lattice. The problem of calculating exactly the number of concepts of an arbitrary context has been proven to be #P-complete by Sergei Kuznetsov in [5]. Therefore, results that limit above the number of concepts (i.e. upper bounds) are of great importance. In [4], there is a reference to an early result in this direction which is due to D. Schütt[7]. There exist a few other results of the same nature, and many of them were obtained using Graph Theory results and were presented in graph-theoretic language. This is due to the connection between formal concepts and maximal bicliques (to be shortly explained in Preliminaries). A short survey of known upper bounds can be seen in [1].

## 2 Preliminaries

In this paper, all formal contexts and graphs will be finite. Graphs will moreover be simple (no loops or multiple edges) and undirected. Given a graph $G$, we denote its vertex and edge sets by $V(G)$ and $E(G)$, respectively. An arbitrary edge $\{u, v\} \in E(G)$ will be denoted by $uv$. Similarly, we will write $gIm$ to denote an incidence in a formal context $(G, M, I)$. For a vertex $v \in V(G)$ of a graph $G$, we define $N(v) = \{w \in V(G) \mid vw \in E(G)\}$ and call it the set of *neighbors* of $v$.

A *path of length* $k$ in a graph $G = (V, E)$ is a sequence $(v_0, v_1, \ldots, v_k)$ of distinct elements in $V$ such that $v_i v_{i+1} \in E$ for each $i \in \{0, \ldots, k-1\}$. A *circuit* is a sequence $(v_0, v_1, \ldots, v_k)$ with $(v_0, v_1, \ldots, v_{k-1})$ is a path, $v_k = v_0$ and $v_{k-1} v_k \in E$. Its *length* is defined to be $k + 1$. A graph is *connected* if there

C.V. Glodeanu, M. Kaytoue, and C. Sacarea (Eds.): ICFCA 2014, LNAI 8478, pp. 44–53, 2014.

exists a path between every pair of vertices; it is *acyclic* if there are no circuits. Graphs which are connected and acyclic are called *trees*. A *subgraph* $H$ of a graph $G = (V, E)$ is a pair $(W, F)$ satisfying $W \subseteq V$ and $F \subseteq E \cap \{w_1 w_2 \mid w_1, w_2 \in W\}$; it is *spanning* if $V = W$. Whenever clear that $H$ and $G$ are graphs and $H$ is a subgraph of $G$, this will be denoted by $H \subseteq G$. It is a basic fact from Graph Theory that every connected graph has a subgraph which is a spanning tree. For a graph $G = (V, E)$ endowed with a weight function $w : E \to \mathbb{R}$, there certainly exists a spanning tree $T \subseteq G$ which maximizes $\sum_{e \in E(T)} w(e)$ (since all graphs considered here are finite), and a tree in this conditions will be called a *maximum spanning tree*.

A formal context $(G, M, I)$ can be thought as a bipartite graph $(V, E)$, simply by creating one vertex $u \in U \subseteq V$ for each object $g \in G$ and one vertex $w \in W \subseteq V$ for each attribute $m \in M$ (one then has $U \cap W = \emptyset$, even if $G \cap M \neq \emptyset$). For each incidence $gIm$ of the formal context, there exists an edge $g, m \in E$ and all edges of $(V, E)$ are obtained in this way. This association between formal contexts and bipartite graphs result in a correspondence between formal concepts and maximal bicliques (maximal complete bipartite subgraphs) of the associated bipartite graph. Indeed, this can be seen after noting that, for example, for a set of objects $A$, the derivation $A'$ corresponds to the intersection of all neighbors sets $N(v)$, with $v$ ranging over all vertices associated to objects in $A$.

For $n \in \mathbb{N}^*$, we denote by $[n]$ the set $\{1, 2, \ldots, n\}$ and define $[0] = \emptyset$. Given $i \in \mathbb{N}$, we denote by $CN(i)$ the contranominal scale $([i], [i], \neq)$. The following Ferrers context will be denoted by $F(i)$: $([i], [i], \{(j, k) \in [i] \times [i] : j < k\})$. When a context $\mathbb{K}_1$ is a subcontext of a context $\mathbb{K}_2$, we shall write $\mathbb{K}_1 \subseteq \mathbb{K}_2$. Given two sets $S$ and $T$, we will write their symmetric difference as $S \triangle T$. If those sets are incomparable with respect to $\subseteq$, this will be denoted by $S \| T$.

We will need the following easy graph theory lemmas, which proofs will only be sketched.

**Lemma 1.** *Let $G$ be a connected graph with weights on edges $w : E(G) \to \mathbb{R}$ and $T$ a maximum spanning tree of $G$. If $(u, v, x, y)$ is a four-vertex circuit of $G$ and $uv \in E(T)$, then $w(uv) \geq \min\{w(vx), w(xy), w(yu)\}$.*

*Proof.* Suppose the opposite. Divide in cases, depending on which vertex among $u, v, x$ is the closest to $y$ in $T$. Except when $u$ is the closest, it is straightforward to substitute one edge of $T$ by a heavier one which is not in $T$, resulting in a spanning tree with larger maximum weight than $T$. For the case where $u$ is the closest, divide in three cases, depending on which vertices the path from $y$ to $x$ goes through.

**Lemma 2.** *Let $G$ be a connected graph with weights on edges $w : E(G) \to \mathbb{R}$ and $T$ a maximum spanning tree of $G$. If $(u, v, x, y)$ is a four-vertex circuit of $G$ and $uv, xy \in E(T)$, then $\max\{w(uv), w(xy)\} \geq \min\{w(uy), w(vx)\}$.*

*Proof.* By contradiction. Certainly one can obtain a path in $T$ that has $uv$ as first edge and $xy$ as last edge. It is easy to verify that such path always implies that one edge from $\{uy, vx\}$ can replace $uv$ or $xy$ in $T$, resulting in a tree with larger maximum weight.

## 3   Improved Bound

We call a formal context $(G, M, I)$ *non-trivial* if neither $G$ nor $M$ are empty sets. In this section, only non-trivial contexts will be considered. For $j \in \mathbb{N}$, we define a context $\mathbb{K}$ to be $CN(j)$-*free* if there is no subcontext of $\mathbb{K}$ which is isomorphic to $CN(j)$.

Using graph theory language, Prisner has proved in [6] the following:

**Theorem 1.** *Let* $\mathbb{K} = (G, M, I)$ *be a* $CN(j)$-*free context. Then,*

$$|\mathfrak{B}(\mathbb{K})| \leq (|G||M|)^{j-1} + 1.$$

Since $c \mapsto (|G||M|)^c$ is a non-decreasing function on $c$, the result above is equivalent to

**Theorem 2.** *Let* $\mathbb{K} = (G, M, I)$ *be a context and define* $c$ *to be the largest* $j \in \mathbb{N}$ *such that* $CN(j) \subseteq \mathbb{K}$. *Then,*

$$|\mathfrak{B}(\mathbb{K})| \leq (|G||M|)^c + 1.$$

Our improvement of the result above is the following:

**Theorem 3.** *Let* $\mathbb{K} = (G, M, I)$ *be a context and define* $c$ *to be the largest* $j \in \mathbb{N}$ *such that* $CN(j) \subseteq \mathbb{K}$. *Then,*

$$|\mathfrak{B}(\mathbb{K})| \leq 3 \min\{|G|, |M|\}^c - 1.$$

*Proof.* We will proceed by induction on $c$. If $c = 0$, then $I = G \times M$ and, clearly, $|\mathfrak{B}(\mathbb{K})| = 1 < 3 \cdot 1 - 1$. If $c = 1$, then $\mathbb{K}$ is a Ferrers context. Therefore, $\{g' \mid g \in G\}$ as well as $\{m' \mid m \in M\}$ are chains. Thus, $|\mathfrak{B}| \leq \min\{|G|, |M|\} + 1 \leq 3 \min\{|G|, |M|\} - 1$. For the step, let $c \in \mathbb{N}$ be such that $c \geq 2$.

Let us suppose, without loss of generality, that $|G| = \min\{|G|, |M|\}$. Note that $c \geq 2 \Rightarrow |G| \geq 2$. We define $\Omega$ to be a graph with vertex set equal to $\mathfrak{B}(\mathbb{K})$ and there exists an edge between vertices $(A_1, B_1), (A_2, B_2)$ if and only if $|A_1 \cap A_2| + |B_1 \cap B_2| \neq 0$. Moreover, every edge of $\Omega$ is weighted with the positive natural number $|A_1 \cap A_2| + |B_1 \cap B_2|$. We call $w : E \to \mathbb{N}^*$ this weight function.

If $\Omega$ is disconnected, then it cannot be the case that there exists $(A, B) \in \mathfrak{B}$ satisfying $A \neq \emptyset$ and $B \neq \emptyset$, since $(G, G')$ and $(M', M)$ are always concepts. Indeed, the concept $(G, G')$ is adjacent to every concept with non-empty extent and $(M', M)$ adjacent to every concept with non-empty intent. Therefore, if $\Omega$ is disconnected, we will have $I = \emptyset$ and $|\mathfrak{B}| = 2 \leq 3|G|^c - 1$, because $|G| \geq 1$.

Therefore, we can assume that $\Omega$ is connected and we will take a maximum spanning tree $T$ of $\Omega$. The rest of the proof consists of two claims, both of which will make use of the following piece of notation. Given two concepts $(A_i, B_i), (A_j, B_j)$, we will denote by $e_{i,j}$ the edge of $\Omega$ between concepts $(A_i, B_i)$ and $(A_j, B_j)$, provided these two concepts are adjacent in $\Omega$.

*Claim #1:* Let $(A_1, B_1) \neq (A_2, B_2)$ be two distinct concepts of $\mathbb{K}$. There exist $g \in A_1 \triangle A_2$ and $m \in B_1 \triangle B_2$ with $g \nmid m$. Also, if $e_{1,2} \in E(T)$, then, for any such choice of $g$ and $m$, the pair $(A_1 \cap A_2, B_1 \cap B_2)$ is a concept of the subcontext $\mathbb{K}_2 = (m', g', I \cap (m' \times g'))$. Moreover, $\mathbb{K}_2$ is $CN(j-1)$-free whenever $\mathbb{K}$ is $CN(j)$-free.

The existence of such a pair $(g, m)$ is clear. For the proof that $(A_1 \cap A_2, B_1 \cap B_2)$ is indeed a concept, let $A_3 = A_1 \cap A_2$, $B_3 = B_1 \cap B_2$. Suppose, by contradiction, that $(A_3, B_3) \notin \mathfrak{B}(\mathbb{K}_2)$. By the definition of $(A_3, B_3)$ and because of the fact that $(A_1, B_1), (A_2, B_2)$ are concepts, it follows that $g^* I m^*$ for every $g^* \in A_3, m^* \in B_3$. Therefore, we can take a concept of $\mathbb{K}_2$, $(A_4, B_4)$, which satisfies $A_4 \supseteq A_3$, $B_4 \supseteq B_3$ but obviously $(A_4, B_4) \neq (A_3, B_3)$. Now, note that $g \notin A_4$ and that $gIn$ for every $n \in B_4$. Moreover, $m \notin B_4$ and $hIm$ for every $h \in A_4$. Thus, we can take $(A_5, B_5) \in \mathfrak{B}(\mathbb{K})$ satisfying $A_5 \supseteq A_4 \cup \{g\}$ and $B_5 \supseteq B_4$. Analogously, we take $(A_6, B_6) \in \mathfrak{B}(\mathbb{K})$ with $A_6 \supseteq A_4$ and $B_6 \supseteq B_4 \cup \{m\}$. Since $g \nmid m$, it follows that $(A_5, B_5)$ and $(A_6, B_6)$ are different concepts. Clearly, $A_5 \supsetneq A_4$ and that $B_6 \supsetneq B_4$.

We calculate:

$$w(e_{1,5}) = |A_1 \cap A_5| + |B_1 \cap B_5| > |A_1 \cap A_4| + |B_1 \cap B_4|$$
$$> |A_1 \cap A_3| + |B_1 \cap B_3| = |A_1 \cap A_2| + |B_1 \cap B_2|$$
$$w(e_{2,6}) = |A_2 \cap A_6| + |B_2 \cap B_6| > |A_2 \cap A_4| + |B_2 \cap B_4|$$
$$> |A_2 \cap A_3| + |B_2 \cap B_3| = |A_1 \cap A_2| + |B_1 \cap B_2|$$
$$w(e_{5,6}) = |A_5 \cap A_6| + |B_5 \cap B_6| \geq |A_4| + |B_4|$$
$$> |A_3| + |B_3| = |A_1 \cap A_2| + |B_1 \cap B_2|.$$

Therefore, $e_{1,5}, e_{2,6}$ and $e_{5,6}$ have heavier edges than $e_{1,2}$. By Lemma 1, this is a contradiction.

Lastly, since $g \nmid m$, $g$ and $m$ can be used to extend any $CN(j-1)$ in $\mathbb{K}_2$ to a $CN(j)$ in $\mathbb{K}$. Therefore, a $CN(j-1)$ found as a subcontext of $\mathbb{K}_2$ implies the existence of a $CN(j)$ in $\mathbb{K}$. Claim #1 is now proved.

*Claim #2:* Let $\{(A_1, B_1), (A_2, B_2)\}$ and $\{(A_3, B_3), (A_4, B_4)\}$ be two different edges of $T$. Then, at least one of the following holds:

1. $(A_1 \triangle A_2) \cap (A_3 \triangle A_4) = \emptyset$
2. $A_1 \cap A_2 \neq A_3 \cap A_4$

We will again proceed by contradiction. Suppose that $(A_1 \triangle A_2) \cap (A_3 \triangle A_4)$ is a non-empty set and, without loss of generality, that there exists an element $u \in A_1 \cap A_3 \setminus (A_2 \cup A_4)$. Still by contradiction, we assume that $A_1 \cap A_2 = A_3 \cap A_4$. By claim #1, $(A_1 \cap A_2, B_1 \cap B_2)$ is a concept (of a subcontext of $\mathbb{K}$) and, therefore, we also have that $B_1 \cap B_2 = B_3 \cap B_4$.

Consider the edges $e_{1,2}$ and $e_{3,4}$. It could be the case that they are adjacent (that is, they share a vertex). This happens precisely when, among the four concepts involved, only three are distinct. It is clear that

$$w(e_{1,2}) = |A_1 \cap A_2| + |B_1 \cap B_2| = |A_3 \cap A_4| + |B_3 \cap B_4| = w(e_{3,4}). \quad (1)$$

It is not hard to verify that, for $i, j, k \in \{1, 2, 3, 4\}$ with $i \neq j \neq k, i \neq k$:

$$A_1 \cap A_2 = A_3 \cap A_4 \Rightarrow A_i \cap A_j \cap A_k = A_1 \cap A_2 = A_3 \cap A_4. \quad (2)$$

In the same way,

$$A_1 \cap A_2 \cap A_3 \cap A_4 = A_3 \cap A_4 = A_1 \cap A_2. \quad (3)$$

For every $i, j \in \{1, 2, 3, 4\}$, it therefore holds that

$$A_i \cap A_j \supseteq A_1 \cap A_2 \cap A_3 \cap A_4 = A_1 \cap A_2 = A_3 \cap A_4. \quad (4)$$

Relations which are dual to (2), (3) and (4) (i.e. relating intersections of $B_1, B_2, B_3, B_4$) can be analogously derived. They will be referred to as (2)', (3)' and (4)'.

In particular, (4) implies the following. Since $A_1 \cap A_3 \supseteq A_1 \cap A_2, B_1 \cap B_3 \supseteq B_1 \cap B_2$ as well as $A_2 \cap A_4 \supseteq A_1 \cap A_2$ and $B_2 \cap B_4 \supseteq B_1 \cap B_2$, it follows that there exist edges $e_{1,3}, e_{2,4} \in E(\Omega)$, as long as the vertices involved are distinct.

Note that, when one combines (2) and (3), one has that for any $i, j, k \in \{1, 2, 3, 4\}$ with $i \neq j \neq k, i \neq k$:

$$A_1 \cap A_2 \cap A_3 \cap A_4 = A_i \cap A_j \cap A_k. \quad (5)$$

We shall now divide into cases.

Case 1 $(A_2, B_2) = (A_4, B_4)$

In this case, $e_{1,2}, e_{3,4}$ and $e_{1,3}$ form a triangle in $\Omega$ and, among these three edges, only $e_{1,2}$ and $e_{3,4}$ belong to $T$. Since $A_1 \cap A_3 \supseteq A_1 \cap A_2$, $B_1 \cap B_3 \supseteq B_1 \cap B_2$ and $u \in A_1 \cap A_3 \setminus (A_2 \cup A_4)$, it follows that $A_1 \cap A_3 \supsetneq A_1 \cap A_2$. Thus,

$$w(e_{1,3}) = |A_1 \cap A_3| + |B_1 \cap B_3| > |A_1 \cap A_2| + |B_1 \cap B_2| = w(e_{1,2}).$$

which is a contradiction,

Case 2 $(A_2, B_2) \neq (A_4, B_4)$ but $A_2, A_4$ are comparable (therefore, $B_2, B_4$ are comparable as well)

Without loss of generality, suppose that $A_2 \supsetneq A_4$. If there exists an element $u^* \in (A_2 \cap A_4) \setminus (A_1 \cup A_3) = A_4 \setminus (A_1 \cup A_3)$, then

$$\begin{aligned} w(e_{2,4}) &= |A_2 \cap A_4| + |B_2 \cap B_4| \\ &> |A_1 \cap A_2| + |B_1 \cap B_2| \text{ (by (4) and } u^* \notin A_1) \\ &= w(e_{1,2}). \end{aligned}$$

Now, if $(A_1, B_1) = (A_3, B_3)$, we will have the same contradiction as in case 1 (a triangle in $\Omega$ having only one edge not in $T$, and such edge being heavier than the other two). On the other hand, if $(A_1, B_1) \neq (A_3, B_3)$, we will have

$$
\begin{aligned}
w(e_{1,3}) &= |A_1 \cap A_3| + |B_1 \cap B_3| \\
&> |A_1 \cap A_2| + |B_1 \cap B_2| \text{ (by (4) and } u \notin A_2) \\
&= w(e_{1,2}),
\end{aligned}
$$

Therefore, the weights of both edges $e_{1,3}$ and $e_{2,4}$ are greater than $w(e_{1,2}) = w(e_{3,4})$, which is a contradiction by Lemma 1. Therefore, we can assume that $A_4 \setminus (A_1 \cup A_3) = \emptyset$. Thus, every element of $A_4$ belongs to $A_3$ or $A_1$, besides belonging to $A_2$ and $A_4$ itself. By equation, (5), it follows that $A_4 \subseteq A_1, A_2, A_3$. We assert that $A_2$ and $A_1$ are incomparable. Indeed, if we had $A_2 \subseteq A_1$, it would have followed that $A_1 \cap A_2 = A_2 \supsetneq A_4 = A_3 \cap A_4$, which contradicts $A_1 \cap A_2 = A_3 \cap A_4$. On the other hand, the containment $A_1 \subseteq A_2$ does not hold since $u \in (A_1 \cap A_3) \setminus (A_2 \cup A_4)$. The relations $A_4 \subseteq A_1, A_2, A_3$ and $A_1 \| A_2$ imply that, when one considers the corresponding intents, it is clear that $B_4 \supseteq B_1$, $B_4 \supseteq B_3$ and $B_1 \| B_2$. Therefore, there exists $w \in B_2 \cap B_4 \setminus (B_1 \cup B_3)$. We calculate:

$$
\begin{aligned}
w(e_{2,4}) &= |A_2 \cap A_4| + |B_2 \cap B_4| \\
&> |A_1 \cap A_2| + |B_1 \cap B_2| \text{ (by (4) and } w \notin B_1) \\
&= w(e_{1,2}),
\end{aligned}
$$

Like before, if $(A_1, B_1) \neq (A_3, B_3)$, we can get a lower bound $w(e_{1,3})$ in exactly the same way and arrive at a contradiction with Lemma 1. On the other hand, if $(A_1, B_1) = (A_3, B_3)$, the concepts form a triangle in $\Omega$ and also result in a contradiction.

Case 3 $(A_2, B_2) \neq (A_4, B_4)$ and $A_2, A_4$ are incomparable (therefore, $B_2, B_4$ are incomparable as well)

Case 3.1 $(A_2, B_2) \neq (A_4, B_4)$, $A_2, A_4$ are incomparable and $(A_1, B_1) = (A_3, B_3)$

Note that, in this case, $(A_1, B_1), (A_2, B_2)$ and $(A_4, B_4)$ are three different concepts. By equation (4), we have that $e_{2,4} \in E(\Omega)$ and, furthermore, $w(e_{2,4}) \geq w(e_{1,2})$. Because of $e_{1,2}, e_{1,4} \in E(T)$, it clearly follows that $e_{2,4} \notin E(T)$. Therefore, we can assume that $w(e_{2,4}) = w(e_{1,2})$, because, otherwise, it would suffice to substitute in $T$ the edge $e_{1,2}$ with the edge $e_{2,4}$ and obtain a contradiction.

Define $(A_5, B_5) = (A_2, B_2) \wedge (A_4, B_4) = (A_2 \cap A_4, (B_2 \cup B_4)'')$. Observe that $u \notin A_5$, since $u \in (A_1 \cap A_3) \setminus (A_2 \cup A_4)$. In particular, $(A_5, B_5) \neq (A_1, B_1)$. Since $A_2$ and $A_4$ are incomparable, it follows that $(A_5, B_5) \neq (A_2, B_2), (A_4, B_4)$. Summing up, we have that $(A_1, B_1), (A_2, B_2), (A_4, B_4)$ and $(A_5, B_5)$ are four different concepts.

We calculate:

$$w(e_{2,5}) = |A_2 \cap A_2 \cap A_4| + |B_2 \cap (B_2 \cup B_4)''|$$
$$= |A_2 \cap A_4| + |B_2|$$
$$> |A_2 \cap A_4| + |B_2 \cap B_4| \text{ (since } B_2 \| B_4)$$
$$= w(e_{2,4}) = w(e_{1,2}).$$

In the same way, we obtain that $w(e_{4,5}) > w(e_{1,2})$.

Let $P$ be the path in $T$ from $(A_5, B_5)$ to $(A_2, B_2)$. We distinguish two cases:

I) The path $P$ goes through $(A_1, B_1)$: in this case, $P$ goes through $(A_1, B_1)$ just before its end-vertex, $(A_2, B_2)$. Therefore $e_{2,5} \notin E(T)$ and we can substitute $e_{1,2}$ with $e_{2,5}$ and obtain a contradiction.

II) The path $P$ does not go through $(A_1, B_1)$: then, clearly, $P$ does not go through $(A_4, B_4)$ as well. Now, we append the length two path having edges $e_{2,1}, e_{1,4}$ to $P$, obtaining a path between $(A_5, B_5)$ and $(A_4, B_4)$. Therefore $e_{4,5} \notin E(T)$ and we can substitute $e_{1,2}$ with $e_{4,5}$ and obtain a contradiction.

Case 3.2 $(A_2, B_2) \neq (A_4, B_4)$, $A_2, A_4$ *are incomparable and* $(A_1, B_1) \neq (A_3, B_3)$

Define $(A_5, B_5) = (A_2, B_2) \wedge (A_4, B_4) = (A_2 \cap A_4, (B_2 \cup B_4)'')$. Since $A_2$ and $A_4$ are incomparable, it follows that $(A_5, B_5) \neq (A_2, B_2), (A_4, B_4)$. Moreover, since it holds that $u \in (A_1 \cap A_3) \setminus (A_2 \cup A_4)$ and $u \notin A_5$, necessarily $(A_5, B_5) \neq (A_1, B_1), (A_3, B_3)$ holds also.

We calculate:

$$w(e_{2,5}) = |A_2 \cap A_2 \cap A_4| + |B_2 \cap (B_2 \cup B_4)''|$$
$$= |A_2 \cap A_4| + |B_2|$$
$$> |A_2 \cap A_4| + |B_2 \cap B_4| \text{ (since } B_2 \| B_4)$$
$$= w(e_{2,4})$$
$$\geq w(e_{1,2}) \text{ (by (4)).}$$

Analogously, it holds that $w(e_{4,5}) > w(e_{1,2})$.

Let $P$ be the path in $T$ from $(A_5, B_5)$ to $(A_1, B_1)$. We will divide again in cases:

I) $P$ does not pass through $(A_2, B_2)$

It suffices to append $e_{1,2}$ to $P$ and substitute edge $e_{1,2}$ with $e_{2,5}$, arriving to a contradiction.

II) $P$ passes through $(A_2, B_2)$

II.I) $P$ passes through $(A_4, B_4)$.

In this case, $P$ passes through the vertex $(A_4, B_4)$ before the vertex $(A_2, B_2)$, since $e_{1,2} \in E(T)$. Clearly, in this situation, there exists a path between $(A_3, B_3)$ and $(A_1, B_1)$ that passes through $(A_4, B_4)$ and $(A_2, B_2)$, in this order. Then, after using the edge $e_{1,3}$ to replace the edge $e_{1,2}$ we will have a contradiction.

II.II) $P$ does not pass through $(A_4, B_4)$

II.II.I) $P$ passes through $(A_3, B_3)$.

Like in subcase II.I, we have that $P$ passes through $(A_3, B_3)$ before the vertex $(A_2, B_2)$. Therefore, there exists a path between $(A_5, B_5)$ and $(A_4, B_4)$ passing through $(A_3, B_3)$. Consequently, $e_{4,5} \notin E(T)$, and using this edge to replace $e_{3,4}$ gives us a contradiction.

II.II.II) $P$ does not pass through $(A_3, B_3)$.

Let $Q$ be the path between $(A_5, B_5)$ and $(A_3, B_3)$. Note that $Q$ can not pass through $(A_1, B_1)$ without passing through $(A_2, B_2)$, otherwise there would be a circuit in $T$ (made of subpaths of $P$ and $Q$). For this reason, there exist six cases with respect to the property of $Q$ passing or not through $(A_1, B_1), (A_2, B_2)$ or $(A_4, B_4)$ before it reaches $(A_3, B_3)$. In each of the six cases, we will show that at least one of two things happen: *A)* there exists a path from $(A_1, B_1)$ to $(A_3, B_3)$ that uses the edge $e_{1,2}$ or the edge $e_{3,4}$ or *B)* there exists a path from $(A_5, B_5)$ to $(A_4, B_4)$ which uses the edge $e_{3,4}$.

II.II.II.I) $Q$ goes through $(A_1, B_1), (A_2, B_2)$ and $(A_4, B_4)$. In this case, the order of relevant vertices visited is $(A_2, B_2), (A_1, B_1), (A_4, B_4), (A_3, B_3)$. This is clearly an A) case.

II.II.II.II) $Q$ goes through $(A_1, B_1)$ and $(A_2, B_2)$ but not through $(A_4, B_4)$. We can clearly append the edge $e_{3,4}$ to $Q$ and get to the B) case.

II.II.II.III) $Q$ goes through $(A_2, B_2)$ and $(A_4, B_4)$ but not through $(A_1, B_1)$. In this case, we can add the edge $e_{1,2}$ to the subpath of $Q$ that starts at $(A_2, B_2)$ and ends at $(A_3, B_3)$, leading to the A) case.

II.II.II.IV) $Q$ goes through $(A_4, B_4)$ but not through $(A_1, B_1)$ and neither $(A_2, B_2)$. The paths $P$ and $Q$ are disjoint in this case and meet at $(A_5, B_5)$. Their concatenation is a path described in the A) case.

II.II.II.V) $Q$ goes through $(A_2, B_2)$ but not through $(A_1, B_1)$ and neither $(A_4, B_4)$. Same as II.II.II.II.

II.II.II.VI) $Q$ does not go through $(A_1, B_1), (A_2, B_2)$ and neither $(A_4, B_4)$. We can clearly append $e_{3,4}$ to $Q$ and get to the B) case.

For the A) cases, we can add edge $e_{1,3}$ to $T$ and remove $e_{1,2}$ or $e_{3,4}$, obtaining a contradiction. For the B) cases, we can add edge $e_{4,5}$ and remove $e_{3,4}$ and also obtain a contradiction.

Claim #2 is therefore proved. By Claim #2, each edge of $T$ is injectively associated with a pair $(g, A)$, where $g \in G$ and $A$ is an extent of a $CN(c)$-free context (by Claim #1).

Therefore, the number of edges of $T$ must be at most the number of such pairs. That is:

$$|E(T)| \leq |G|(3(|G|)^{c-1} - 1) = 3|G|^c - |G|.$$

Hence, since every tree has one edge less than its number of vertices:

$$|\mathfrak{B}(\mathbb{K})| \leq 3|G|^c - |G| + 1$$
$$\leq 3|G|^c - 1 \text{ (since } |G| \geq 2).$$

$\square$

## 4   Lower Bound

In this section we will make use of the commonly used asymptotic notation symbols $O(f), o(f), \omega(f)$ and $\Theta(f)$. Their definitions can be seen in [2]. We will, however, denote membership in those classes using the $\in$ symbol instead of the equal sign. As an application of Prisner's upper bound, one obtains that, if every context in a sequence $(\mathbb{K}_i)_{i \in \mathbb{N}^*}$ has at most a constant-sized contranominal scale subcontext, then the associated sequence $(|\mathfrak{B}(\mathbb{K}_i)|)_{i \in \mathbb{N}^*}$ grows polynomially. In this section, we show that the condition $c \in O(1)$ is necessary to establish the polynomial growth of $(|\mathfrak{B}(\mathbb{K}_i)|)_{i \in \mathbb{N}^*}$.

Let $n, k \in \mathbb{N}$ with $k \le n$ and let $l$ be the remainder of the division of $n$ by $k$. As usual, we will denote by $\mathbb{K}_1 + \mathbb{K}_2$ the direct sum between two contexts and, more generally, we will use $\Sigma$ to denote the generalized $n$-ary sum. We define the following formal context:

$$Z(n,k) = \left( \sum_{i=1}^{\lfloor \frac{n}{k} \rfloor} F(k) \right) + F(l).$$

Note that $Z(n,n) = F(n)$ and that $Z(n,1) = CN(n)$. The following proposition calculates the size of the biggest contranominal scale found as a subcontext in $Z(n,k)$:

**Proposition 1.** *Let $n, k \in \mathbb{N}$ with $k \le n$. Then, $c(Z(n,k)) = \lceil n/k \rceil$.*

*Proof.* Consider the complementary context $\mathbb{K}^c$ of $\mathbb{K} = Z(n,k)$. A subcontext of $\mathbb{K}$ which is a contranominal scale then corresponds to a subcontext of $\mathbb{K}^c$ which is a nominal scale. Note that $\mathbb{K}^c$ is the disjoint union of Ferrers contexts, since the complement of a Ferrers context is again a Ferrers context. Now, consider a subcontext $(H, N, J)$ of $\mathbb{K}^c$ which is a nominal scale. For every attribute $m \in N$, at most one object $g$ belonging to the same term as $m$ may belong to $H$. Indeed, since, otherwise, the restricted subcontext $(\{g, h\}, N, J \cap (\{g, h\} \times N))$ would be Ferrers and therefore free of nominal scales of size two. Therefore, the size of a nominal scale $(H, N, J)$ is upper bounded by $\lceil n/k \rceil$. One can clearly obtain a nominal scale of size $\lceil n/k \rceil$ by choosing one attribute and one corresponding object from each Ferrers term.

The content of the next theorem is our lower bound relating $|\mathfrak{B}(\mathbb{K})|$ and $c(\mathbb{K})$.

**Theorem 4.** *For every function $f(n) \in o(1) \cap O(\log n)$, there exists a sequence of contexts $(\mathbb{K}_i)_{i \in \mathbb{N}}$ with $c(\mathbb{K}) \in \Theta(f)$ for which the corresponding sequence of numbers of concepts is not bounded above by any polynomial function.*

*Proof.* By the definition of $Z(n,k)$ and by the property relating concepts lattices of direct sums [4], we have that $|\mathfrak{B}(Z(n,k))| = (k+1)^{\lfloor n/k \rfloor} \cdot (l+1)$, where $l$ is the remaining of the division of $n$ by $k$.

Let $f(n) \in o(1) \cap O(\log n)$ and define $\mathbb{K}_n = Z(n, \lfloor n/f(n) \rfloor)$. Note that $n/f(n) \in \omega(n^{1-\epsilon}) \cap o(n)$ for every $\epsilon > 0$, because of $f(n) \in O(\log n)$. Applying Proposition 1, we obtain that $c(\mathbb{K}_n) = \lceil n/\lfloor n/f(n) \rfloor \rceil \in \Theta(f)$.

Substituting:

$$|\mathfrak{B}(\mathbb{K}_n)| = (\lfloor n/f(n)\rfloor + 1)^{\lfloor n/\lfloor n/f(n)\rfloor\rfloor} \cdot (l+1)$$
$$> (\lfloor n/f(n)\rfloor)^{\lfloor n/\lfloor n/f(n)\rfloor\rfloor} \in \Theta((n/f(n))^{f(n)}),$$

which is superpolynomial since $n/f(n) \in \omega(n^{1-\epsilon})$ and $f(n) \in o(1)$. $\qquad\square$

## 5  Related Work

The problem of calculating $c$ is polynomially equivalent to that of finding the maximum induced matching (MIM) of a bipartite graph. Indeed, given a context $\mathbb{K}$, $c(\mathbb{K})$ equals the size of the biggest induced matching of the complementary context $\mathbb{K}^c$, when $\mathbb{K}^c$ is viewed as a bipartite graph. A connection with the largest independent set problem is provided by the following fact: given a bipartite graph $B$, an independent set of $L^2(B)$ (that is, the square of its line graph) corresponds to an induced matching of $B$.

Stockmeyer and Vazirani introduced in [8] the $\delta$-separated-matching problem. The particular case $\delta = 2$ is MIM, and the authors call it the "risk-free marriage problem". In the same paper, the authors prove that MIM is NP-hard, even when one restricts its instances to graphs of maximum degree four. In terms of approximability, Duckworth, Manlove and Zito proved in [3] that, for every $r \geq 3$, the decision version of MIM restricted to $r$-regular graphs is a problem that belongs to APX but is also APX-complete.

**Acknowledgements.** The author is deeply grateful for the valuable and patient work of the anonymous reviewers.

## References

1. Albano, A., do Lago, A.P.: A convexity upper bound for the number of maximal bicliques of a bipartite graph. Discrete Applied Mathematics 165, 12–24 (2011); 10th Cologne/Twente Workshop on Graphs and Combinatorial Optimization (CTW 2011)
2. Cormen, T.H., Leiserson, C.E., Rivest, R.L., Stein, C.: Introduction to Algorithms, 2nd edn. The MIT Press and McGraw-Hill Book Company (2001)
3. Duckworth, W., Manlove, D., Zito, M.: On the approximability of the maximum induced matching problem. J. Discrete Algorithms 3(1), 79–91 (2005)
4. Ganter, B., Wille, R.: Formal Concept Analysis: Mathematical Foundations. Springer, Heidelberg (1999)
5. Kuznetsov, S.O.: On computing the size of a lattice and related decision problems. Order 18(4), 313–321 (2001)
6. Prisner, E.: Bicliques in graphs I: Bounds on their number. Combinatorica 20(1), 109–117 (2000)
7. Schütt, D.: Abschätzungen für die Anzahl der Begriffe von Kontexten. Master's thesis, TH Darmstadt (1987)
8. Stockmeyer, L.J., Vazirani, V.V.: NP-completeness of some generalizations of the maximum matching problem. Inf. Process. Lett. 15(1), 14–19 (1982)

# Algebraicity and the Tensor Product
# of Concept Lattices

Bogdan Chornomaz

Department of Computer Science
V.N. Karazin Kharkiv National University
Kharkiv, Ukraine
markyz.karabas@gmail.com

**Abstract.** In this paper we prove that the tensor product of complete lattices, as it is defined in formal concept analysis, preserves algebraicity. The proof of this fact is based on the compactness of propositional logic. We use this property to show that the box product of $(0, \vee)$-semilattices, introduced by G.Grätzer and F.Wehrung in 1999, can be obtained from the tensor product of concept lattices in a manner similar to how it is done in the definition of tensor product in "general" lattice theory.

## 1 Introduction

Traditionally, the tensor product in lattice theory is defined on $(0, \vee)$-semilattices as a join-semilattice of compact *bi-ideals* in the direct product of the corresponding lattices, see [4]. The formal concept analysis provides a different (and nonequivalent) approach toward the concept of tensor product. Namely, the tensor product of concept lattices is defined as the concept lattice of the direct product of their formal contexts. Theorem 14 of [3] proves that the resulting lattice is independent of the choice of formal contexts, thus justifying the definition. However, the concept lattices are exactly the complete lattices, and so thus defined tensor product has a narrower scope than that from [4].

In Section 2 we argue that we can define a *complete tensor product* of complete lattices, denoted $\overset{\text{bi}}{\otimes}$, as a set of *complete bi-ideals* in their direct product, in much the same way as it is done for the tensor product from [4], which we call *finite tensor product* and denote by $\overset{\text{bi}}{\boxtimes}$. It can be easily verified that $\overset{\text{bi}}{\otimes}$ preserves algebraicity, thus enabling the following construction: for any $(0, \vee)$-semilattices $A$ and $B$ we take the complete tensor product of complete algebraic lattices $\operatorname{Id} A$ and $\operatorname{Id} B$, and then take the $(0, \vee)$-semilattice of compact elements of $\operatorname{Id} A \overset{\text{bi}}{\otimes} \operatorname{Id} B$. Unsurprisingly, we get

$$A \overset{\text{bi}}{\boxtimes} B = \operatorname{Cp}\left(\operatorname{Id} A \overset{\text{bi}}{\otimes} \operatorname{Id} B\right). \tag{1}$$

Further on, we will omit the word "complete" whenever the context is clear.

C.V. Glodeanu, M. Kaytoue, and C. Sacarea (Eds.): ICFCA 2014, LNAI 8478, pp. 54–66, 2014.
© Springer International Publishing Switzerland 2014

The results of Section 2 are rather trivial and mainly given without the proof. As suggested by the section title, they serve as a motivation for introducing similar construction for the tensor product defined via formal contexts, which we call *complete fc-tensor product* and denote $\overset{\text{fc}}{\otimes}$. The key property of this construction is the preservation of algebraicity, which is trivial for $\overset{\text{bi}}{\otimes}$, but takes some effort to prove in case of $\overset{\text{fc}}{\otimes}$.

As a prerequisite for this proof, in Section 3 we argue that for lattices $A$ and $B$ the tensor product $A \overset{\text{fc}}{\otimes} B$ can be represented as a lattice of *closed* complete bi-ideals in $A \times B$, and that it is a complete meet-subsemilattice of $A \overset{\text{bi}}{\otimes} B$. Also, we give two characterizations of closed bi-ideals.

In Section 4 we prove the key result of the paper that the tensor product $\overset{\text{fc}}{\otimes}$ preserves algebraicity. In fact, we prove that this problem can be reduced to the compactness of propositional logic, see Corollary 1.2.12 in [1]. Thus, we can define *finite fc-tensor product* of $(0, \vee)$-semilattices, denoted $\overset{\text{fc}}{\boxtimes}$, by the formula

$$A \overset{\text{fc}}{\boxtimes} B = \mathrm{Cp}\left( \mathrm{Id}\, A \overset{\text{fc}}{\otimes} \mathrm{Id}\, B \right). \tag{2}$$

Notice that while the tensor products $\overset{\text{bi}}{\otimes}$ and $\overset{\text{bi}}{\boxtimes}$ are defined in their own right, and (1) simply establishes a relation between them, the alike formula (2) is used as the definition for $\overset{\text{fc}}{\boxtimes}$. Also notice that the "general" tensor product from [4] in our terminology is called *finite* tensor product, while the tensor product defined in the formal concept analysis is the *complete* fc-tensor product.

Finally, in Section 5 we identify finite fc-tensor product as the lattice tensor product, introduced by G.Grätzer and F.Wehrung in [5]. The key concept on which this definition is based is the *box product*, thus we will call this tensor product *box tensor product*. In that paper the authors notice the resemblance of their construction with Wille's construction, in particular that this tensor products coincide in case of finite lattices. Our construction thus can be used to back this resemblance and establish a parallel between the definition of the box tensor product and the finite tensor product.

## 2   Motivational Example

We start with the definition of a tensor product $\overset{\text{bi}}{\otimes}$, almost literally repeating the definition of $\overset{\text{bi}}{\boxtimes}$ given in [4].

For a lattice $A$ we call a set $X \subseteq L$ *hereditary* if $x \in X$ and $y \leq x$ implies $y \in X$. For complete lattices $K$ and $L$ we define the *complete lateral join* as a partial function $\bigvee_{CL} : 2^{K \times L} \to K \times L$, given by

$$\bigvee_{CL} \langle x_\alpha, y_\alpha \rangle = \begin{cases} \langle x, \bigvee y_\alpha \rangle, & \forall \alpha : x_\alpha = x; \\ \langle \bigvee x_\alpha, y \rangle, & \forall \alpha : y_\alpha = y; \end{cases}.$$

A subset $I$ of $K \times L$ is called a *complete bi-ideal* if it is hereditary, it contains the set

$$\perp_{K,L} = (\{0_K\} \times L) \cup (K \times \{0_L\})$$

and it is closed under complete lateral joins. We say that $J \subseteq K \times L$ is a *complete dual bi-ideal*, if $J$ is a bi-ideal in $K^d \times L^d$.

The *complete tensor product* of $K$ and $L$, denoted $K \overset{\text{bi}}{\otimes} L$, is the set of complete bi-ideals in $K \times L$ ordered by set inclusion. Obviously, $K \overset{\text{bi}}{\otimes} L$ is a complete lattice where the meet coincides with set intersection.

Let $A$ and $B$ be $(0, \vee)$-semilattices, $x \in A$ and $y \in B$. We adopt the conventional notation $(x]$ and $[x)$ for the principal ideal and the principal filter of $x$ in $A$ correspondingly. Also, by $(x, y]$ and $[x, y)$ we denote the principal ideal and the principal filter of $(x, y)$ in $A \times B$. The same notation is also used for complete lattices.

We recall that for $(0, \vee)$-semilattices $A$ and $B$, the complete algebraic lattice of all bi-ideals in $A \times B$ is called in [4] an *extended tensor product* and is denoted $A \overline{\otimes} B$, and $A \boxtimes B$ is then defined as the join-semilattice of compact elements of $A \overline{\otimes} B$.

**Lemma 1.** *For $(0, \vee)$-semilattices $A$ and $B$ the complete lattices $\operatorname{Id} A \overset{\text{bi}}{\otimes} \operatorname{Id} B$ and $A \overline{\otimes} B$ are isomorphic, and the isomorphism $\varepsilon \colon \operatorname{Id} A \overset{\text{bi}}{\otimes} \operatorname{Id} B \to A \overline{\otimes} B$ is given by*

$$\varepsilon(\mathcal{I}) = \{(x, y) \in A \times B \mid ((x], (y]) \in \mathcal{I}\},$$

*for every complete bi-ideal $\mathcal{I}$ in $\operatorname{Id} A \times \operatorname{Id} B$. The inverse mapping takes form*

$$\varepsilon^{-1}(\mathcal{J}) = \{(I_x, I_y) \in \operatorname{Id} A \times \operatorname{Id} B \mid I_x \times I_y \subseteq \mathcal{J}\},$$

*for every bi-ideal $\mathcal{J}$ in $A \times B$.*

**Proof.** We left to the reader the proof of an easy fact that $\varepsilon(\mathcal{I})$ is a bi-ideal and $\varepsilon^{-1}(\mathcal{J})$ is a complete bi-ideal, for any $\mathcal{I} \in \operatorname{Id} A \overset{\text{bi}}{\otimes} \operatorname{Id} B$ and $\mathcal{J} \in A \overline{\otimes} B$. Let us now prove that $\varepsilon \circ \varepsilon^{-1}$ and $\varepsilon^{-1} \circ \varepsilon$ are identity mappings.

Indeed, for $\mathcal{I} \in \operatorname{Id} A \overset{\text{bi}}{\otimes} \operatorname{Id} B$ we get

$$(I_x, I_y) \in \varepsilon^{-1} \circ \varepsilon(\mathcal{I}) \Leftrightarrow I_x \times I_y \subseteq \varepsilon(\mathcal{I})$$
$$\Leftrightarrow \forall x \in I_x, y \in I_y \colon (x, y) \in \varepsilon(\mathcal{I})$$
$$\Leftrightarrow \forall x \in I_x, y \in I_y \colon ((x], (y]) \in \mathcal{I}$$
$$\Leftrightarrow \forall x \in I_x \colon \left((x], \bigvee_{y \in I_y} (y]\right) \in \mathcal{I}$$
$$\Leftrightarrow \left(\bigvee_{x \in I_x} (x], \bigvee_{y \in I_y} (y]\right) = (I_x, I_y) \in \mathcal{I}.$$

And for $\mathcal{J} \in A \overline{\otimes} B$ we get

$$(x, y) \in \varepsilon \circ \varepsilon^{-1}(\mathcal{J}) \Leftrightarrow ((x], (y]) \in \varepsilon^{-1}(\mathcal{J})$$
$$\Leftrightarrow (x] \times (y] \subseteq \mathcal{J} \Leftrightarrow (x, y) \in \mathcal{J}. \quad \blacksquare$$

Let us recall the notions of compactness and algebraicity. An element $x$ in a complete lattice $K$ is *compact* if $x \leq \bigvee S$ for some $S$ implies $x \leq \bigvee T$ for some finite $T \subseteq S$. The set $\mathbf{C}(K)$ of all compact elements in a complete lattice $K$ is a $(\vee, 0)$-semilattice of $K$. A complete lattice $A$ is called *algebraic* if every element is the join of compact elements.

The fact that $A \overline{\otimes} B$ is an algebraic lattice for any $(0, \vee)$-semilattices $A$ and $B$ is thus equivalent to the fact that $\overset{\text{bi}}{\otimes}$ preserves algebraicity.

**Proposition 1.** *If $K$ and $L$ are complete algebraic lattices then the lattice $K \overset{\text{bi}}{\otimes} L$ is also algebraic.*

Now, using Lemma 1 we get

$$A \overset{\text{bi}}{\boxtimes} B = \mathrm{Cp}\left(\mathrm{Id}\, A \overset{\text{bi}}{\otimes} \mathrm{Id}\, B\right),$$
$$K \overset{\text{bi}}{\otimes} L = \mathrm{Id}\left(\mathrm{Cp}\, K \overset{\text{bi}}{\boxtimes} \mathrm{Cp}\, L\right),$$

for all $(0, \vee)$-semilattices $A$ and $B$ and all complete algebraic lattices $K$ and $L$.

It is shown in [4] that the bi-ideals can be represented by join-homomorphisms. Below we introduce similar technique for complete bi-ideals.

For complete lattices $K$ and $L$ let us define $K \overset{\rightarrow}{\otimes} L$ as

$$K \overset{\rightarrow}{\otimes} L = \mathrm{Hom}\left(\left(K; \bigvee; 0\right), \left(L; \bigwedge; 1\right)\right),$$

that is, $K \overset{\rightarrow}{\otimes} L$ is a lattice of complete dual join-homomorphisms from $K$ to $L$ sending 0 to 1.

**Proposition 2.** *For complete lattices $K$ and $L$ the mapping $\eta\colon K \overset{\rightarrow}{\otimes} L \to K \overset{\text{bi}}{\otimes} L$ defined by*

$$\eta(\varphi) = \{(x, y) \in K \times L \mid y \leq \varphi(x)\},$$

*for any $\varphi \in K \overset{\rightarrow}{\otimes} L$, establishes an isomorphism between $K \overset{\rightarrow}{\otimes} L$ and $K \overset{\text{bi}}{\otimes} L$. And the inverse mapping is given by*

$$\eta^{-1}(H)(a) = \bigvee \{x \in L \mid (a, x) \in H\},$$

*for any $H \in K \overset{\text{bi}}{\otimes} L$ and $a \in L$.*

# 3   Properties of fc-tensor Product

The definition of the tensor product in formal concept analysis stems from two papers of R. Wille [9,10]. However, in this paper to introduce this tensor product we are following the observational paper of B.Ganter and R.Wille [3]; same results but presented with proofs can be found in sections 4.4 and 5.4 of the monography by the same authors [2].

In formal concept analysis the tensor product of complete lattices $K$ and $L$ is defined as the concept lattice

$$\mathfrak{B}(K \times L, K \times L, \nabla),$$

where $\nabla \subseteq (K \times L) \times (K \times L)$ is a relation defined by

$$\nabla = \left\{ ((x_1, y_1), (x_2, y_2)) \mid x_1 \leq x_2 \text{ or } y_1 \leq y_2 \right\},$$

see [3]. This concept lattice can be represented as a lattice of subsets of $K \times L$, closed under the closure operator

$$X \mapsto X^{*+}, \tag{3}$$

where

$$X^* := \{ b \in K \times L \mid \forall a \in X \colon a \nabla b \},$$
$$Y^+ := \{ a \in K \times L \mid \forall b \in Y \colon a \nabla b \},$$

for all $X, Y \subseteq K \times L$. We call such sets simply *closed* when the closure operator is clear from the context. Note that the mappings $X \mapsto X^*$ and $Y \mapsto Y^+$ are antitone, and thus the mapping $X \mapsto X^{*+}$ is isotone. Also notice that the mapping $X \mapsto X^{+*}$ is also isotone and is a dual closure in $K \times L$. We will take the representation by closed sets as a definition for fc-tensor product, which we denote by $\overset{\text{fc}}{\otimes}$. By the properties of the closure operator, the complete meet in $K \overset{\text{fc}}{\otimes} L$ coincides with set intersection, that is, $K \overset{\text{fc}}{\otimes} L$ is a complete meet-subsemilattice in the powerset of $K \times L$.

The following easily verified proposition gives a necessary condition for a set to be closed.

**Proposition 3.** *For complete lattices $K$ and $L$ and $X \subseteq K \times L$, the set $X^{*+}$ is a complete bi-ideal, and $X^{*+}$ is a complete dual bi-ideal.*

Thus, $K \overset{\text{fc}}{\otimes} L$ is the lattice of *closed complete bi-ideals*. Further on, we will omit the word "complete" and call them simply closed bi-ideals.

Now we will investigate how the closure operator (3) acts on bi-ideals represented by homomorphisms.

Let $P$ be a poset and $X \subseteq P$. Then we define a *hereditary closure* of $X$ as a smallest hereditary set containing $X$. One can easily verify that if $Y$ is a hereditary closure of $X$ then it can be represented as

$$Y = \{ y \in P \mid \exists x \in X \colon y \leq x \}.$$

**Lemma 2.** *For complete lattices $K$ and $L$ and a set $X \subseteq K$ let $Y$ be a hereditary closure of $X$. Then $X^{*+} = Y^{*+}$.*

**Proof.** As $X \subseteq Y$ we get $X^{*+} \subseteq Y^{*+}$. On the other hand, the set $X^{*+}$ is hereditary and contains $X$, thus, it contains its hereditary closure, that is, $Y \subseteq X^{*+}$. But then $Y^{*+} \subseteq X^{*+*+} = X^{*+}$, which proves our claim. ∎

**Lemma 3.** *For complete lattices $K$ and $L$ let $I \subseteq K \times L$ be a set defined as*

$$I = \{(x, y) \mid y \le f(x)\},$$

*for some $f : K \to L$, and let $f^{*+} = \eta^{-1}(I^{*+})$. Then*

$$f^{*+}(x) = \bigwedge_{y \in K-[x]} \bigvee_{w \in K-(y]} f(w)$$

Note that, in particular, this lemma is applicable in case when $I$ is a complete bi-ideal and $f = \eta^{-1}(I)$

**Proof.** Let us define the mapping $f^* : K \to L$ as

$$f^*(y) = \bigvee_{w \in K-(y]} f(w),$$

for all $y \in K$.

Let $I_b = \{(x, f(x)) \mid x \in K\}$, then $I$ is a hereditary closure of $I_b$ and by Lemma 2 we get $I^* = I_b^*$. Now

$$
\begin{aligned}
I^* = I_b^* &= \{(x, y) \mid \forall x' : x' \le x \text{ or } f(x') \le y\} \\
&= \{(x, y) \mid \forall x' : x' \in K - (x] \text{ implies } f(x') \le y\} \\
&= \left\{(x, y) \mid \bigvee_{x' \in K-(x]} f(x') \le y\right\} \\
&= \{(x, y) \mid f^*(x) \le y\}.
\end{aligned}
$$

By Proposition 3, the set $I^*$ is a complete dual bi-ideal. Taking into account that the mapping $A \mapsto A^{+*}$ is a dual closure, by the same argument as above we get $I^{*+} = I_c^+$ where $I_c = \{(x, f^*(x)) \mid x \in K\}$. Then

$$
\begin{aligned}
I^{*+} = I_c^+ &= \{(x, y) \mid \forall x' : x \le x' \text{ or } y \le f^*(x')\} \\
&= \{(x, y) \mid \forall x' : x' \in K - [x) \text{ implies } y \le f^*(x')\} \\
&= \left\{(x, y) \mid y \le \bigwedge_{x' \in K-[x)} f^*(x')\right\}.
\end{aligned}
$$

This easily yields

$$f^{*+}(x) = \bigwedge_{x' \in K-[x)} f^*(x'),$$

which proves the claim of the theorem.    ∎

For complete lattices $K$ and $L$ let us notice that the set $(x, y] \cup \perp$ is a closed bi-ideal, for any $x \in K$ and $y \in L$. Following [4], we call it a *pure tensor* and denote it by $x \otimes y$. We also introduce a set $[x, y] \subseteq K \times L$ defined by

$$[x, y] = \{(x', y') \mid x' \leq x \text{ or } y' \leq y\}.$$

Obviously, $[x, y]$ is also a closed bi-ideal, for every $x \in K$ and $y \in L$.

For a set $A \subseteq K \times L$ we define the sets $A|_K \subseteq K$ and $A|_L \subseteq L$ as

$$A|_K = \{x \mid \exists y, (x, y) \in A\},$$
$$A|_L = \{y \mid \exists x, (x, y) \in A\}.$$

We use the symbol $\bigsqcup$ to denote the disjoint union of sets.

Now we give another characterization of the closure operator and, correspondingly, of closed bi-ideals

**Lemma 4.** *For complete lattices $K$ and $L$ and a set $I \subseteq K \times L$ the closure $I^{*+}$ is given by*

$$I^{*+} = \bigcap_{X \sqcup Y = I} \left[ \bigvee X|_K, \bigvee Y|_L \right] \tag{4}$$

**Proof.**

$$I^{*+} = \{(x, y) \mid \forall (x', y') \in I^* : x \leq x' \text{ or } y \leq y'\}$$
$$= \bigcap \{[x', y'] \mid (x', y') \in I^*\}$$
$$= \bigcap \{[x', y'] \mid \forall (x'', y'') \in I : x'' \leq x' \text{ or } y'' \leq y'\}$$
$$= \bigcap \left\{ [x', y'] \mid \exists X \subseteq I : \bigvee X|_K \leq x' \text{ and } \bigvee (I - X)|_L \leq y' \right\}$$
$$= \bigcap \left\{ \left[ \bigvee X|_K, \bigvee Y|_L \right] \mid X \sqcup Y = I \right\}.$$

    ∎

**Corollary 1.** *The family of closed bi-ideals of $K \times L$ is the minimal family of sets which contains all sets $[x, y]$ and is closed under $\bigcap$.*

## 4    Algebraicity

For a $(0, \vee)$-semilattices $A$ and $B$, the extended tensor product $A \overline{\otimes} B$ is an algebraic lattice, see [4]. By Lemma 1 this means that $\overset{bi}{\otimes}$ preserves algebraicity. The goal of this section is to prove similar property for $\overset{fc}{\otimes}$.

**Lemma 5.** *If $K$ and $L$ are complete lattices and $x_0 \in K$ and $y_0 \in L$ are compact elements, then $x_0 \otimes y_0$ is a compact element of $K \overset{fc}{\otimes} L$.*

**Proof.** Arguing by contradiction, suppose that $x_0 \otimes y_0$ is not compact. Then there is an infinite family of closed bi-ideals $\{I_\alpha\}_{\alpha \in \mathfrak{A}} \subseteq K \overset{fc}{\otimes} L$ such that

$$(x_0, y_0] \leq \bigvee \{I_\alpha \mid \alpha \in \mathfrak{A}\}$$

and for every finite subfamily $A \subset \mathfrak{A}$ holds

$$(x_0, y_0] \not\leq \bigvee_{\alpha \in A} I_\alpha.$$

As every bi-ideal from this family can be represented as an infinite join of pure tensors, then, without losing generality, we may assume that every bi-ideal $I_\alpha$ is a pure tensor, that is $I_\alpha = x_\alpha \otimes y_\alpha$, for every $\alpha \in \mathfrak{A}$.

As $\bigcup \{(x_\beta, y_\beta] \mid \beta \in \mathfrak{B}\}$ is the hereditary closure of the set $\{(x_\beta, y_\beta) \mid \beta \in \mathfrak{B}\}$, by Lemma 2 we infer

$$\left( \bigcup \{(x_\beta, y_\beta] \mid \beta \in \mathfrak{B}\} \right)^{*+} = \left( \{(x_\beta, y_\beta) \mid \beta \in \mathfrak{B}\} \right)^{*+},$$

for every $\mathfrak{B} \subseteq \mathfrak{A}$. Then

$$(x_0, y_0] \leq \bigvee_{\alpha \in \mathfrak{A}} (x_\alpha, y_\alpha] = \left( \bigcup_{\alpha \in \mathfrak{A}} [x_\alpha, y_\alpha) \right)^{*+} = \{(x_\alpha, y_\alpha) \mid \alpha \in \mathfrak{A}\}^{*+},$$

and using (4) we get

$$(x_0, y_0] \leq \bigcap_{\mathfrak{B} \subseteq \mathfrak{A}} \left[ \bigvee_{\beta \in \mathfrak{B}} x_\beta, \bigvee_{\gamma \in \mathfrak{A} - \mathfrak{B}} y_\gamma \right]. \tag{5}$$

And similarly

$$(x_0, y_0] \not\leq \bigcap_{B \subseteq A} \left[ \bigvee_{\beta \in B} x_\beta, \bigvee_{\gamma \in A - B} y_\gamma \right], \tag{6}$$

for every finite $A \subseteq \mathfrak{A}$.

Let us define two families $\mathfrak{X}$ and $\mathfrak{Y}$ of finite subsets of $\mathfrak{A}$ by

$$\mathfrak{X} = \left\{ A \subseteq \mathfrak{A}, |A| < \infty \;\middle|\; x_0 \leq \bigvee_{\alpha \in A} x_a \right\},$$

$$\mathfrak{Y} = \left\{ A \subseteq \mathfrak{A}, |A| < \infty \;\middle|\; y_0 \leq \bigvee_{\alpha \in A} y_a \right\}.$$

That is, $\mathfrak{X}$ and $\mathfrak{Y}$ are the families of all sets of indexes in $\mathfrak{A}$, defining the finite covers of $x_0$ and $y_0$ correspondingly.

We now need to use some tools from propositional logic, namely the compactness theorem. We are following the terminology of H.J.Keisler and C.C.Chang, see Section 1.2 in [1]. Let us consider the set $\mathfrak{A}$ as the set of *simple statements* and build a set $\Sigma$ of *propositional sentences* over it

$$\Sigma := \Sigma_{\mathfrak{X}} \bigsqcup \Sigma_{\mathfrak{Y}},$$

$$\Sigma_{\mathfrak{X}} := \left\{ \neg \left( \bigwedge_{\alpha \in A} \alpha \right) \middle| A \in \mathfrak{X} \right\},$$

$$\Sigma_{\mathfrak{Y}} := \left\{ \neg \left( \bigwedge_{\alpha \in A} \neg \alpha \right) \middle| A \in \mathfrak{Y} \right\}.$$

Notice that the symbol $\bigwedge$ in the definition above is used not as join, but as a *connective* in the *propositional language*. Its usage is justified by the fact that all considered "joins" are finite. The *models* of our language are simply subsets of $\mathfrak{A}$. For a model $\mathfrak{B} \subseteq \mathfrak{A}$ a simple statement $\beta$ is true in $\mathfrak{B}$ iff $\beta \in \mathfrak{B}$.

We claim that (5) is equivalent to the following statement: The set $\Sigma$ of sentences is not *satisfiable*, that is, $\Sigma$ has no model. Indeed, (5) can be restated as: For any model $\mathfrak{B} \subseteq \mathfrak{A}$, either $x_0 \leq \bigvee \{x_\beta \mid \beta \in \mathfrak{B}\}$ or $y_0 \leq \bigvee \{y_\gamma \mid \gamma \in \mathfrak{A} - \mathfrak{B}\}$. As $x_0$ is compact, it follows that $x_0 \leq \bigvee \{x_\beta \mid \beta \in \mathfrak{B}\}$ iff there is $A_0 \in \mathfrak{X}$ such that $A_0 \subseteq \mathfrak{B}$, in which case the propositional sentence

$$\left( \neg \bigwedge \{a \mid a \in A_0\} \right) \in \Sigma_{\mathfrak{X}}$$

is not satisfied in $\mathfrak{B}$. Similarly, $y_0 \leq \bigvee \{y_\gamma \mid \gamma \in \mathfrak{A} - \mathfrak{B}\}$ iff there is a set $B_0 \in \mathfrak{Y}$ such that $B_0 \subseteq \mathfrak{A} - \mathfrak{B}$, and consequently the propositional sentence

$$\left( \neg \bigwedge \{\neg a \mid a \in B_0\} \right) \in \Sigma_{\mathfrak{Y}}$$

is not satisfied in $\mathfrak{B}$. Combined together, these observations prove our claim.

Similarly, (6) is equivalent to the statement: The set $\Sigma$ of sentences is *finitely satisfiable*, that is, every finite subset of $\Sigma$ has a model. However, by the compactness theorem for propositional calculus, see Corollary 1.2.12 in [1], $\Sigma$ is satisfiable if it is finitely satisfiable, a contradiction. ∎

Now as an easy corollary we get

**Theorem 1.** *If $K$ and $L$ are complete algebraic lattices then $K \overset{fc}{\otimes} L$ is algebraic.*

**Proof.** Notice that by Lemma 5, $x \otimes y$ is a compact element of $K \overset{fc}{\otimes} L$, for any $x \in \operatorname{Cp} K$ and $y \in \operatorname{Cp} L$. Any closed bi-ideal $I \subseteq K \times L$ can be represented as

$$I = \bot \cup \bigcup \{(x, y] \mid (x, y) \in I, x \in \operatorname{Cp} K, y \in \operatorname{Cp} L\}$$
$$= \bigvee \{x \otimes y \mid (x, y) \in I, x \in \operatorname{Cp} K, y \in \operatorname{Cp} L\}.$$

Thus, every element of $K \overset{fc}{\otimes} L$ can be represented as a join of compact elements, so $K \overset{fc}{\otimes} L$ is algebraic. ∎

Now, using Theorem 1, for $(0, \vee)$-semilattices $A$ and $B$ we can define *finite fc-tensor product* as

$$A \overset{\text{fc}}{\boxtimes} B = \mathrm{Cp}\left(\mathrm{Id}\, A \overset{\text{fc}}{\otimes} \mathrm{Id}\, B\right).$$

and observe that, just as for "regular" tensor product, holds

$$K \overset{\text{fc}}{\otimes} L = \mathrm{Id}\left(\mathrm{Cp}\, K \overset{\text{fc}}{\boxtimes} \mathrm{Cp}\, L\right),$$

for all complete algebraic lattices $K$ and $L$.

## 5   The Box Tensor Product

Now we are going to show that, thus defined, finite fc-tensor product coincides with the *box tensor product*, introduced in [5].

For lattices with zero $A$ and $B$, $a \in A$ and $b \in B$ we define the *box tensor product* of $A$ and $B$, denoted $A \boxtimes B$, as the set of all finite intersections of the form

$$H = \bigcap \{[a_i, b_i] \mid i < n\}, \tag{7}$$

satisfying

$$\bigwedge \{a_i \mid i < n\} = 0_A, \tag{8}$$

$$\bigwedge \{b_i \mid i < n\} = 0_B, \tag{9}$$

where $n > 0$, $(a_i, b_i) \in A \times B$, for all $i < n$.

Let us point out that in making this definition we have skipped few intermediate steps as compared to [5]; the definition now corresponds to Lemma 3.8 of the mentioned paper.

Now, let us identify $A$ and $B$ with the sets of principal ideals in $\mathrm{Id}\, A$ and $\mathrm{Id}\, B$ correspondingly, using canonical embeddings $\pi_A \colon x \mapsto (x]$ and $\pi_B \colon y \mapsto (y]$; and let us extend this embeddings to the embedding $\pi \colon A \boxtimes B \to A \overset{\text{fc}}{\otimes} B$ defined by

$$\pi \colon \bigcap \{[a_i, b_i] \mid i < n\} \mapsto \bigcap \{[\pi_A(a_i), \pi_B(b_i)] \mid i < n\}. \tag{10}$$

Notice that by Corollary 1, all elements of $\pi[H]$ are valid elements of $\mathrm{Id}\, A \overset{\text{fc}}{\otimes} \mathrm{Id}\, B$, that is, valid closed bi-ideals in $\mathrm{Id}\, A \times \mathrm{Id}\, B$.

For a lattice $C$ and a set $X \subseteq C$ we denote the lattice generated by $X$ in $C$ by $\langle X \rangle_C$, or simply by $\langle X \rangle$ if the underlying lattice is clear from the context.

**Proposition 4.** *For lattices with zero $C$ and $D$, let $c_i \in C$ and $d_i \in D$, for $i = 1, \ldots, n$. Then*

$$\bigcap_{i=1,\ldots,n} [c_i, d_i] = \perp_{c_i, d_i} \cup \bigcup_{i=1,\ldots,m} (z_i, w_i],$$

*where*

$$\perp_{c_i, d_i} = \left( \left( \bigwedge c_i \right] \times D \right) \cup \left( C \times \left( \bigwedge d_i \right] \right)$$

*for some $m$, and $z_i \in \langle c_j \mid j = 1, \ldots, n \rangle$ and $w_i \in \langle d_j \mid j = 1, \ldots, n \rangle$, for all $i$.*

Using Proposition 4 we can easily prove the desired result.

**Theorem 2.** *For lattices with zero $A$ and $B$, the mapping $\pi$ defined by (10) establishes an isomorphism between $A \boxtimes B$ and $A \overset{fc}{\boxtimes} B$, that is*

1. *for any $H \in A \boxtimes B$, $\pi(H)$ is a compact closed bi-ideal in $\mathrm{Id}\, A \times \mathrm{Id}\, B$,*
2. *any compact compact closed bi-ideal in $\mathrm{Id}\, A \times \mathrm{Id}\, B$ can be represented as $\pi(H)$, for some $H \in A \boxtimes B$.*

**Proof.** (1): Let us take $H$ as in (7), then

$$\pi(H) = \bigcap \left\{ [\pi_A(a_i), \pi_B(b_i)] \mid i < n \right\},$$

and, by Proposition 4

$$\pi(H) = \perp_{\pi_A(a_i), \pi_B(b_i)} \cup \bigcup_{i=1, \ldots, m} (z_i, w_i],$$

where $z_i \in \langle \pi_A(a_j) \mid j = 1, \ldots, n \rangle \subseteq \pi_A[A]$ and $w_i \in \langle \pi_B(b_j) \mid i = 1, \ldots, n \rangle \subseteq \pi_B[B]$.

Also from (8) and (9) it follows that $\perp_{\pi_A(a_i), \pi_B(b_i)} = \perp_{\mathrm{Id}\, A, \mathrm{Id}\, B}$, which yields

$$\pi(H) = \bigcup_{i=1, \ldots, m} z_i \otimes w_i = \bigvee_{i=1, \ldots, m} z_i \otimes w_i,$$

where $z_i$ and $w_i$ are compact elements of $\mathrm{Id}\, A$ and $\mathrm{Id}\, B$ correspondingly, for all $i \leq n$. Thus, by Lemma 5 the elements $z_i \otimes w_i$ are compact elements of $\mathrm{Id}\, A \overset{fc}{\otimes} \mathrm{Id}\, B$ and $\pi(H)$ is a finite join of compact elements, thus, it is also compact.

(2): Let $\mathcal{H}$ be a closed compact bi-ideal in $\mathrm{Id}\, A \times \mathrm{Id}\, B$. Definitely, $\mathcal{H}$ can be represented as

$$\mathcal{H} = \bigcup \{ a \otimes b \mid a \in \pi_A[A], b \in \pi_B[B], (\pi_A(a), \pi_B(b)) \in H \}$$

$$= \bigvee \{ a \otimes b \mid a \in \pi_A[A], b \in \pi_B[B], (\pi_A(a), \pi_B(b)) \in H \}.$$

Using compactness of $\mathcal{H}$, we get

$$\mathcal{H} = \bigvee \{ a_i \otimes b_i \mid i < n \} = \left( \bigcup \{ a_i \otimes b_i \mid i < n \} \right)^{*+},$$

for some $n > 0$ and $a_i \in \pi_A[A]$ and $b_i \in \pi_B[B]$ ,for all $i < n$. By Lemma 2 the latter gives

$$\mathcal{H} = \left( \bigcup \{ (a_i, b_i) \mid i < n \} \right)^{*+},$$

and the application of Lemma 4 gives us the desired representation of $\mathcal{H}$.    ∎

# 6   Conclusion

The paper [5] contains a very profound discussion on the similarities between various kinds of tensor products and their properties, as well as a list of open problems. In this list we would like to single out two problems that explicitly deal with the connection between the tensor box product and finite tensor product. Problem 1 asks for a characterization of a situation when $A \boxtimes B = A \overset{bi}{\boxtimes} B$, for lattices with zero $A$ and $B$, and Problem 6 seeks for another tensor products "between" $\boxtimes$ and $\overset{bi}{\boxtimes}$.

A problem in comparing these tensor products is that the finite tensor product is defined on $(0, \vee)$-semilattices, while the box tensor product is defined on lattices with zero. This situation is natural in the following sense: for lattices with zero $A$ and $B$, $A \boxtimes B$ is always a lattice, while $A \overset{bi}{\boxtimes} B$ in general is only a $(0, \vee)$-semilattice, see, for example, Corollary 8.2 in [6]. Now, our construction of $\overset{fc}{\boxtimes}$ enables to extend the definition of the box tensor product to $(0, \vee)$-semilattices, and thus enables to compare these constructions on some "natural" domain. Again, correspondences (1) and (2) enable to characterize the connection between $\overset{bi}{\boxtimes}$ with $\overset{fc}{\boxtimes}$ by comparing $\overset{bi}{\otimes}$ with $\overset{fc}{\otimes}$.

The author would also like to draw the parallel to the paper of M. Krötzsch and G. Malik [8]. For complete lattices $K$ and $L$, the space of regular Galois connections described in this paper is $F \overset{fc}{\otimes} L$; indeed, *regular* Galois connections are deliberately defined this way. In the same time one can show, not without some effort, that the space of all Galois connections will be exactly $F \overset{bi}{\otimes} L$. The large part of [8] is dedicated to describing the situation when $K$ and $L$ have only regular Galois connections between them, and in particular to the case when $K$ have only regular Galois connections to any complete lattice $L$.

Regarding the latter case, the author believe that he has the proof that this holds iff $K$ satisfies complete infinite distributive identity (CIDI). The one direction of this statement is provided by Theorem 4 of [8]. For the other direction, when $K$ does not satisfy CIDI, the counterexample is provided by the identity mapping from $K$ to $K^{\mathrm{op}}$, which would be an irregular Galois connection. However, the complete proof of this fact requires efforts which fall beyond the scope of the present article.

# References

1. Chang, C.C., Keisler, H.J.: Model Theory, 3rd edn. North-Holland, Amsterdam (1990)
2. Ganter, B., Wille, R.: Formal concept analysis - mathematical foundations. Springer (1999)
3. Ganter, B., Wille, R.: Applied lattice theory: formal concept analysis, Appendix H in [7], pp. 591–605

4. Grätzer, G., Wehrung, F.: A survey of tensor products and related constructions in two lectures. Algebra Universalis 45, 117–134 (2001)
5. Grätzer, G., Wehrung, F.: A new lattice construction: the box product. J. Algebra 221, 315–344 (1999)
6. Grätzer, G., Wehrung, F.: Tensor products and transferability of semilattices. Canad. J. Math. 51, 792–815 (1999)
7. Grätzer, G.: General Lattice Theory, 2nd edn. Birkhäuser, Basel (1998)
8. Krötzsch, M., Malik, G.: The Tensor Product as a Lattice of Regular Galois Connections. In: Missaoui, R., Schmidt, J. (eds.) Formal Concept Analysis. LNCS (LNAI), vol. 3874, pp. 89–104. Springer, Heidelberg (2006)
9. Wille, R.: Tensorial decompositions of concept lattices. Order 2, 81–95 (1985)
10. Wille, R.: Tensor products of complete lattices as closure systems. Contributions to General Algebra 7, 381–386 (1991)

# On the Existence of Isotone Galois Connections between Preorders

Francisca García-Pardo, Inma P. Cabrera, Pablo Cordero,
Manuel Ojeda-Aciego, and Francisco J. Rodríguez-Sanchez

Universidad de Málaga, Spain*

**Abstract.** Given a mapping $f\colon A \to B$ from a preordered set $A$ into an unstructured set $B$, we study the problem of defining a suitable preordering relation on $B$ such that there exists a mapping $g\colon B \to A$ such that the pair $(f, g)$ forms an adjunction between preordered sets.

## 1 Introduction

Galois connections were introduced by Ore [30] in 1944 as a pair of antitone mappings aimed at generalizing Birkhoff's theory of polarities to the framework of complete lattices. Later, in 1958, Kan [23] introduced the notion of pair of adjoint functors in a categorical context. It is not surprising to find a plethora of examples of adjunction in several disparate research areas, ranging from the most theoretical to the most applied. It is remarkable to note that the importance of adjunctions quickly increased to an extent that, for instance, the interest of category theorists moved from universal mapping properties and natural transformations to adjointness.

When instantiating an adjunction to categories of ordered sets, it can be seen that both constructions, adjunctions and Galois connections, are fairly similar and, to some extent, are interdefinable: in some sense, an adjunction between $A$ and $B$ is a Galois connection in which the order relation on $B$ is reversed (this leads to the use of the term *isotone Galois connection* which is exactly that of adjunction between ordered structures).

Nowadays, one can often find publications concerning Galois connections, both isotone and antitone, focused on either theoretical developments or theoretical applications [7,9,24]. Another term for adjunction, frequently used in the context of ordered sets, is that of pair of residuated mappings [5].

Concerning applications to informatics, we can find a first survey [28] on computer science applications published back in 1986. Of course, a number of more specific references on certain topics can be found, for instance, to programming [29], data analysis [34], logic [12, 21]. It is specially remarkable that the research topic of approximate reasoning using rough sets has benefitted specially from the use of the theory of Galois connections [13, 20, 31, 32].

* This work is partially supported by the Spanish research projects TIN2009-14562-C05-01, TIN2011-28084 and TIN2012-39353-C04-01, and Junta de Andalucía project P09-FQM-5233.

C.V. Glodeanu, M. Kaytoue, and C. Sacarea (Eds.): ICFCA 2014, LNAI 8478, pp. 67–79, 2014.

It is worth to recall that many recent works on Galois connections use them in the framework of Formal Concept Analysis (FCA), either theoretically or applicatively.This is not surprising, since the operators used to build concepts form a Galois connection. In [33] one can find an extension of conceptualization modes, [1] describes a general approach to fuzzy FCA, [6] studies two previously existing frameworks and proved them equivalent, [10] use them for solving multi-adjoint relation equations, [27] provides new generalizations for FCA, [11] relates FCA and possibility theory, [3] stress on the "duality" between isotone and antitone Galois connections in showing a case of mutual reducibility of the concept lattices generated by using each type of connection, etcetera.

Being able to define a Galois connection between two ordered structures is a matter of major importance, and not only for FCA. For instance, [8] establishes a Galois connection between valued constraint languages and sets of weighted polymorphisms in order to develop an algebraic theory of complexity for valued constraint languages.

Browsing the related literature, one can find several publications concerning sufficient or necessary conditions for the existence of Galois connections between ordered structures. The main results of this paper are related to the existence and construction of the adjoint pair to a given mapping $f$, but *in a more general framework*.

Our initial setting is to consider a mapping $f \colon A \to B$ from a preordered set $A$ into an unstructured set $B$, and then characterize those situations in which the set $B$ can be preordered and an isotone mapping $g \colon B \to A$ can be built such that the pair $(f, g)$ is an adjunction. (*Note that hereafter we will use exclusively this term since is shorter than isotone Galois connection*).

The structure of the paper is as follows: in Section 2, given $f \colon A \to B$ we introduce the preliminary definitions, and recall the necessary and sufficient conditions for the existence of a unique partial ordering on $B$ and a mapping $g$ such that $(f, g)$ is an adjunction; then, in Section 3 we study the existence of preordering in $B$ and the existence of $g$ such that $(f, g)$ is an adjunction between preordered structures; at this point, the absence of antisymmetry makes that both the statements and the proofs of the results to be much more involved. Finally, in Section 5, we draw some conclusions and discuss future work.

## 2 Preliminary Definitions and Results

We assume basic knowledge of the properties and constructions related to a partially ordered and preordered sets. Anyway, for the sake of self-completion, we include below the formal definitions of the main concepts to be used in this work.

**Definition 1.** *Given a partially ordered set* $\mathbb{A} = (A, \leq_A)$, $X \subseteq A$, *and* $a \in A$.

- *An element* $u$ *is said to be an* upper bound *of* $X$, *if* $x \leq u$ *for all* $x \in X$. *We write* $UB(X)$ *to refer to the set of upper bounds of* $X$.

- *An element $a$ is said to be the* maximum *of $X$, denoted $\max X$, if $a \in X$ and $x \leq a$ for all $x \in X$.*
- *The* downset $a^{\downarrow}$ *of $a$ is defined as $a^{\downarrow} = \{x \in A \mid x \leq_A a\}$.*
- *The* upset $a^{\uparrow}$ *of $a$ is defined as $a^{\uparrow} = \{x \in A \mid x \geq_A a\}$.*

*A mapping $f : (A, \leq_A) \to (B, \leq_B)$ between partially ordered sets is said to be*

- isotone *if $a_1 \leq_A a_2$ implies $f(a_1) \leq_B f(a_2)$, for all $a_1, a_2 \in A$.*
- antitone *if $a_1 \leq_A a_2$ implies $f(a_2) \leq_B f(a_1)$, for all $a_1, a_2 \in A$.*

*In the particular case in which $A = B$,*

- *$f$ is* inflationary *(also called extensive) if $a \leq_A f(a)$ for all $a \in A$.*
- *$f$ is* deflationary *if $f(a) \leq_A a$ for all $a \in A$.*

As we are including the necessary definitions for the development of the construction of adjunctions, we state below the definition of adjunction we will be working with.

**Definition 2.** *Let $\mathbb{A} = (A, \leq_A)$ and $\mathbb{B} = (B, \leq_B)$ be posets, $f : A \to B$ and $g : B \to A$ be two mappings. The pair $(f, g)$ is said to be an* adjunction *between $\mathbb{A}$ and $\mathbb{B}$, denoted by $(f, g) : \mathbb{A} \leftrightharpoons \mathbb{B}$, whenever for all $a \in A$ and $b \in B$ we have that*

$$f(a) \leq_B b \qquad \text{if and only if} \qquad a \leq_A g(b)$$

*The mapping $f$ is called* left adjoint *and $g$ is called* right adjoint.

As we will not be working with partially ordered sets but with preordered sets, some of the previous notions have to be adapted to this more general setting.

The definitions of downset (resp. upset) of an element in a preordered set, and those of isotone, antitone, inflationary and deflationary mapping between preordered sets are exactly the same as those given for posets.

The notion of maximum or minimum element of a subset of a preordered set is defined as usual. Note, however, that due to the absence of antisymmetry, these elements need not be unique. This is an important difference which justifies the introduction of special terminology in this context.

**Definition 3.** *Given a preordered set $(A, \leq_A)$ and a subset $X \subseteq A$, an element $a \in A$ is said to be a* p-maximum *(resp., p-minimum) of $X$ if $a \in X$ and $x \leq_A a$ (resp., $a \leq_A x$) for all $x \in X$. The set of p-maxima (resp., p-minima) of $X$ will be denoted as* p-max$(X)$ *(resp., p-min$(X)$).*

Notice that p-max$(X)$ (resp., p-min$(X)$) need not be a singleton. In the event that, say $a, b \in$ p-max$(X)$, then the two relations $a \leq b$ and $b \leq a$ hold. As this situation will repeat several times, we introduce the equivalence relation $\approx_A$ in any preordered set $(A, \leq_A)$, defined as follows for $a_1, a_2 \in A$:

$$a_1 \approx_A a_2 \quad \text{if and only if} \quad a_1 \leq_A a_2 \text{ and } a_2 \leq_A a_1 \tag{1}$$

The equivalence class of element $a$ wrt an equivalence relation $R$ will be written, as usual, as $[a]_R$. If there is no risk of ambiguity, the subscript will be omitted.

In this work we will assume a mapping $f\colon A \to B$ such that the original set is preordered. In order to study the existence of adjoints in this framework, we will need to use the previously defined relation $\approx_A$, together with the kernel relation $\equiv_f$, defined as $a \equiv_f b$ if and only if $f(a) = f(b)$.

The two relations above are used together in the definition of the *p-kernel* relation defined below:

**Definition 4.** *Let* $\mathbb{A} = (A, \leq_A)$ *be a preordered set, and* $f\colon A \to B$ *a mapping. The* p-kernel *relation* $\cong_A$ *is the equivalence relation obtained as the transitive closure of the union of the relations* $\approx_A$ *and* $\equiv_f$.

It is well-known that the transitive closure in the definition above can be described as follows: given $a_1, a_2 \in A$, we have that $a_1 \cong_A a_2$ if and only if there exists a finite chain $\{x_i\}_{i \in \{1,\dots,n\}} \subseteq A$ such that $x_1 = a_1$, $x_n = a_2$ and, for all $i \in \{1, \dots, n-1\}$, either $x_i \equiv_f x_{i+1}$ or $x_i \approx_A x_{i+1}$.

The following theorem [17] states different equivalent characterizations of the notion of adjunction *between preordered sets* that will be used in the main construction of the right adjoint. As expected, the general structure of the definitions is preserved, but those concerning the actual definition of the adjoints have to be modified by using the notions of p-maximum and p-minimum.

**Theorem 1.** *Let* $\mathbb{A} = (A, \leq_A), \mathbb{B} = (B, \leq_B)$ *be two preordered sets, and* $f\colon \mathbb{A} \to \mathbb{B}$ *and* $g\colon \mathbb{B} \to \mathbb{A}$ *be two mappings. The following statements are equivalent:*

1. $(f, g)\colon \mathbb{A} \leftrightharpoons \mathbb{B}$.
2. $f$ *and* $g$ *are isotone maps, and* $g \circ f$ *is inflationary map,* $f \circ g$ *is deflationary map.*
3. $f(a)^{\uparrow} = g^{-1}(a^{\uparrow})$ *for all* $a \in A$.
4. $g(b)^{\downarrow} = f^{-1}(b^{\downarrow})$ *for all* $b \in B$.
5. $f$ *is isotone and* $g(b) \in \text{p-max}\, f^{-1}(b^{\downarrow})$ *for all* $b \in B$.
6. $g$ *is isotone and* $f(a) \in \text{p-min}\, g^{-1}(a^{\uparrow})$ *for each* $a \in A$.

Once again, the absence of antisymmetry leads to slight modifications of some well-known properties of adjunctions, as that stated in the result below and its corollary.

**Theorem 2.** *Let* $\mathbb{A} = (A, \leq), \mathbb{B} = (B, \leq)$ *be two preordered sets, and* $f\colon A \to B$ *and* $g\colon B \to A$ *be two mappings. If* $(f, g)\colon \mathbb{A} \leftrightharpoons \mathbb{B}$ *then,* $(f \circ g \circ f)(a) \approx_B f(a)$ *for all* $a \in A$, *and* $(g \circ f \circ g)(b) \approx_A g(b)$ *for all* $b \in B$.

**Corollary 1.** *Let* $\mathbb{A} = (A, \leq_A), \mathbb{B} = (B, \leq_B)$ *be two preordered sets, and* $f\colon A \to B$ *and* $g\colon B \to A$ *be two mappings. If* $(f, g)\colon \mathbb{A} \leftrightharpoons \mathbb{B}$ *then,* $(g \circ f \circ g \circ f)(a) \approx_A (g \circ f)(a)$ *for all* $a \in A$, *and* $(f \circ g \circ f \circ g)(b) \approx_B (f \circ g)(b)$ *for all* $b \in B$.

The following definition recalls the notion of Hoare ordering between subsets of a preordered set, and then introduces an alternative statement in the subsequent lemma.

**Definition 5.** *Let* $(A, \leq)$ *be a preordered set, and consider* $X, Y \subseteq A$.

- *We will denote by* $\sqsubseteq_H$ *the Hoare relation,* $X \sqsubseteq_H Y$ *if and only if, for all* $x \in X$, *there exists* $y \in Y$ *such that* $x \leq y$.
- *We define* $X \sqsubseteq Y$ *if and only if there exist* $x \in X$ *and* $y \in Y$ *such that* $x \leq y$.

**Lemma 1.** *Let* $(A, \leq)$ *be a preordered set, and consider* $X, Y \subseteq A$ *such that* p-min$(X) \neq \varnothing$ *and* p-min$(Y) \neq \varnothing$. *The following statements are equivalent:*

1. p-min$(X) \sqsubseteq_H$ p-min$(Y)$
2. p-min$(X) \sqsubseteq$ p-min$(Y)$
3. *For all* $x \in$ p-min$(X)$ *and for all* $y \in$ p-min$(Y)$, $x \leq y$.

*Proof.* The implications 1) $\Rightarrow$ 2) and 3) $\Rightarrow$ 1) are straightforward. Let us prove, 2) $\Rightarrow$ 3). For this, consider any $x \in$ p-min$(X)$ and $y \in$ p-min$(Y)$. Using the hypothesis and $x \in$ p-min$(X)$, we have that, there exists $y_1 \in$ p-min$(Y)$ such that $x \leq y_1$. Since $y_1 \in$ p-min$(Y)$, we have that $y_1 \leq y$ for all $y \in Y$. Therefore, $x \leq y$ for all $x \in$ p-min$(X)$ and $y \in$ p-min$(Y)$. $\qquad\square$

We finish this preliminary section by stating the characterization theorem of existence of a suitable *partial ordering* on $B$ so that a right adjoint exists. The core of this work is to develop a generalized version of the theorem below:

**Theorem 3 ( [18]).** *Given a poset* $(A, \leq_A)$ *and a map* $f \colon A \to B$, *let* $\equiv_f$ *be the kernel relation. Then, there exists an ordering* $\leq_B$ *in* $B$ *and a map* $g \colon B \to A$ *such that* $(f, g) \colon A \leftrightharpoons B$ *if and only if*

1. *There exists* max$([a])$ *for all* $a \in A$.
2. *For all* $a_1, a_2 \in A$, $a_1 \leq_A a_2$ *implies* max$([a_1]) \leq_A$ max$([a_2])$.

Roughly speaking, the proof of the previous theorem is done by using the canonical decomposition theorem via the quotient set $A_f$ wrt the kernel relation, and building right adjoints to any of the arrows in the path.

$$g = \max \circ \varphi^{-1} \circ j_m$$

$$
\begin{array}{ccc}
A & \xrightarrow{\;\;f\;\;} & B \\
{\scriptstyle \max} \uparrow \downarrow {\scriptstyle \pi} & & {\scriptstyle i} \uparrow \downarrow {\scriptstyle j_m} \\
A_f & \underset{\varphi^{-1}}{\overset{\varphi}{\rightleftarrows}} & f(A)
\end{array}
$$

The tricky part of the proof was to extend the ordering on $f(A)$ to the whole set $B$ so that it is still compatible with the existence of right adjoint $j_m$, obviously when $f$ is not surjective. The underlying idea here is related to the definition of an order-embedding of the image into the codomain set; more generally, the idea is to extend a partial ordering defined just on a subset of a set to the whole set.

**Definition 6.** *Given a subset* $X \subseteq B$, *and a fixed element* $m \in X$, *any pre-ordering* $\leq_X$ *in* $X$ *can be extended to a preordering* $\leq_m$ *on* $B$, *defined as the reflexive and transitive closure of the relation* $\leq_X \cup \{(m, y) \mid y \notin X\}$.

Note that the relation above can be described as, for all $x, y \in B$, $x \leq_m y$ if and only if some of the following holds:

(a) $x, y \in X$ and $x \leq_X y$
(b) $x \in X, y \notin X$ and $x \leq_X m$
(c) $x, y \notin X$ and $x = y$

## 3    Building Adjunctions between Preordered Sets

Given a mapping $f \colon \mathbb{A} \to B$ from a preordered set $\mathbb{A} = (A, \leq)$ to an unstructured set $B$, our first goal is to find sufficient conditions to define a suitable preordering on $B$ such that a right adjoint exists. Notice that there is much more than a mere adaptation of the result for posets.

**Lemma 2.** *Let $\mathbb{A} = (A, \leq_A)$ be a preordered set and $f \colon \mathbb{A} \to B$ a surjective map. Let $S \subseteq \bigcup_{a \in A} \text{p-max}[a]_{\cong_A}$ such that the following conditions hold:*

- *$\text{p-min}(UB[a]_{\cong_A} \cap S) \neq \varnothing$, for all $a \in A$.*
- *If $a_1 \leq_A a_2$, then $\text{p-min}(UB[a_1]_{\cong_A} \cap S) \sqsubseteq \text{p-min}(UB[a_2]_{\cong_A} \cap S)$.*

*Then, there exists a preorder $\leq_B$ in $B$ and a map $g$ such that $(f, g) \colon \mathbb{A} \leftrightharpoons \mathbb{B}$.*

*Proof.* The definition of the preorder $\leq_B$ in $B$, given $b_1, b_2 \in B$, is as follows:
$b_1 \leq_B b_2$ if and only if there exist $a_1 \in f^{-1}(b_1)$ and $a_2 \in f^{-1}(b_2)$ such that

$$\text{p-min}(UB[a_1]_{\cong_A} \cap S) \sqsubseteq \text{p-min}(UB[a_2]_{\cong_A} \cap S).$$

Let us prove that it is a preordering:

**Reflexivity:** By the first hypothesis, we have that $\text{p-min}(UB[a]_{\cong_A} \cap S) \neq \varnothing$.
Now, trivially, $\text{p-min}(UB[a]_{\cong_A} \cap S) \sqsubseteq \text{p-min}(UB[a]_{\cong_A} \cap S)$ holds for any $a \in f^{-1}(b)$. Therefore, $b \leq_B b$ for any $b \in B$.
**Transitivity:** Assume $b_1 \leq_B b_2$ and $b_2 \leq_B b_3$.
From $b_1 \leq_B b_2$, there exist $a_i \in f^{-1}(b_i)$, and $c_i \in \text{p-min}(UB[a_i]_{\cong_A} \cap S)$ for $i = 1, 2$ such that $c_1 \leq_A c_2$.
From $b_2 \leq_B b_3$, there exist $a_i' \in f^{-1}(b_i)$, and $c_i' \in \text{p-min}(UB[a_i']_{\cong_A} \cap S)$ for $i = 2, 3$ such that $c_2' \leq_A c_3'$.
As $a_2, a_2' \in f^{-1}(b_2)$, we have that $[a_2]_{\cong_A} = [a_2']_{\cong_A}$, which implies that $c_2 \approx c_2'$. Therefore, $c_1 \leq_A c_2 \approx_A c_2' \leq_A c_3'$ and, as a result, $b_1 \leq_B b_3$.

In order to define $g \colon B \to A$, firstly notice that, as $f$ is onto, given $b \in B$ there exists $x_b \in A$ with $f(x_b) = b$. By hypothesis, $\text{p-min}(UB[x_b]_{\cong_A} \cap S) \neq \varnothing$ for all $b \in B$ and, therefore, there exists a choice function. Any of these functions can be used to define $g$, in such a manner that $g(b) \in \text{p-min}(UB[x_b]_{\cong_A} \cap S)$.
To finish the proof, we have just to check that $(f, g) \colon (A, \leq_A) \leftrightharpoons (B, \leq_B)$.
Assume $f(a) \leq_B b$, then there exist $a_1 \in f^{-1}(f(a))$, $a_2 \in f^{-1}(b)$, $c_1 \in \text{p-min}(UB[a_1]_{\cong_A} \cap S)$ and $c_2 \in \text{p-min}(UB[a_2]_{\cong_A} \cap S)$ with $c_1 \leq_A c_2$; as $[a_1]_{\cong_A} = [a]_{\cong_A}$, and $c_1 \in UB[a_1]$, we also have $a \leq_A c_1$. By definition, we have that $g(b) \in$

p-min$(UB[x]_{\cong_A} \cap S)$ for $x \in f^{-1}(b)$, then $[a_2]_{\cong_A} = [x]_{\cong_A}$, and p-min$(UB[a_2]_{\cong_A} \cap S) =$ p-min$(UB[x]_{\cong_A} \cap S)$. Thus, $c_2 \approx g(b)$ and, as $a \leq_A c_1 \leq_A c_2$, then $a \leq_A g(b)$.

Assuming now that $a \leq_A g(b)$, let us prove $f(a) \leq_B b$. For this, consider $a \in f^{-1}(f(a))$ and $x \in f^{-1}(b)$ where $x$ is the element in $f^{-1}(b)$ used in the definition of $g(b)$, and let us prove that p-min$(UB[a]_{\cong_A} \cap S) \sqsubseteq$ p-min$(UB[x]_{\cong_A} \cap S)$. For this, it is enough to see that for all $z \in$ p-min$(UB[a]_{\cong_A} \cap S)$ the inequality $z \leq_A g(b)$ holds, since obviously $g(b) \in$ p-min$(UB[x]_{\cong_A} \cap S)$.

Fixed $z \in$ p-min$(UB[a]_{\cong_A} \cap S)$, firstly consider that from $g(b) \in S$, using the hypothesis on $S$, we have that $g(b) \in$ p-max$[g(b)]_{\cong_A}$, which means that $g(b) \in UB[g(b)]_{\cong_A}$ as well; that is, $g(b) \in (UB[g(b)]_{\cong_A} \cap S)$. On the other hand, from $a \leq_A g(b)$ and the second hypothesis we have p-min$(UB[a]_{\cong_A} \cap S) \sqsubseteq$ p-min$(UB[g(b)]_{\cong_A} \cap S)$. By Lemma 1, we have that $z \leq_A t$ for all $t \in$ p-min$(UB[g(b)]_{\cong_A} \cap S)$. Since, obviously $t \leq_A g(b)$, we obtain $z \leq_A g(b)$.  $\square$

The following lemma gets rid of the condition of $f$ being surjective, and will be used in the proof of the main theorem of this work, stated as Theorem 4.

**Lemma 3.** *Consider $(A, \leq_A)$ a preordered set, $B$ a set, and $f \colon A \to B$. Then, there exist both a preorder $\leq_B$ and an adjunction $(f, g) \colon (A, \leq_A) \leftrightharpoons (B, \leq_B)$ if and only if there exist a preorder $\leq_{f(A)}$ and an adjunction $(f, g') \colon (A, \leq_A) \leftrightharpoons (f(A), \leq_{f(A)})$.*

*Proof.* The direct implication is trivial, by considering $\leq_{f(A)}$ and $g'$ as the restrictions to $f(A)$ of $\leq_B$ and $g$, respectively.

Conversely, consider the adjunction $(f, g') \colon (A, \leq_A) \leftrightharpoons (f(A), \leq_{f(A)})$, fix $m \in f(A)$, and choose $\leq_B$ to be its associated preorder, as introduced in Definition 6. It is just a matter of straightforward computation to check that we have an adjunction $(f, g) \colon (A, \leq_A) \leftrightharpoons (B, \leq_B)$ where $g$ is the extension of $g'$ defined as follows:

$$g(x) = \begin{cases} g'(x) & \text{if } x \in f(A) \\ g'(m) & \text{if } x \notin f(A) \end{cases}$$

$\square$

The corresponding version of Theorem 3 is a twofold extension of the statement of Lemma 2 in that, firstly, the mapping $f$ need not be onto and, secondly, gives a necessary and sufficient condition for the existence of adjunction.

**Theorem 4.** *Given any preordered set $\mathbb{A} = (A, \leq_A)$ and a mapping $f \colon \mathbb{A} \to B$, there exists a preorder $\mathbb{B} = (B, \leq_B)$ and $g \colon B \to A$ such that $(f, g) : \mathbb{A} \leftrightharpoons \mathbb{B}$ if and only if there exists $S \subseteq \bigcup\limits_{a \in A}$ p-max$[a]_{\cong_A}$ such that*

1. *p-min$(UB[a]_{\cong_A} \cap S) \neq \varnothing$, for all $a \in A$.*
2. *If $a_1 \leq_A a_2$, then p-min$(UB[a_1]_{\cong_A} \cap S) \sqsubseteq$ p-min$(UB[a_2]_{\cong_A} \cap S)$.*

*Proof.* Assume the existence of the preordering in $B$ and the mapping $g$ such that $(f, g) : \mathbb{A} \leftrightarrows \mathbb{B}$, and let us prove the three properties in the statement.

Define $S = g(f(A))$, consider $g(f(a)) \in S$, and let us show that $g(f(a)) \in$ p-max$[g(f(a))]_{\cong_A}$. Consider $x \in [g(f(a))]_{\cong_A}$, by a straightforward induction argument we obtain $f(x) \approx_B f(g(f(a)))$; now, using $f(g(f(a))) \approx_B f(a)$ we have $f(x) \approx_B f(a)$. Since $f(x) \leq_B f(a)$, by using the adjunction, we obtain $x \leq_A g(f(a))$, hence $g(f(a)) \in$ p-max$[g(f(a))]_{\cong_A}$.

For property 1, we will check that $g(f(a)) \in$ p-min$(UB[a]_{\cong_A} \cap S)$. To begin with, by definition $g(f(a)) \in S$; then, we will prove that $g(f(a)) \in UB[a]_{\cong_A}$. Given $x \in [a]_{\cong_A}$ we have to prove $x \leq_A g(f(a))$; the argument follows by induction on the length of the chain connecting $x$ and $a$

- For $n = 0$, we have $a \leq_A g(f(a))$ by properties of adjunction.
- Assume the result is true for any chain of length $n$, and consider $a \cong_A a_2 \cong_A \ldots a_n \cong_A x$, then, by induction hypothesis, $a_n \leq_A g(f(a))$. Now, as $a_n \cong_A x$, there are two possibilities:
  - $a_n \approx_A x$ and, trivially $x \leq_A g(f(a))$.
  - $f(a_n) = f(x)$, using the properties of adjunction twice we firstly obtain $f(x) \leq_A f(g(f(a)))$ and, then, $x \leq_A g(f(g(f(a)))) \approx_A g(f(a))$.

We have just proved that $g(f(a)) \in UB[a]_{\cong_A} \cap S$, the remaining point is to prove that it is a p-minimum element. Consider $x \in UB[a]_{\cong_A} \cap S$; then $z \leq_A x$ for all $z \in [a]_{\cong_A}$ and, by definition of $S$, $x = g(f(a_1))$. Particularly, for $z = a$ we have that, $a \leq_A g(f(a_1))$, by properties of adjunction, $g(f(a)) \leq_A g(f(g(f(a_1)))) \approx_A g(f(a_1)) = x$, i.e. $g(f(a)) \leq_A x$.

For Property 2, assume $a_1 \leq_A a_2$, by adjunction, $f$ and $g$ are isotone maps, then $g(f(a_1)) \leq_A g(f(a_2))$. From this, we directly obtain p-min$(UB[a_1]_{\cong_A} \cap S) \sqsubseteq$ p-min$(UB[a_2]_{\cong_A} \cap S)$ since we just proved above that for all $a \in A$ $g(f(a)) \in$ p-min$(UB[a]_{\cong_A} \cap S)$.

Conversely, if we assume properties 1 and 2, then by Lemma 2 and Lemma 3, there exist a preorder $\mathbb{B} = (B, \leq_B)$ and a map $g$ such that $(f, g) : \mathbb{A} \leftrightarrows \mathbb{B}$. $\square$

## 4    On the Uniqueness of Right Adjoints and the Inherited Ordered Structure in the Codomains

The unicity of the right adjoint between posets is well-known. Specifically, given two posets $\mathbb{A} = (A, \leq_A)$ and $\mathbb{B} = (B, \leq_B)$ and a mapping $f \colon A \to B$, if there exists $g \colon B \to A$ such that the pair $(f, g)$ is an adjunction, then it is unique.

This behavior was further analyzed in [18], where the uniqueness property was extended, in the case of surjective mappings, not only to the right adjoint, but also to the ordering relation in the codomain: namely, there exists just one partial ordering on the codomain $B$ such that a right adjoint exists, that is, given a surjective mapping $f$ from a poset $\mathbb{A}$ to an unstructured set $B$, we introduced necessary and sufficient conditions to ensure the existence of an ordering $\leq_B$ in $B$ and a mapping $g\colon B \to A$ such that $(f, g)$ is an adjunction. Moreover, both $\leq_B$ and $g$ are uniquely determined by $\leq_A$ and $f$.

Contrariwise to the partially ordered case, given two preordered sets $\mathbb{A} = (A, \leq_A)$ and $\mathbb{B} = (B, \leq_B)$ and a mapping $f\colon A \to B$, the unicity of the mapping $g\colon B \to A$ satisfying $(f, g)\colon \mathbb{A} \leftrightarrows \mathbb{B}$, when it exists, cannot be guaranteed. However, it is well known that if $g_1$ and $g_2$ are right adjoints, then $g_1(b) \approx_A g_2(b)$ for all $b \in B$, and one usually says that the right adjoint is *essentially unique*. This scenario is much more similar to what occurs in category theory: if one functor $F$ has two right adjoints $G$ and $G'$, then $G$ and $G'$ are naturally isomorphic.

However, and this is the interesting part, the unicity cannot be extended to the case in which the codomain is unstructured. In this section we introduce several examples supporting this statement.

**Examples.** Let $A = \{a, b, c, d\}$, $B = \{o, p, q\}$ be two sets and $f\colon A \to B$ defined as $f(a) = f(c) = p$, $f(b) = o$ and $f(d) = q$. Consider $(A, \leq_A)$ ordered by $a \leq_A b \leq_A c \leq_A d$. We have $[a]_{\cong_A} = [c]_{\cong_A} = \{a, c\}$, $[b]_{\cong_A} = \{b\}$ and $[d]_{\cong_A} = \{d\}$ and $\bigcup_{x \in A} \text{p-max}[x]_{\cong_A} = \{b, c, d\}$.

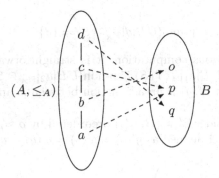

Notice that $f$ is surjective, and does not fulfill the conditions in Theorem 3, specifically the second one. Thus, there does not exist any *partial ordering* relation in $B$ for which some $g\colon B \to A$ would be a right adjoint to $f$. Notice, however, that if we relax the requirement to be an adjunction *between preordered* sets, then there exist a preordering (actually more than one) which generates a right adjoint to $f$. Some examples are worked out below to illustrate the previous situation.

*Example 1.* Consider $\mathbb{B} = (B, \leq_B)$ preordered with $o \approx_B p$, $o \leq_B q$ and $p \leq_B q$, and the mapping $g\colon B \to A$ defined as $g(o) = g(p) = c$ and $g(q) = d$.

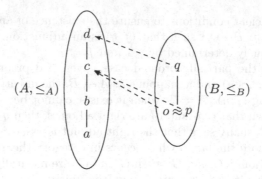

To begin with, we have that $S = gf(A) = \{c, d\}$ is a subset of $\bigcup\limits_{x \in A}$ p-max$[x]_{\cong_A}$
and, then, check the two conditions in Theorem 4.

It is not difficult to check that p-min$(UB[x]_{\cong_A} \cap S) \neq \varnothing$ for all $x \in A$. Specifically, we have

$$\text{p-min}(UB[a]_{\cong_A} \cap S) = \text{p-min}(UB[b]_{\cong_A} \cap S) = \text{p-min}(UB[c]_{\cong_A} \cap S) = \{c, d\}$$

and

$$\text{p-min}(UB[d]_{\cong_A} \cap S) = \{d\}$$

Finally, with the previous computation, it is straightforward to check that if $a_1 \leq_A a_2$ then p-min$(UB[a_1]_{\cong_A} \cap S) \sqsubseteq$ p-min$(UB[a_2]_{\cong_A} \cap S)$.

As a result, the pair $(f, g)$ is an adjunction between $\mathbb{A}$ and $\mathbb{B}$.    □

*Example 2.* Now, consider $\mathbb{B}' = (B, \leq'_B)$ preordered by $o \approx'_B p$ and $p \approx'_B q$, and the mapping $g' \colon B \to A$ defined as $g'(o) = g'(p) = g'(q) = d$.

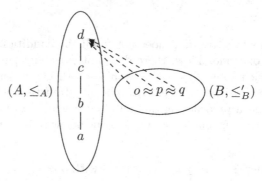

Again we will check the conditions in Theorem 4.

In this case, $S = g'f(A) = \{d\}$ which is a subset of $\bigcup_{x \in A}$ p-max$[x]_{\cong_A} = \{b, c, d\}$. The first condition holds since p-min$(UB[a]_{\cong_A} \cap S) = $ p-min$(UB[b]_{\cong_A} \cap S) = $ p-min$(UB[c]_{\cong_A} \cap S) = $ p-min$(UB[d]_{\cong_A} \cap S) = \{d\}$. As all the previous sets coincide, the second condition follows trivially.

As a result, the pair $(f, g')$ is an adjunction between the preorders $\mathbb{A}$ and $\mathbb{B}'$. □

# 5    Conclusions

Given a mapping $f\colon A \to B$ from a preordered set $A$ into an unstructured set $B$, we have obtained necessary and sufficient conditions which allow us for defining a suitable preordering relation on $B$ such that there exists mapping $g\colon B \to A$ such that the pair of mappings $(f, g)$ forms an adjunction between preordered sets.

Whereas the results in the partially ordered case followed more or less the intuition of what should be expected (Theorem 3), the description of the conditions on the preordered case is much more involved (Theorem 4). A first piece of future work should be to consider alternative approaches to this problem in order to obtain, if possible, a simpler alternative characterization.

Concerning potential applications of the present work, let us recall that the Galois connections used in FCA are given between the Boole algebras of the powersets of objects and the powerset of attributes. There exist several generalizations in FCA which weaken the structure on which a Galois connection is defined: for instance, in fuzzy FCA the residuated structure of the powerset of fuzzy sets is used. In [16], a general approach called *pattern structures* was proposed, which allows for extending FCA techniques to arbitrary partially ordered data descriptions. Using pattern structures, one can compute taxonomies, ontologies, implications, implication bases, association rules, concept-based (or JSM-) hypotheses in the same way it is done with standard concept lattices [26].

In this generalization, instead of associating each object with the set of attributes it satisfies, a pattern is given, which can be either a graph, or a sequence or an interval, and the semantics of these patterns can be different in each case. For instance, [15] represents scenarios of conflict between human agents, or [22] use gene expression data. These sets of patterns are provided with a partial ordering relation such as *"being a subgraph of"* or *"being a subchain of"*.

The results obtained in this work are aimed at not only extending these results to sets in which there is a preordering previously defined but, more specifically, to the problem of knowledge discovery on the existing structure between the patterns. The scenario in which this work could be applied is as follows: we start from a set of objects each one related to the set of patterns it satisfies, ignoring whether there exists some (pre-)ordering relation between patterns, but assuming that the semantics of the problem guarantees the existence of a Galois connection between them, the goal would be to obtain as much information as possible about the relation existing in the set of patterns.

To finish with the future work, it is remarkable the number of papers on fuzzy Galois connections have been written since its introduction in [2]; consider for instance [4, 14, 19, 25] for some recent ones. As future work in the short term, we would like to extend the results in this work to the fuzzy case, for instance to the framework of fuzzy posets and fuzzy preorders, and study the potential relationship to other approaches based on generalized structures.

# References

1. Antoni, L., Krajči, S., Krídlo, O., Macek, B., Pisková, L.: On heterogeneous formal contexts. Fuzzy Sets and Systems 234, 22–33 (2014)
2. Bělohlávek, R.: Fuzzy Galois connections. Mathematical Logic Quarterly 45(4), 497–504 (1999)
3. Bělohlávek, R., Konečný, J.: Concept lattices of isotone vs. antitone Galois connections in graded setting: Mutual reducibility revisited. Information Sciences 199, 133–137 (2012)
4. Bělohlávek, R., Osička, P.: Triadic fuzzy Galois connections as ordinary connections. In: IEEE Intl Conf. on Fuzzy Systems (2012)
5. Blyth, T.S.: Lattices and Ordered Algebraic Structures. Springer (2005)
6. Butka, P., Pócs, J., Pócsová, J.: On equivalence of conceptual scaling and generalized one-sided concept lattices. Information Sciences 259, 57–70 (2014)
7. Castellini, G., Koslowski, J., Strecker, G.: Closure operators and polarities. Annals of the New York Academy of Sciences 704, 38–52 (1993)
8. Cohen, D.A., Creed, P., Jeavons, P.G., Živný, S.: An algebraic theory of complexity for valued constraints: Establishing a Galois connection. In: Murlak, F., Sankowski, P. (eds.) MFCS 2011. LNCS, vol. 6907, pp. 231–242. Springer, Heidelberg (2011)
9. Denecke, K., Erné, M., Wismath, S.L.: Galois connections and applications, vol. 565. Springer (2004)
10. Díaz, J.C., Medina, J.: Multi-adjoint relation equations: Definition, properties and solutions using concept lattices. Information Sciences 253, 100–109 (2013)
11. Dubois, D., Prade, H.: Possibility theory and formal concept analysis: Characterizing independent sub-contexts. Fuzzy Sets and Systems 196, 4–16 (2012)
12. Dzik, W., Järvinen, J., Kondo, M.: Intuitionistic propositional logic with Galois connections. Logic Journal of the IGPL 18(6), 837–858 (2010)
13. Dzik, W., Järvinen, J., Kondo, M.: Representing expansions of bounded distributive lattices with Galois connections in terms of rough sets. International Journal of Approximate Reasoning 55(1), 427–435 (2014)
14. Frascella, A.: Fuzzy Galois connections under weak conditions. Fuzzy Sets and Systems 172(1), 33–50 (2011)
15. Galitsky, B.A., Kovalerchuk, B., Kuznetsov, S.O.: Learning Common Outcomes of Communicative Actions Represented by Labeled Graphs. In: Priss, U., Polovina, S., Hill, R. (eds.) ICCS 2007. LNCS (LNAI), vol. 4604, pp. 387–400. Springer, Heidelberg (2007)
16. Ganter, B., Kuznetsov, S.O.: Pattern Structures and Their Projections. In: Delugach, H.S., Stumme, G. (eds.) ICCS 2001. LNCS (LNAI), vol. 2120, pp. 129–142. Springer, Heidelberg (2001)
17. García-Pardo, F., Cabrera, I.P., Cordero, P., Ojeda-Aciego, M.: On Galois Connections and Soft Computing. In: Rojas, I., Joya, G., Cabestany, J. (eds.) IWANN 2013, Part II. LNCS, vol. 7903, pp. 224–235. Springer, Heidelberg (2013)
18. García-Pardo, F., Cabrera, I.P., Cordero, P., Ojeda-Aciego, M., Rodríguez, F.J.: Generating isotone Galois connections on an unstructured codomain. In: Proc. of Information Processing and Management of Uncertainty in Knowledge-based Systems, IPMU (to appear, 2014)
19. Guo, L., Zhang, G.-Q., Li, Q.: Fuzzy closure systems on $L$-ordered sets. Mathematical Logic Quarterly 57(3), 281–291 (2011)
20. Järvinen, J.: Pawlak's information systems in terms of Galois connections and functional dependencies. Fundamenta Informaticae 75, 315–330 (2007)

21. Järvinen, J., Kondo, M., Kortelainen, J.: Logics from Galois connections. Int. J. Approx. Reasoning 49(3), 595–606 (2008)
22. Kaytoue, M., Kuznetsov, S.O., Napoli, A., Duplessis, S.: Mining gene expression data with pattern structures in formal concept analysis. Information Sciences 181(10), 1989–2001 (2011)
23. Kan, D.M.: Adjoint functors. Transactions of the American Mathematical Society 87(2), 294–329 (1958)
24. Kerkhoff, S.: A general Galois theory for operations and relations in arbitrary categories. Algebra Universalis 68(3), 325–352 (2012)
25. Konecny, J.: Isotone fuzzy Galois connections with hedges. Information Sciences 181(10), 1804–1817 (2011)
26. Kuznetsov, S.O.: Fitting Pattern Structures to Knowledge Discovery in Big Data. In: Cellier, P., Distel, F., Ganter, B. (eds.) ICFCA 2013. LNCS, vol. 7880, pp. 254–266. Springer, Heidelberg (2013)
27. Medina, J.: Multi-adjoint property-oriented and object-oriented concept lattices. Information Sciences 190, 95–106 (2012)
28. Melton, A., Schmidt, D.A., Strecker, G.E.: Galois connections and computer science applications. In: Poigné, A., Pitt, D.H., Rydeheard, D.E., Abramsky, S. (eds.) Category Theory and Computer Programming. LNCS, vol. 240, pp. 299–312. Springer, Heidelberg (1986)
29. Mu, S.-C., Oliveira, J.N.: Programming from Galois connections. The Journal of Logic and Algebraic Programming 81(6), 680–704 (2012)
30. Ore, Ø.: Galois connections. Transactions of the American Mathematical Society 55, 493–513 (1944)
31. Poelmans, J., Ignatov, D.I., Kuznetsov, S.O., Dedene, G.: Fuzzy and rough formal concept analysis: a survey. Intl Journal of General Systems 43(2), 105–134 (2014)
32. Restrepo, M., Cornelis, C., Gómez, J.: Duality, conjugacy and adjointness of approximation operators in covering-based rough sets. Intl. Journal of Approximate Reasoning 55(1), 469–485 (2014)
33. Valverde-Albacete, F.J., Peláez-Moreno, C.: Extending conceptualisation modes for generalised formal concept analysis. Information Sciences 181, 1888–1909 (2011)
34. Wolski, M.: Galois connections and data analysis. Fundamenta Informaticae 60, 401–415 (2004)

# Directed Tree Decompositions

Sebastian Kerkhoff and Friedrich Martin Schneider

Technische Universität Dresden, 01062 Dresden, Germany
{Sebastian.kerkhoff,martin.schneider}@tu-dresden.de

**Abstract.** In the problem session of the ICFCA 2006, Sándor Radeleczki asked for the meaning of the smallest integer $k$ such that a given poset can be decomposed as the union of $k$ directed trees. The problem also asks for the connection of this number to the order dimension. Since it was left open what kind of decomposition might be used, there is more than one reading of this problem. In the paper, we discuss different versions and give some answers to this open problem.

**Keywords:** directed tree, poset decomposition, order dimension.

## 1  Introduction

Motivated by research surrounding CD-bases [CHS09, HR12], Sándor Radeleczki posed the following open problem at the ICFCA 2006 in Dresden[1]:

> A finite poset $(P, \leq)$ is a directed tree-order if it has a greatest element and for all $x \in P$, $\uparrow x$ is a chain. Each finite poset can be decomposed as unions of tree-orders. Let $k$ be the minimal number. What is the mean[ing] of $k$? What is the relation to the order dimension of $(P, \leq)$?

The term "decomposed as unions" may be interpreted in different ways. For instance, we may or may not require that the order relation of each of the directed tree-orders (we will simply call them *directed trees* from now on) is the restriction of $\leq$ to the carrier set of the tree (instead of just having to be a subset of $\leq$). Also, it is not specified whether the transitive closure of the union of the trees' order relations has to be the entire relation $\leq$ or just a subset. This leaves us with (at least) four versions of Radeleczki's problem. We decided to write this paper about all four of them. That is, we study the numbers defined in the next four definitions, discuss their interconnections and their relation to the order dimension.

**Definition 1.** *For a poset $\mathbb{P} = (P, \leq)$, denote by $\kappa_1(\mathbb{P})$ the smallest integer $k$ such that there exist $k$ directed trees $(T_1, \leq_1), \ldots, (T_k, \leq_k)$ with $\bigcup_{i=1}^{k} T_i = P$ and $\leq_i$ being contained in $\leq$ for all $i \in \{1, \ldots, k\}$.*

---

[1] http://www.upriss.org.uk/fca/problems06.pdf

C.V. Glodeanu, M. Kaytoue, and C. Sacarea (Eds.): ICFCA 2014, LNAI 8478, pp. 80–95, 2014.

**Definition 2.** *For a poset* $\mathbb{P} = (P, \leq)$*, denote by* $\kappa_2(\mathbb{P})$ *the smallest integer* $k$ *such that there exist* $k$ *directed trees* $(T_1, \leq_1), \ldots, (T_k, \leq_k)$ *such that* $\leq$ *is the transitive closure of* $\bigcup_{i=1}^{k} \leq_i$.

Note that the condition in Definition 2 implies $\bigcup_{i=1}^{l} T_i = P$, so we have $\kappa_2(\mathbb{P}) \geq \kappa_1(\mathbb{P})$. Before we continue, let us clarify that we call a poset $\mathbb{Q} = (Q, \leq_{\mathbb{Q}})$ a *sub-poset* of $\mathbb{P} = (P, \leq_{\mathbb{P}})$ if $\mathbb{Q}$ is a relational substructure of $\mathbb{P}$, that is, we have $Q \subseteq P$, and $\leq_{\mathbb{Q}} = \leq_{\mathbb{P}} \cap Q^2$. In this case, we also say that $Q$ *induces* $\mathbb{Q}$.

**Definition 3.** *For a poset* $\mathbb{P} = (P, \leq)$*, denote by* $\kappa_3(\mathbb{P})$ *the smallest integer* $k$ *such that there exist* $k$ *directed trees* $(T_1, \leq_1), \ldots, (T_k, \leq_k)$ *with* $\bigcup_{i=1}^{l} T_i = P$ *and* $(T_i, \leq_i)$ *being a sub-poset of* $\mathbb{P}$ *for all* $i \in \{1, \ldots, k\}$.

**Definition 4.** *For a poset* $\mathbb{P} = (P, \leq)$*, denote by* $\kappa_4(\mathbb{P})$ *the smallest integer* $k$ *such that there exist* $k$ *directed trees* $(T_1, \leq_1), \ldots, (T_k, \leq_k)$ *with* $(T_i, \leq_i)$ *being a sub-poset of* $\mathbb{P}$ *for all* $i \in \{1, \ldots, k\}$ *and* $\leq$ *being the transitive closure of* $\bigcup_{i=1}^{k} \leq_i$.

The paper is structured as follows. After the preliminaries and some basic examples for the purpose of illustration, the heart of the paper are Sections 4–7, where we will discuss each version of the problem. That is, for $i \in \{1, \ldots, 4\}$, Section i+3 will discuss the meaning of $\kappa_i(\mathbb{P})$, elaborate on some of its properties, describe its connection to the other three numbers, and study its relation to the order dimension. The final section summarizes the results.

## 2    Preliminaries

In this section, we set some notation and introduce the order dimension in the amount of detail that we need for the results of our study. Throughout the paper, we denote by $\mathbb{N}$ the set of natural numbers not including zero, that is, $\mathbb{N} = \{1, 2, \ldots\}$. Moreover, all posets in this paper are assumed to be finite, and if we say that a poset is bounded, then we mean that it has a greatest and a least element.

**Definition 5.** *The order dimension of a poset* $\mathbb{P} = (P, \leq)$*, denoted by* $\dim \mathbb{P}$*, is the least integer* $t$ *for which there exists a family of linear orders* $<_1, \ldots, <_t$ *on* $P$ *such that* $\leq$ *is the intersection* $\bigcap_{i=1}^{t} <_i$.

The choice of the term "dimension" is justified, because we have $\dim \mathbb{P} = t$ if and only if $t$ is the least integer such that $\mathbb{P}$ can be embedded into the direct product of $t$ chains. In particular, a directed tree has order dimension at most 2. For a detailed study of the order dimension, we refer to the monograph [Tro92] and the references therein.

*Example 6.* For $k \in \mathbb{N}$, consider the $2k$-element poset $\mathbb{H}_k = (H_k, \leq_{\mathbb{H}_k})$ defined by setting $H_k := \{a_1, \ldots, a_k, b_1, \ldots, b_k\}$ and defining $\leq_{\mathbb{H}_k}$ to be the reflexive closure of $\{(a_i, b_j) \mid i \neq j\}$.

It is the standard example of the poset with order dimension $k$.

The example shows that $\dim \mathbb{P}$ is not bounded by the height of $\mathbb{P}$, which we denote by height $\mathbb{P}$ and define as the number of points in a maximum chain. It is, however, known to be bounded by the width of $\mathbb{P}$, which we denote by width $\mathbb{P}$ and define as the number of points in a maximum antichain.

**Definition 7.** *Let $\mathbb{P} = (P, \leq)$ be a poset and $Q \subseteq P$. We define the poset $\mathbb{P} - Q$ as the sub-poset of $\mathbb{P}$ that is induced by $P \setminus Q$.*

It is well-known that the order dimension is in a reasonable sense continuous. For instance, it was shown in [Hir55] that removing a single point from an at least two-element poset decreases the order dimension by at most 1. The following theorem can also be found in [Hir55] and is in a similar spirit:

**Theorem 8.** *Let $\mathbb{P} = (P, \leq)$ be a poset and let $C \subseteq P$ be a chain with $P \setminus C \neq \emptyset$. Then $\dim \mathbb{P} \leq 2 + \dim (\mathbb{P} - C)$.*

Note that an analogon of this theorem cannot hold for any of the numbers $\kappa_1(\mathbb{P}), \dots, \kappa_4(\mathbb{P})$. For instance if $\mathbb{P}$ is the poset obtained by adding a greatest element $1_\mathbb{P}$ to a $k$-element antichain, then we have $\kappa_i(\mathbb{P}) = 1$, but $\kappa_i(\mathbb{P} - \{1_\mathbb{P}\}) = k$ for all $i \in \{1, \dots, 4\}$.

**Definition 9.** *Let $\mathbb{P}$ and $\mathbb{Q}$ be posets with disjoint carrier sets such that $\mathbb{P}$ has a greatest element $1_\mathbb{P}$ and $\mathbb{Q}$ has a least element $0_\mathbb{Q}$. We denote by $\mathbb{P} \oplus \mathbb{Q}$ the poset on $P \cup Q$ whose order relation is the transitive closure of $\leq_\mathbb{P} \cup \leq_\mathbb{Q} \cup \{(1_\mathbb{P}, 0_\mathbb{Q})\}$.*

Observe that we have $\dim (\mathbb{P} \oplus \mathbb{Q}) = \max\{\dim \mathbb{P}, \dim \mathbb{Q}\}$. The following diagram illustrates the construction of $\mathbb{P} \oplus \mathbb{Q}$:

**Definition 10.** *For an order relation $\leq$, we define the relation $\lessdot$ by setting $x \lessdot y$ if and only if $y$ is an upper neighbour of $x$ with respect to $\leq$.*

## 3   Examples

*Example 11.* For $k \geq 2$, let $\mathbb{M}_k = (M_k, \leq_{\mathbb{M}_k})$ denote the $(k + 2)$-element poset with $M_k := \{0, a_1, \dots, a_k, 1\}$ and $\leq_{\mathbb{M}_k}$ being the reflexive closure of the binary relation $(\{0\} \times M_k) \cup (M_k \times \{1\})$. That is, $(M_k, \leq)$ is the poset given by the following diagram:

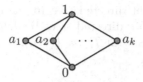

We have

- $\kappa_1(\mathbb{M}_k) = 1$, where the single directed tree that yields the decomposition may for instance be given by $(T_1, \leq_1)$ with $T_1 := M_k$ and $\leq_1$ being the reflexive-transitive closure of $\{(0, a_1), (a_1, 1), (a_2, 1), \ldots, (a_k, 1)\}$;
- $\kappa_2(\mathbb{M}_k) = k$, because we can clearly decompose $\mathbb{M}_k$ into $k$ chains but any two of the pairs $(0, a_1), \ldots, (0, a_k)$ have to be in the order relation of a different directed tree;
- $\kappa_3(\mathbb{M}_k) = 2$, where the two directed trees in the decomposition may for instance be given by the two sub-posets induced by $\{a_1, \ldots, a_k, 1\}$ and $\{0\}$;
- $\kappa_4(\mathbb{M}_k) = k$ for the same reason as in the second case.

*Example 12.* For $k \in \mathbb{N}$, consider the $2k$-element poset $\mathbb{L}_k = (L_k, \leq_{L_k})$ defined by setting $L_k := \{1, \ldots, k\} \times \{a, b\}$ and

$$(x_1, x_2) \leq (y_1, y_2) :<=> x_1 \leq y_1 \wedge (x_2 = y_2 \vee y_2 = b).$$

That is, $(L_k, \leq_{L_k})$ is the poset given by the following diagram:

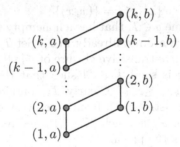

We have

- $\kappa_1(\mathbb{L}_k) = 1$, where the single directed tree that yields the decomposition is given by $(T_1, \leq_1)$ with $T_1 = L_k$ and

$$(x_1, x_2) \leq_1 (y_1, y_2) :\Longleftrightarrow (x_1 \leq y_1 \wedge x_2 = y_2) \vee (y_1, y_2) = (k, b);$$

- $\kappa_2(\mathbb{L}_k) = 2$, where the two directed trees may for instance be given by the chain $(T_1, \leq_1)$ induced by $T_1 := \{1, \ldots, k\} \times \{a\}$ and $(T_2, \leq_2)$ with $T_2 := L_k$ and

$$(x_1, x_2) \leq_2 (y_1, y_2) :\Longleftrightarrow (x_1 \leq y_1 \wedge y_2 = b) \vee (x_1, x_2) = (y_1, y_2);$$

- $\kappa_3(\mathbb{L}_k) = 2$, where the two directed trees in the decomposition may for instance be given by the two chains induced by $T_1 = \{1, \ldots, k\} \times \{a\}$ and $T_2 = \{1, \ldots, k\} \times \{b\}$;

- $\kappa_4(\mathbb{L}_k) = k$, because any two of the $k$ pairs $((1,a),(1,b)),\ldots,((k,a),(k,b))$ must belong to a different directed tree in the decomposition and we can find a decomposition into $k$ directed trees by defining $(T_1,\leq_1),\ldots,(T_k,\leq_k)$ to be the induced chains with $T_i := \{(1,a),\ldots,(i,a),(i,b),\ldots,(k,b)\}$ for $i \in \{1,\ldots,k\}$.

## 4   First Version: $\kappa_1$

First, let us note the obvious fact that we have $\kappa_1(\mathbb{P}) \leq \kappa_2(\mathbb{P}), \kappa_3(\mathbb{P}), \kappa_4(\mathbb{P})$ for each poset $\mathbb{P}$. Moreover, we will see in this section that $\kappa_1(\mathbb{P})$ is not a very interesting quantity. In fact, it is simply the number of maximal elements in $\mathbb{P}$.

**Proposition 13.** *Let $\mathbb{P}$ be a poset. Then $\kappa_1(\mathbb{P})$ is the number of maximal elements in $\mathbb{P}$.*

*Proof.* It is obvious that $\kappa_1(\mathbb{P})$ is at least the number of maximal elements of $\mathbb{P}$. Since $P$ is the union of the downsets of all maximal elements, the other direction follows if we can show that each of these downsets can be covered with a single directed tree. To this end, let $x$ be a maximal element. Doing essentially the same thing as in the standard proof of the basic fact that every maximal (graph theoretic) tree in a connected graph contains all vertices of that graph, we construct a series of directed trees $(T_0,\leq_0),(T_1,\leq_1),(T_2,\leq_2),\ldots$ recursively as follows:

Recursion start: Set $T_0 = \{x\}$, $\leq_0 := \{(x,x)\}$.

Recursive step: Take some $y \in T_i$ that has a nonempty set of those lower neighbours $y_1,\ldots,y_k$ in $\mathbb{P}$ which are not already in $T_i$. Set $T_{i+1} := T_i \cup \{y_1,\ldots,y_k\}$ and define $\leq_i$ as the reflexive-transitive closure of $\leq_i \cup\{(y_1,y),\ldots,(y_k,y)\}$.

Since $P$ is finite, there is some $k \in \mathbb{N}$ such that the recursion stops, that is, we have $(T_k,\leq_k) = (T_{k+1},\leq_{k+1})$. Clearly, $T_k$ contains every element of the downset of $x$ since there exists a path from $x$ to every element in its downset. By construction, it is obvious that each $(T_i,\leq_i)$ is a directed tree and that $\leq_i$ is a subset of $\leq$. This finishes the proof.                    □

Of course, this also means that $\kappa_1(\mathbb{P})$ has virtually no connection to the order dimension of $\mathbb{P}$.

## 5   Second Version: $\kappa_2$

In this section, we discuss the second version of the problem. We give an easy characterization of the number $\kappa_2(\mathbb{P})$ for the case that $\mathbb{P}$ has a greatest element, and we provide a tight interval for the case that $\mathbb{P}$ has multiple maximal elements. Besides showing some other properties of $\kappa_2(\mathbb{P})$ on the way, we also prove that $\kappa_2(\mathbb{P})$ is bounded by the width of $\mathbb{P}$ and that it bears essentially no relation to the order dimension of $\mathbb{P}$.

Let us start with the following easy observation (recall that, for a given order relation $\leq$, we write $x \lessdot y$ if $y$ is an upper neighbour of $x$, see Definition 10):

**Lemma 14.** *Let* $\mathbb{P} = (P, \leq)$ *be a poset. Then*

$$\kappa_2(\mathbb{P}) \geq \max_{x \in P} |\{y \in P \mid x \lessdot y\}|.$$

*In other words, the maximum number of upper neighbours that an element* $x \in P$ *can have is a lower bound on* $\kappa_2(\mathbb{P})$.

*Proof.* Assume that some $x \in P$ has $k$ upper neighbours $y_1, \ldots, y_k$ and that $(T_1, \leq_1), \ldots, (T_l, \leq_l)$ is a directed tree decomposition of $(P, \leq)$ in the sense of $\kappa_2$. For $i \in \{1, \ldots, k\}$, each pair $(x, y_i)$ must be in at least one of the relations $\leq_1, \ldots, \leq_l$. However, for $i \neq j$, the tuples $(x, y_i)$ and $(x, y_j)$ cannot be in the order relation of the same directed tree. Hence, $k \geq l$ and thus $k \geq \kappa_2(\mathbb{P})$. □

**Proposition 15.** *Let* $\mathbb{P} = (P, \leq)$ *be a poset with a greatest element. Then*

$$\kappa_2(\mathbb{P}) = \max_{x \in P} |\{y \in P \mid x \lessdot y\}|$$

*Proof.* We need to show "$\leq$". Set $k := \max_{x \in P} |\{y \in P \mid x \lessdot y\}|$. We apply the construction from Proposition 13 and obtain a directed tree $(P, \leq_1)$ with $\leq_1$ being contained in $\leq$. Let $x_1, \ldots, x_l \in P$ be all elements of $P$ that have more than one upper neighbour with respect to $\leq$. By construction of $\leq_1$, there exists exactly one $y_i \in P$ for each $i \in \{1, \ldots, l\}$ such that $x_i \lessdot_1 y_i$. Define $\preccurlyeq$ to be the reflexive-transitive closure of $\lessdot \setminus \{(x_1, y_1), \ldots, (x_k, y_k)\}$ on $P$. Note that $(P, \preccurlyeq)$ has the same greatest element as $\mathbb{P}$. We repeat the entire procedure, that is, we apply Proposition 13 to $(P, \preccurlyeq)$ to obtain a directed tree $(P, \leq_2)$ with $\leq_2$ being contained in $\preccurlyeq$ and each $x \in P$ having at most $k - 1$ upper neighbours with respect to $\preccurlyeq$. Applying this technique exactly $k$ times gives us a series of directed trees $(P, \leq_1), \ldots, (P, \leq_k)$. By construction, each $(x, y) \in \lessdot$ is contained in at least one of the order relations. Hence, the trees cover $\mathbb{P}$ in the sense of $\kappa_2$. □

If $\mathbb{P}$ contains more than one maximal element, then $\kappa_2(\mathbb{P})$ will also be influenced by how the downsets of the maximal elements intersect. Indeed, in this case, we can easily prove the following Proposition:

**Proposition 16.** *Let* $\mathbb{P}$ *be a poset with precisely* $l$ *maximal elements, and let* $(P_1, \leq_1), \ldots, (P_l, \leq_l)$ *be the downsets induced by these elements. Set*

$$k_i := \max_{x \in P_i} |\{y \in P_i \mid x \lessdot_i y\}|$$

*for all* $i \in \{1, \ldots, l\}$. *Then, we have*

$$\max_{i \in \{1, \ldots, l\}} k_i \leq \kappa_2(\mathbb{P}) \leq \sum_{i=1}^{l} k_i,$$

*where both bounds are tight in the sense that equality occurs for some posets* $\mathbb{P}$ *with* $l$ *maximal elements.*

*Proof.* The inequality is obvious by Lemma 14 and Proposition 15. Equality on the left hand side of the inequality occurs, for instance, if $\mathbb{P}$ is chosen to be $\mathbb{M}_k \oplus \mathbb{M}_l - \{1_{\mathbb{M}_l}\}$. Equality on the right hand side of the inequality occurs, for instance, if the induced downsets are disjoint. □

Let us now show that there is another upper bound on $\kappa_2(\mathbb{P})$ that might be of combinatorial interest.

**Proposition 17.** *For each poset $\mathbb{P}$, we have* width $\mathbb{P} \geq \kappa_2(\mathbb{P})$.

*Proof.* Let $\mathbb{P} = (P, \leq)$ and set $k := $ width $\mathbb{P}$. By Dilworth's Theorem [Dil50], this means that the poset $\mathbb{P}$ can be covered (in terms of points) by $k$ chains $\mathbb{C}_1 = (C_1, \leq_{\mathbb{C}_1}), \ldots, \mathbb{C}_k = (C_k, \leq_{\mathbb{C}_k})$. Set $M := \lessdot \setminus \bigcup_i \leq_{C_i}$. For each $s \in \{1, \ldots, k\}$, we define a poset $\mathbb{T}_s = (T_s, \leq_s)$ by setting $T_s := C_s \cup \{x \mid \exists y \in C_s : (x, y) \in M\}$ and defining $\leq_s$ to be the reflexive-transitive closure of $\leq_{C_s} \cup \{(x, y) \in M \mid y \in C_s\}$. Note that this makes $\mathbb{T}_s$ a directed tree: the greatest element is the maximum from $\mathbb{C}_s$ and each upset is a chain, because our construction ensures that, for each $x \in T_s$, there is at most one $y \in T_s$ with $(x, y) \in \lessdot_s$. What is more, for each $(x, y) \in M$, there exists some $j \in \{1, \ldots, k\}$ such that $y \in C_j$ and hence $(x, y) \in \leq_j$. Therefore, $\leq$ is the transitive closure of $\bigcup_{i=1}^k \leq_i$. Thus, $\mathbb{T}_1, \ldots, \mathbb{T}_k$ decompose $\mathbb{P}$ is the sense of $\kappa_2$. □

It is easy to see that there is no such result for the height of $\mathbb{P}$, even if $\mathbb{P}$ is assumed to be bounded. Indeed, there exist posets $\mathbb{P}_1, \mathbb{P}_2$ such that $\kappa_2(\mathbb{P}_1) > $ height $\mathbb{P}_1$ and $\kappa_2(\mathbb{P}_2) < $ height $\mathbb{P}_2$ (take, for instance $\mathbb{P}_1 := \mathbb{M}_k$ for any $k \geq 3$ and define $\mathbb{P}_2$ to be any chain with at least two elements).

Let us now turn our attention to the relation between $\kappa_2(\mathbb{P})$ and the order dimension of $\mathbb{P}$. As it turns out, they have virtually no connection, even for bounded posets (except for the obvious cases in which one of the numbers is 1).

In order to show this, we will need the following lemma.

**Lemma 18.** *Let $\mathbb{P} = (P, \leq_\mathbb{P})$ be a poset. Then there exists a poset $\mathbb{Q} = (Q, \leq_\mathbb{Q})$ and some embedding $\varphi \colon \mathbb{P} \to \mathbb{Q}$ such that*

$$\max_{x \in Q} |\{y \in Q \mid x \lessdot_\mathbb{Q} y\}| \leq 2.$$

*Furthermore, if $\mathbb{P}$ is bounded, then $\mathbb{Q}$ may be chosen to be bounded as well.*

*Proof.* The proof proceeds by induction on the number $|P|$. Of course, the statement is obvious for $|P| \leq 2$. Let us elaborate the inductive step. Suppose the claim to be true for all posets of cardinality less than $|P|$. Choose a minimal element $p$ in $\mathbb{P}$, and define $\mathbb{P}'$ to be the sub-poset of $\mathbb{P}$ that is induced by $P' := P \setminus \{p\}$. By induction hypothesis, there is a poset $\mathbb{Q}' = (Q', \leq_{\mathbb{Q}'})$ and some embedding $\varphi' \colon \mathbb{P}' \to \mathbb{Q}'$ such that $\max_{x \in Q'} |\{y \in Q' \mid x \lessdot_{\mathbb{Q}'} y\}| \leq 2$. Moreover, let $N$ denote the set of all upper neighbours of $p$ in $\mathbb{P}$. Choose a finite sequence $\pi_1, \ldots, \pi_n$ of partitions of $N$ such that $\pi_n = \{N\}$ and, for every $i \in \{1, \ldots, n-1\}$, the following three conditions are met:

(1) $\pi_{i+1} \neq \pi_i$,

(2) $\pi_{i+1}$ coarsens $\pi_i$, i.e., $\forall M \in \pi_i \, \exists M' \in \pi_{i+1} : M \subseteq M'$,

(2) each element of $\pi_{i+1}$ is the union of at most two members of $\pi_i$.

Let us define $J := \bigcup\{\{i\} \times \pi_i \mid i \in \{1, \ldots, n\}\}$. Choose a family $(x_{(i,M)})_{(i,M) \in J}$ of pairwise distinct elements not occurring in $Q'$. Define a poset $\mathbb{Q} := (Q, \leq_\mathbb{Q})$, where $Q := Q' \cup \{x_{(i,M)} \mid (i,M) \in J\}$ and $\leq_\mathbb{Q}$ is defined to be the reflexive-transitive closure of

$$\leq_{\mathbb{Q}'} \cup \{(x_{(i+1,M')}, x_{(i,M)}) \mid i \in \{1, \ldots, n-1\}, M \in \pi_i, M' \in \pi_{i+1}, M \subseteq M'\}$$
$$\cup \{(x_{(1,M)}, y) \mid M \in \pi_1, y \in M\}.$$

Evidently, $\mathbb{Q}$ has got the desired property, i.e., $\max_{x \in Q} |\{y \in Q \mid x <_\mathbb{Q} y\}| \leq 2$. Moreover, we obtain an embedding $\varphi \colon \mathbb{P} \to \mathbb{Q}$ by defining $\varphi(p) := x_{(n,N)}$ and $\varphi(z) := \varphi'(z)$ for all $z \in P'$. This proves the claim for $\mathbb{P}$. Finally, we observe that, if $\mathbb{P}$ is bounded, then $\varphi[P]$ is contained in some interval of $\mathbb{Q}$, which clearly inherits the desired property from $\mathbb{Q}$. □

**Proposition 19.** (i) For all $k \geq 2$ and $l \geq k$, there exists a bounded poset $\mathbb{P}$ such that $\dim \mathbb{P} = k$ and $\kappa_2(\mathbb{P}) \geq l$.

(ii) For all $k \geq 2$ and $l \geq k$, there exists a bounded poset $\mathbb{P}$ such that $\kappa_2(\mathbb{P}) = k$ and $\dim \mathbb{P} \geq l$.

*Proof.* (i) Let $k \geq 2$ and $l \geq k$. Let $\mathbb{Q}$ be the $k$-dimensional cube and let $\mathbb{M}_l = (M_l, \leq_{\mathbb{M}_l})$ be defined as in Example 11. Set $\mathbb{P} := \mathbb{M}_l \oplus \mathbb{Q}$ (see Definition 9). We have $\dim \mathbb{P} = \max\{\dim \mathbb{M}_l, \dim \mathbb{Q}\} = \max\{2, k\} = k$ and, by Proposition 15, $\kappa_2(\mathbb{P}) = \max\{l, k\} = l$.

(ii) Let $k \geq 2$ and $l \geq k$. By Lemma 18, it is possible to construct a bounded poset $\mathbb{Q} = (Q, \leq_\mathbb{Q})$ such that the $l$-dimensional cube can be embedded into $\mathbb{Q}$ and each element $x \in Q$ has at most 2 upper neighbours. Define $\mathbb{P} := \mathbb{M}_k \oplus \mathbb{Q}$. By Proposition 15 and $k \geq 2$, we have $\kappa_2(\mathbb{P}) = k$, and since the $l$-dimensional cube can be embedded into $\mathbb{P}$, we have $\dim \mathbb{P} \geq l$. □

It remains the question of how the value $\kappa_2(\mathbb{P})$ relates to $\kappa_1(\mathbb{P}), \kappa_3(\mathbb{P}), \kappa_4(\mathbb{P})$. For any poset $\mathbb{P}$, it is of course obvious that we have $\kappa_1(\mathbb{P}) \leq \kappa_2(\mathbb{P})$, and it is an easy consequence of Proposition 13 and Proposition 15 that the difference between $\kappa_2(\mathbb{P})$ and $\kappa_1(\mathbb{P})$ can be arbitrarily large. Indeed, for all $k \in \mathbb{N}$ and $l \geq k$, there exists some poset $\mathbb{P}$ such that $\kappa_1(\mathbb{P}) = k$ and $\kappa_2(\mathbb{P}) \geq l$. However, the connection between $\kappa_2(\mathbb{P})$ and $\kappa_3(\mathbb{P})$ is perhaps not entirely obvious, and while we of course know $\kappa_2(\mathbb{P}) \leq \kappa_4(\mathbb{P})$, one may still ask whether the difference between $\kappa_2(\mathbb{P})$ and $\kappa_4(\mathbb{P})$ can be arbitrarily high. Among other things, we will discuss these questions in the next sections.

## 6  Third Version: $\kappa_3$

In this section, we will discuss the meaning of the value $\kappa_3(\mathbb{P})$. Among other things, we will elaborate its relation to the other three quantities, show its connection to the width and the height of $\mathbb{P}$, and show that it has a slight yet nontrivial connection to the order dimension of $\mathbb{P}$.

In the introductory examples from Section 3, we have only seen posets $\mathbb{P}$ with $\kappa_3(\mathbb{P}) = 2$. Hence, let us start this section by quickly showing that there is indeed, for each $k \in \mathbb{N}$, some nontrivial example of a (bounded) poset $\mathbb{P}$ with $\kappa_3(\mathbb{P}) = k$.

**Lemma 20.** *For each $k \in \mathbb{N}$, there exists a bounded poset $\mathbb{P}$ with $\kappa_3(\mathbb{P}) = k$.*

*Proof.* Let $k \in \mathbb{N}$. We define the $(k^2 - k + 2)$-element poset $\mathbb{D}_k = (D_k, \leq_{\mathbb{D}_k})$ by setting $D_k := \{0_{\mathbb{D}_k}, 1_{\mathbb{D}_k}\} \cup (\{1, \ldots, k-1\} \times \{1, \ldots, k\})$ and $x \leq_{\mathbb{D}_k} y$ if and only if $x = 0_{\mathbb{D}_k}$ or $y = 1_{\mathbb{D}_k}$ or there exist $x_1, y_1 \in \{1, \ldots, k-1\}$ and $x_2, y_2 \in \{1, \ldots, k\}$ such that $x = (x_1, x_2)$, $y = (y_1, y_2)$ and $x_1 \leq y_1$.

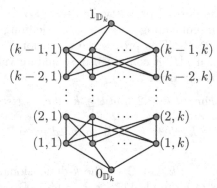

It is obvious that we have $\kappa_3(\mathbb{D}_k) \leq k$. To prove $\kappa_3(\mathbb{D}_k) \geq k$, assume for contradiction that the directed trees $\mathbb{T}_1 = (T_1, \leq_1), \ldots, \mathbb{T}_{k-1} = (T_{k-1}, \leq_{k-1})$ decompose $\mathbb{D}_k$ in the sense of $\kappa_3$. Evidently, there must exist some $i \in \{1, \ldots, k-1\}$, such that $T_i$ contains at least two points from $\{k-1\} \times \{1, \ldots, k\}$. Without loss of generality, assume $i = k-1$. Since $\mathbb{T}_{k-1}$ is a sub-poset of $\mathbb{D}_k$, we have

$$T_{k-1} \cap (\{0_{\mathbb{D}_k}\} \cup (\{1, \ldots, k-2\} \times \{1, \ldots, k\})) = \emptyset.$$

Thus,

$$\{0_{\mathbb{D}_k}\} \cup (\{1, \ldots, k-2\} \times \{1, \ldots, k\}) \subseteq T_1 \cup \ldots \cup T_{k-2}.$$

Proceeding similarly, there must be some $i \in \{1, \ldots, k-2\}$, such that $T_i$ contains at least two points from $\{k-2\} \times \{1, \ldots, k\}$. Assuming this set is $T_{k-2}$ and applying the same arguments as above, we may conclude

$$\{0_{\mathbb{D}_k}\} \cup (\{1, \ldots, k-3\} \times \{1, \ldots, k\}) \subseteq T_1 \cup \ldots \cup T_{k-3}.$$

Using this argument $k-2$ times, we finally obtain $\{0_{\mathbb{D}_k}\} \cup (\{1\} \times \{1, \ldots, k\}) \subseteq T_1$, which clearly contradicts that $\mathbb{T}_1$ is a directed tree and a sub-poset of $\mathbb{D}_k$.     $\square$

In Section 5, we have seen that, for any bounded poset $\mathbb{P}$, the integer $\kappa_2(\mathbb{P})$ is bounded by the width of $\mathbb{P}$ but not by the height. Let us now show that, for $\kappa_3(\mathbb{P})$, both claims are true.

**Proposition 21.** *For any poset $\mathbb{P} = (P, \leq)$ with a greatest element and $|P| \geq 2$, we have $\kappa_3(\mathbb{P}) \leq \min\{\text{width}\,\mathbb{P}, \text{height}\,\mathbb{P} - 1\}$.*

*Proof.* Let $1_\mathbb{P}$ be the greatest element of $\mathbb{P}$. If we have $k = $ width $\mathbb{P}$, then we can cover $\mathbb{P}$ with $k$ chains by Dilworth's Theorem [Dil50]. Since every chain necessarily forms a sub-poset of $\mathbb{P}$, this means $\kappa_3(\mathbb{P}) \le k$. If we have $k = $ height $\mathbb{P}$ (note that this implies $k \ge 2$ since $\mathbb{P}$ has a maximum and at least two elements), then we can cover $\mathbb{P}$ with $k$ antichains $A_1, \ldots, A_k$ by Mirsky's Theorem [Mir71]. One of these antichains must be $\{1_\mathbb{P}\}$, say $A_k$. Now, define the $k - 1$ directed trees $\mathbb{T}_1, \ldots, \mathbb{T}_{k-1}$ by setting $\mathbb{T}_i$ to be the sub-poset of $\mathbb{P}$ that is induced by $A_i \cup \{1_\mathbb{P}\}$. Since $\{1_\mathbb{P}\} \cup A_1 \cup \ldots \cup A_{k-1} = P$, this proves $\kappa_3(\mathbb{P}) \le k - 1$.    $\square$

While the argument for $\kappa_3(\mathbb{P}) \le $ width $\mathbb{P}$ holds for any poset, the inequality $\kappa_3(\mathbb{P}) \le $ height $\mathbb{P}-1$ can only be ensured if $\mathbb{P}$ has a greatest element. In particular, requiring a least element is not enough (as a counterexample, take for instance an antichain with at least two elements and add a least element).

Let us now discuss the connection of the integers $\kappa_3(\mathbb{P})$ and $\kappa_2(\mathbb{P})$ for a given poset $\mathbb{P}$. It will turn out that these numbers say practically nothing about each other (except the obvious equivalence between $\kappa_2(\mathbb{P}) = 1$ and $\kappa_3(\mathbb{P}) = 1$). We start with a small lemma that we will need to show the desired result.

**Lemma 22.** *Let $\mathbb{P} = (P, \le_\mathbb{P})$ be a sub-poset of $\mathbb{Q} = (Q, \le_\mathbb{Q})$ and assume that $\mathbb{P}$ has a greatest element. Then we have $\kappa_3(\mathbb{Q}) \ge \kappa_3(\mathbb{P})$.*

*Proof.* Denote the greatest element of $\mathbb{P}$ by $1_\mathbb{P}$ and let $(T_1, \le_1), \ldots, (T_k, \le_k)$ be a directed tree decomposition of $\mathbb{Q}$ in the sense of $\kappa_3$. For each $i \in \{1, \ldots, k\}$, let $(S_i, \le_{S_i})$ be the sub-poset of $\mathbb{P}$ with $S_i := (T_i \cap P) \cup \{1_\mathbb{P}\}$. We are done if we can show that each $(S_i, \le_{S_i})$ is a directed tree. By definition, each $S_i$ has a greatest element. Take some $x \in S_i$ and consider its upset in $S_i$. Assuming that it is not a chain gives us $x_1, x_2 \in S_i$ with $x_1, x_2 \ge_{S_i} x$ and $x_1$ and $x_2$ being incomparable. This implies $1_\mathbb{P} \notin \{x, x_1, x_2\}$. Thus, $\{x, x_1, x_2\} \subseteq T_i$, which yields a contradiction to $(T_i, \le_i)$ being a directed tree.    $\square$

**Proposition 23.** *(i) For all $k \ge 2$ and $l \ge k$, there exists a bounded poset $\mathbb{P}$ such that $\kappa_2(\mathbb{P}) = k$ and $\kappa_3(\mathbb{P}) \ge l$.*
*(ii) For all $k \ge 2$ and $l \ge k$, there exists a bounded poset $\mathbb{P}$ such that $\kappa_3(\mathbb{P}) = k$ and $\kappa_2(\mathbb{P}) \ge l$.*

*Proof.* (i) Let $k \ge 2$ and $l \ge k$. Let us show that, given a bounded poset $\mathbb{P}$ with $\kappa_2(\mathbb{P}) \ge 2$, we can construct a bounded poset $\mathbb{P}'$ with $\kappa_2(\mathbb{P}') = \kappa_2(\mathbb{P})$ and $\kappa_3(\mathbb{P}') > \kappa_3(\mathbb{P})$. Define $\mathbb{P}'$ as indicated by the following diagram.

Written in detail, we define $\mathbb{P}' = (P', \le')$ by setting $P' := \{0_{\mathbb{P}'}, 1_{\mathbb{P}'}\} \cup (\{a, b\} \times P)$, and

$$x \le' y :\Longleftrightarrow x = 0_{\mathbb{P}'} \lor y = 1_{\mathbb{P}'} \lor \exists i \in \{a, b\}, x_2 \le y_2 : x = (i, x_2), y = (i, y_2).$$

Evidently, $\mathbb{P}'$ is bounded. By Proposition 15 and $\kappa_2(\mathbb{P}) \geq 2$, we can infer $\kappa_2(\mathbb{P}') = \kappa_2(\mathbb{P})$. Assume that we have a decomposition $(T_1, \leq_1), \ldots, (T_k, \leq_k)$ the sense of $\kappa_3$. The element $0_{\mathbb{P}'}$ belongs to at least one of the sets $T_1, \ldots, T_k$, say $T_1$. Since $(T_1, \leq_1)$ is a sub-poset of $\mathbb{P}'$ and contains $0_{\mathbb{P}'}$, it must be a chain. But now, a chain in $\mathbb{P}'$ will be disjoint with at least one sub-poset of $\mathbb{P}'$ that is isomorphic to $\mathbb{P}$ (the left or right side of the diagram given above). Hence, $(T_2, \leq_2), \ldots, (T_k, \leq_k)$ provide a decomposition (in the sense of $\kappa_3$) of some poset which has that isomorphic copy of $\mathbb{P}$ as a sub-poset. Hence, by Lemma 22, $k - 1$ directed trees are enough to decompose $\mathbb{P}$ in the sense of $\kappa_3$. Thus, $\kappa_3(\mathbb{P}') > \kappa_3(\mathbb{P})$. Now, take any bounded poset $\mathbb{P}_0$ with $\kappa_2(\mathbb{P}_0) = k$ (for instance, we may choose $\mathbb{P}_0 := \mathbb{M}_k$ as defined in Example 11) and construct a series of posets $\mathbb{P}_1, \ldots, \mathbb{P}_l$ by setting $\mathbb{P}_{i+1} := \mathbb{P}'_i$ for all $i \in \{0, \ldots, l-1\}$. By the arguments from above, we have $\kappa_2(\mathbb{P}_l) = \kappa_2(\mathbb{P}_0) = k$ but $\kappa_3(\mathbb{P}_l) \geq \kappa_3(\mathbb{P}_0) + l > l$.

(ii) Let $k \geq 2$ and $l \geq k$. Take some bounded poset $\mathbb{Q} = (Q, \leq_{\mathbb{Q}})$ with greatest element $1_{\mathbb{Q}}$ and $\kappa_3(\mathbb{Q}) = k - 1$ (its existence is guaranteed by Lemma 20). Set $\mathbb{P} := \mathbb{Q} \oplus \mathbb{M}_l$.

By Proposition 15, we have $\kappa_2(\mathbb{P}) \geq l$. It remains to show $\kappa_3(\mathbb{P}) = k$. First, let us show $\kappa_3(\mathbb{P}) \leq k$. By assumption, there exists a directed tree decomposition $(T_1, \leq_1), \ldots, (T_{k-1}, \leq_{k-1})$ of $\mathbb{Q}$ in the sense of $\kappa_3$. Some set $T_1, \ldots, T_{k-1}$ must contain $1_{\mathbb{Q}}$, say $T_1$. Define the tree $(T'_1, \leq'_1)$ by setting $T'_1 := T_1 \cup \{0_{\mathbb{M}_l}\}$ and taking $\leq'_1$ to be the reflexive-transitive closure of $\leq_1 \cup \{(1_{\mathbb{Q}}, 0_{\mathbb{M}_l})\}$. Also, define the directed tree $(T_k, \leq_k)$ as the sub-poset of $\mathbb{P}$ induced by $T_k := \mathbb{M}_l \setminus \{0_{\mathbb{M}_l}\}$. But now, the directed trees $(T'_1, \leq'_1), (T_2, \leq_2), \ldots, (T_k, \leq_k)$ decompose $\mathbb{P}$ in the sense of $\kappa_3$. Hence, $\kappa_3(\mathbb{P}) \leq k$. To finish the proof by showing $\kappa_3(\mathbb{P}) \geq k$, assume for contradiction that some directed trees $(T_1, \leq_1), \ldots, (T_{k-1}, \leq_{k-1})$ give a decomposition of $\mathbb{P}$ in the sense of $\kappa_3$. Let $a_1, \ldots, a_l$ be the elements from the $l$-element antichain in $\mathbb{M}_l$. Since $l \geq k$, some set $T_1, \ldots, T_{k-1}$ must contain at least two elements from this antichain. Assume that $T_1$ is this set. Then we have $T_1 \cap Q = \emptyset$, because $T_1$ must also be a sub-poset of $\mathbb{P}$. But now, this implies that the directed trees $(T_2, \leq_2), \ldots, (T_{k-1}, \leq_{k-1})$ decompose some poset (in the sense of $\kappa_3$) of which $\mathbb{Q}$ is a sub-poset. Hence, by Lemma 22, we obtain $\kappa_3(\mathbb{Q}) \leq k-2$. Contradiction.    □

The next few propositions will illustrates that the number $\kappa_3(\mathbb{P})$ has almost no connection to the order dimension of $\mathbb{P}$, except for one (nontrivial) case.

**Proposition 24.** *(i) For all $k \geq 2$ and $l \geq k$, there exists a bounded poset $\mathbb{P}$ such that $\dim \mathbb{P} = k$ and $\kappa_3(\mathbb{P}) \geq l$.*
*(ii) For all $k \geq 3$ and $l \geq k$, there exists a bounded poset $\mathbb{P}$ such that $\kappa_3(\mathbb{P}) = k$ and $\dim \mathbb{P} \geq l$.*

*Proof.* (i) Assume $k \geq 2$ and $l \geq k$. Let $\mathbb{Q} = (Q, \leq_{\mathbb{Q}})$ be any bounded poset with order dimension $k$ (for instance, take the $k$-dimensional cube). Set $\mathbb{P} := \mathbb{D}_l \oplus \mathbb{Q}$, where $\mathbb{D}_l$ is defined as in Lemma 20. By Lemma 22 and Lemma 20, we have $\kappa_3(\mathbb{P}) \geq \kappa_3(\mathbb{D}_l) = l$. Moreover, it is easy to check that the order dimension of $\mathbb{D}_l$ is 2. Hence, we end up with $\dim \mathbb{P} = \max\{\dim \mathbb{D}_l, \dim \mathbb{Q}\} = \max\{2, k\} = k$.

(ii) Let $k \geq 3$ and $l \geq k$. Take the poset $\mathbb{H}_l = (H_l, \leq_{\mathbb{H}_l})$ as introduced in Example 6 and define $\mathbb{Q}$ as the poset that arises from $\mathbb{H}_l$ by adding a greatest element $1_{\mathbb{Q}}$ and a least element $0_{\mathbb{Q}}$. That is, the poset $\mathbb{Q}$ is illustrated as follows:

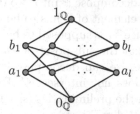

Take the poset $\mathbb{D}_k$ as defined in Lemma 20 and set $\mathbb{P} := \mathbb{Q} \oplus \mathbb{D}_k$. Since we have $\dim \mathbb{Q} = l$, we have $\dim \mathbb{P} \geq l$. Moreover, $\kappa_3(\mathbb{P}) \geq k$ follows immediately from $\kappa_3(\mathbb{D}_k) = k$ and Lemma 22. It remains to show $\kappa_3(\mathbb{P}) \leq k$. Observe that $\mathbb{D}_k$ can be decomposed (in the sense of $\kappa_3$) into $k$ chains $\mathbb{C}_1, \ldots, \mathbb{C}_k$. Moreover, $\mathbb{Q}$ can be decomposed into three directed trees, namely the three sub-posets $\mathbb{S}_1 = (S_1, \leq_1)$, $\mathbb{S}_2 = (S_2, \leq_2)$, $\mathbb{S}_3 = (S_3, \leq_3)$ with $S_1 = \{1_{\mathbb{Q}}, b_1, \ldots, b_l\}$, $S_2 = \{1_{\mathbb{Q}}, a_1, \ldots, a_l\}$, $S_3 = \{1_{\mathbb{Q}}, 0_{\mathbb{Q}}\}$. Thus, we obtain the required $k$-element decomposition of $\mathbb{P}$ into directed trees $\mathbb{T}_1, \ldots, \mathbb{T}_k$ by setting $\mathbb{T}_1 := \mathbb{S}_1 \oplus \mathbb{C}_1$, $\mathbb{T}_2 := \mathbb{S}_2 \oplus \mathbb{C}_2$, $\mathbb{T}_3 := \mathbb{S}_3 \oplus \mathbb{C}_3$, $\mathbb{T}_4 := \mathbb{C}_4, \ldots, \mathbb{T}_k := \mathbb{C}_k$. $\square$

Case (ii) leaves us with two questions:

- If we weaken the condition of $\mathbb{P}$ being bounded, does there then, for each $l \geq 2$, exist some poset $\mathbb{P}$ such that $\kappa_3(\mathbb{P}) = 2$ and $\dim \mathbb{P} \geq l$?
- If we have $\kappa_3(\mathbb{P}) = 2$ for some bounded poset $\mathbb{P}$, what does this tell us about the order relation of $\dim \mathbb{P}$?

The next two proposition answer both of these questions.

**Proposition 25.** *For all $k \geq 2$ and $l \geq k$, there exists a poset $\mathbb{P}$ with a greatest element such that $\kappa_3(\mathbb{P}) = k$ and $\dim \mathbb{P} \geq l$.*

*Proof.* In view of Proposition 24, we need to show the claim for $k = 2$. To this end, let $l \geq k$ and take $\mathbb{H}_l$ as defined in Example 6. Let $\mathbb{P} = (P, \leq)$ be the poset that arises by adding a greatest element $1_{\mathbb{P}}$ to $\mathbb{H}_l$.

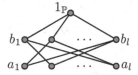

We have $\dim \mathbb{P} = l$ and it is obvious that we have $\kappa_3(\mathbb{P}) \geq 2$, because $\mathbb{P}$ is not a tree. Indeed, we have $\kappa_3(\mathbb{P}) = 2$ because the two sub-posets induced by

$\{1_\mathbb{P}, a_1, \ldots, a_l\}$ and $\{1_\mathbb{P}, b_1, \ldots, b_l\}$ are directed trees and decompose $\mathbb{P}$ in the sense of $\kappa_3$. $\qquad\square$

The next proposition is the first nontrivial connection between our numbers and the order dimension.

**Proposition 26.** *Let $\mathbb{P}$ be a poset with a least element. If $\kappa_3(\mathbb{P}) = 2$, then $\dim \mathbb{P} \leq 4$.*

*Proof.* By $\kappa_3(\mathbb{P}) = 2$, we can decompose $\mathbb{P}$ into two directed trees $\mathbb{T}_1 = (T_1, \leq_1)$ and $\mathbb{T}_2 = (T_2, \leq_2)$. The least element of $\mathbb{P}$ has to be contained in at least one of the two sets $T_1, T_2$, say $T_1$. Since $\mathbb{T}_1$ is supposed to be a sub-poset of $\mathbb{P}$, this means that it is a chain. Observe that the sub-poset of $\mathbb{T}_2$ that is induced by $T_2 \setminus T_1$ is still a directed tree. Hence, we can decompose $\mathbb{P}$ into a chain $\mathbb{C} = (C, \leq_\mathbb{C})$ and a directed tree $\mathbb{T}$ that does not intersect $\mathbb{C}$. By the removal theorem for the order dimension stated in the preliminaries (see Theorem 8), this establishes $\dim \mathbb{P} \leq 2 + \dim (\mathbb{P} - C) = 2 + \dim \mathbb{T} \leq 4$. $\qquad\square$

## 7  Fourth Version: $\kappa_4$

Finally, let us take a look at the meaning of $\kappa_4(\mathbb{P})$. Of the four quantities we have defined in Definitions 1–4, it is obviously the greatest integer, and we have the trivial equivalence $\kappa_2(\mathbb{P}) = 1 \iff \kappa_3(\mathbb{P}) = 1 \iff \kappa_4(\mathbb{P}) = 1$. Except that, however, we will see that knowing the numbers $\kappa_1(\mathbb{P}), \kappa_2(\mathbb{P})$ or $\kappa_3(\mathbb{P})$ tells us nothing about how big $\kappa_4(\mathbb{P})$ can be (observe that this is evident for $\kappa_1(\mathbb{P})$). Besides showing this result, we also discuss the relation of $\kappa_4(\mathbb{P})$ with the order dimension. We will show some partial results, give some nontrivial examples and state an open problem.

Let us start by showing that $\kappa_4(\mathbb{P})$ can take every value that is greater than or equal to $\kappa_2(\mathbb{P})$ and $\kappa_3(\mathbb{P})$, even for bounded posets.

**Proposition 27.**
*(i) For all $k \geq 2$ and $l \geq k$, there exists a bounded poset $\mathbb{P}$ such that $\kappa_2(\mathbb{P}) = k$ and $\kappa_4(\mathbb{P}) = l$.*
*(ii) For all $k \geq 2$ and $l \geq k$, there exists a bounded poset $\mathbb{P}$ such that $\kappa_3(\mathbb{P}) = k$ and $\kappa_4(\mathbb{P}) = l$.*

*Proof.* (i) Let $k \geq 2$ and $l \geq k$. Take the poset $\mathbb{M}_k$ as defined in Example 11, and the poset $\mathbb{L}_l$ as defined in Example 12. Set $\mathbb{P} := \mathbb{M}_k \oplus \mathbb{L}_l$.

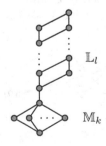

By Proposition 15 and $k \geq 2$, we have $\kappa_2(\mathbb{P}) = k$. Furthermore, $\kappa_4(\mathbb{P}) \geq l$ follows from $\kappa_4(\mathbb{L}_l) \geq l$. In order to show $\kappa_4(\mathbb{P}) \leq l$, let $\mathbb{T}_1, \ldots, \mathbb{T}_l$ be the $l$ chains that are constructed in Example 12 and decompose $\mathbb{L}_l$ in the sense of $\kappa_4$. Moreover, let $\mathbb{T}'_1, \ldots, \mathbb{T}'_k$ be the obvious choice of $k$ chains decomposing $\mathbb{M}_k$. But now, the $l$ chains $\mathbb{T}'_1 \oplus \mathbb{T}_1, \ldots, \mathbb{T}'_k \oplus \mathbb{T}_k, \mathbb{T}_{k+1}, \cdots, \mathbb{T}_l$ decompose $\mathbb{P}$ in the sense of $\kappa_4$. Thus, $\kappa_4(\mathbb{P}) = l$.

(ii) Again, let $k \geq 2$ and $l \geq k$. This time, we define $\mathbb{P} = (P, \leq_{\mathbb{P}})$ by setting $P := \{1, \ldots, k-1\} \times \mathbb{M}_l$, with $(i, x) \leq_{\mathbb{P}} (j, y) :\Longleftrightarrow i < j \vee (i = j \wedge x \leq_{\mathbb{M}_l} y)$.

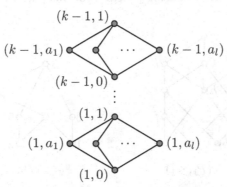

$\kappa_4(\mathbb{P}) \geq l$ follows immediately from $\kappa_4(\mathbb{P}) \geq \kappa_2(\mathbb{P}) = l$, where the latter equality is given by Proposition 15. In fact, we have $\kappa_4(\mathbb{P}) = l$, because $\mathbb{P}$ can obviously be decomposed into $l$ chains. We have $\kappa_3(\mathbb{P}) \leq k$, because the $k$ sub-posets $\mathbb{T}_1, \ldots, \mathbb{T}_k$ that are induced by $T_i := \{(i, 1), (i, a_1), \ldots, (i, a_l)\}$ for $i \in \{1, \ldots, k-1\}$ and $T_k := \{(k-1, 0), (k-2, 0), \ldots, (1, 0)\}$ decompose $\mathbb{P}$ in the sense of $\kappa_3$. It remains to show $\kappa_3(\mathbb{P}) \geq k$, which we are going to prove by a very similar technique as in the proof of Lemma 20. For contradiction, assume that there are directed trees $\mathbb{T}_1 = (T_1, \leq_1), \ldots, \mathbb{T}_{k-1} = (T_{k-1}, \leq_{k-1})$ that decompose $\mathbb{P}$ in the sense of $\kappa_3$. There must be some $i \in \{1, \ldots, k-1\}$ such that $T_i$ contains at least two points from $\{k-1\} \times \{a_1, \ldots, a_l\}$. Without loss of generality, assume $i = k-1$. Since $\mathbb{T}_{k-1}$ is a sub-poset of $\mathbb{P}$, this establishes $\{1, \ldots, k-2\} \times \mathbb{M}_l \subseteq T_1 \cup \ldots \cup T_{k-2}$. Using this argument $k-2$ times, we eventually obtain $\{1\} \times \mathbb{M}_l \subseteq T_1$, which clearly contradicts that $\mathbb{T}_1$ is a directed tree and a sub-poset of $\mathbb{P}$.  $\square$

Also, an analogon of Proposition 21 fails for $\kappa_4(\mathbb{P})$. In fact, $\kappa_4(\mathbb{P})$ is not connected the height or width of $\mathbb{P}$. This is readily demonstrated by Examples 11 and 12.

Finally, let us discuss the connection between $\kappa_4(\mathbb{P})$ and the order dimension of $\mathbb{P}$. First, note that the following two corollaries are direct consequences of $\kappa_4(\mathbb{P}) \geq \kappa_3(\mathbb{P})$ and Propositions 24–26:

**Corollary 28.** *For all $k \geq 3$ and $l \geq k$, there exists a bounded poset $\mathbb{P}$ such that $\dim \mathbb{P} = k$ and $\kappa_4(\mathbb{P}) \geq l$.*

**Corollary 29.** *Let $\mathbb{P}$ be a poset that has a least element. If $\kappa_4(\mathbb{P}) = 2$, then $\dim \mathbb{P} \leq 4$.*

However, we do not know more than that, and the following open problem remains:

*Problem 30.* Is there a fixed function $\psi \colon \mathbb{N} \to \mathbb{N}$ such that $\kappa_4(\mathbb{P}) \leq \psi(\dim \mathbb{P})$ for all (bounded) posets $\mathbb{P}$?

Let us close this section by giving two nontrivial examples that at least contradict the perhaps conceivable claim that we have $\kappa_4(\mathbb{P}) \geq \dim \mathbb{P}$ for all posets $\mathbb{P}$ (note that all previously appearing posets from this paper meet this condition).

*Example 31.*

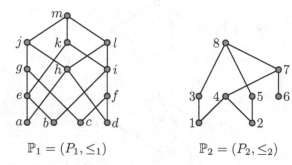

$$\mathbb{P}_1 = (P_1, \leq_1) \qquad\qquad \mathbb{P}_2 = (P_2, \leq_2)$$

It is straightforward to check that we have $\dim \mathbb{P}_1 = 4$ and $\dim \mathbb{P}_2 = 3$. But now, also we have $\kappa_4(\mathbb{P}_1) = 3$ (take the three directed trees given by the sub-posets induced by $\{m, j, g, h, e, b, c, d\}$, $\{m, l, h, i, f, a, b, c\}$ and $\{m, k, i, e, f, a, d\}$) and $\kappa_4(\mathbb{P}_2) = 2$ (take the two directed trees given by the two sub-posets induced by $\{8, 3, 5, 1, 2\}$ and $\{8, 7, 4, 6, 2, 1\}$).

## 8    Summary of Results

We have discussed four versions of Radeleczki's problem by studying the numbers $\kappa_1(\mathbb{P}), \ldots, \kappa_4(\mathbb{P})$ for a poset $\mathbb{P} = (P, \leq)$.

- $\kappa_1(\mathbb{P})$ simply counts the number of maximal elements in $\mathbb{P}$ and hence has almost no connection to the order dimension or the other three numbers (except that, by definition, it is less than or equal to each of them).
- $\kappa_2(\mathbb{P})$ equals $\max_{x \in P} |\{y \mid x \lessdot y\}|$ if $\mathbb{P}$ has a greatest element. In the general case, it depends on how the downsets of the maximal elements intersect, which still provides us with a tight interval for $\kappa_2(\mathbb{P})$. Moreover, $\kappa_2(\mathbb{P})$ is bounded by the width of $\mathbb{P}$ (but not by the height), and has virtually no connection the order dimension. Indeed, even if we only consider bounded posets (to avoid trivial examples), then the numbers $\kappa_2(\mathbb{P})$ and $\dim \mathbb{P}$ say absolutely nothing about each other, except for the trivial case that one of them equals 1.
- $\kappa_3(\mathbb{P})$ has no connection to $\kappa_2(\mathbb{P})$ (except that $\kappa_2(\mathbb{P}) = 1$ is equivalent to $\kappa_3(\mathbb{P}) = 1$). It is bounded by $\min\{\text{width}\,\mathbb{P}, \text{height}\,\mathbb{P} - 1\}$ if $\mathbb{P}$ has a greatest element, but only bounded by width $\mathbb{P}$ if we drop the assumption that there is a maximum. While the order dimension of a (bounded) poset $\mathbb{P}$ does

not provide us with any information about $\kappa_3(\mathbb{P})$ (except if it equals 1), the other direction is only partly true. If $\mathbb{P}$ is required to be bounded, then $\kappa_3(\mathbb{P})$ tells us nothing about the order dimension if it is greater than or equal to 3 (except, of course, that the order dimension cannot be 1 in that case). However, if $\kappa_3(\mathbb{P})$ is 2 for a bounded poset $\mathbb{P}$ (in fact, it is enough to require that $\mathbb{P}$ has a least element), then the order dimension is at most 4. This is the only nontrivial relation between $\kappa_3(\mathbb{P})$ and $\dim \mathbb{P}$, and it becomes false if we do not require $\mathbb{P}$ to have a minimum. Indeed, for each $l \in \mathbb{N}$ there is already a connected poset $\mathbb{P}$ with $\kappa_3(\mathbb{P}) = 2$ and $\dim \mathbb{P} \geq l$.

– $\kappa_4(\mathbb{P})$ is of course greater than or equal to each of the other three values. However, this is all that we can say about their relation. In fact, even if we require $\mathbb{P}$ to be bounded, then knowing $\kappa_2(\mathbb{P})$ or $\kappa_3(\mathbb{P})$ tells us absolutely nothing about $\kappa_4(\mathbb{P})$ beyond the fact that it is at least as large. Moreover, $\kappa_4(\mathbb{P})$ is neither bounded by the width nor by the height of $\mathbb{P}$. For every $k \geq 2$ and $l \geq k$ there exists some poset (even a bounded one) such that we have $\dim \mathbb{P} = k$ and $\kappa_4(\mathbb{P}) \geq l$. However, we do not know whether the converse direction is also true or there exists a function $\psi$ such that $\kappa_4(\mathbb{P}) \leq \psi(\dim \mathbb{P})$ for all (bounded) posets $\mathbb{P}$. All we know is that $\kappa_4(\mathbb{P}) = 2$ implies $\dim \mathbb{P} \leq 4$ if it has a least element and that there are nontrivial examples of posets $\mathbb{P}$ with $\kappa_4(\mathbb{P}) < \dim \mathbb{P}$.

# References

[CHS09]  Czédli, G., Hartmann, M., Schmidt, E.T.: CD-independent subsets in distributive lattices. Publ. Math. Debrecen 74(1-2), 127–134 (2009)

[Dil50]  Dilworth, R.P.: A decomposition theorem for partially ordered sets. Ann. of Math. (2)51, 161–166 (1950)

[Hir55]  Hiraguti, T.: On the dimension of orders. Sci. Rep. Kanazawa Univ. 4(1), 1–20 (1955)

[HR12]  Horváth, E.K., Radeleczki, S.: Notes on CD-independent subsets. Acta Sci. Math (Szeged) 78(1-2), 3–24 (2012)

[Mir71]  Mirsky, L.: A dual of Dilworth's decomposition theorem. Amer. Math. Monthly 78, 876–877 (1971)

[Tro92]  Trotter, W.T.: Combinatorics and partially ordered sets, Dimension theory. Johns Hopkins University Press, Baltimore (1992) (Dimension theory. Johns Hopkins Series in the Mathematical Sciences)

# A Proposition for Combining Pattern Structures and Relational Concept Analysis

Víctor Codocedo and Amedeo Napoli

LORIA - CNRS - INRIA - Université de Lorraine, BP 239, 54506 Vandœuvre-les-Nancy, France
{victor.codocedo,amedeo.napoli}@loria.fr

**Abstract.** In this paper we propose an adaptation of the RCA process enabling the relational scaling of pattern structures. In a nutshell, this adaptation allows the scenario where RCA needs to be applied in a relational context family composed by pattern structures instead of formal contexts. To achieve this we define the heterogeneous pattern structures as a model to describe objects in a combination of spaces, namely the original object description space and the set of relational attributes derived from the RCA scaling process. We frame our approach in the problem of characterizing latent variables (LV) in a latent variable model of documents and terms. LVs are used as compact and improved dataset representations. We approach the problem of LV characterization missing from the original LV-model, through the application of the adapted RCA process using pattern structures. Finally, we discuss the implications of our proposition.

## 1 Introduction

Relational Concept Analysis (RCA) [10] is an extension of Formal Concept Analysis (FCA) [4] based on a scaling process. RCA enables the application of FCA algorithms over a relational context family (RCF) which models the situation where different object sets, in different formal contexts ($\mathcal{K}_1$ and $\mathcal{K}_2$) are associated by a binary relation $r \subseteq G_1 \times G_2$ (e.g. people and their professions liking movies with different genres). In this paper we present an adaptation of RCA which enables its application when one of the object sets cannot be described by set of attributes as usual, but rather by complex descriptions (and thus calling for pattern structures [3] for taking into account these complex descriptions). Particularly, we consider the case when the "domain context" (i.e. the context where the object set is the domain of relation $r$) is a pattern structure of the form $\mathcal{K}_1 = (G_1, (D, \sqcap), \delta)$. To achieve this adaptation, we define the heterogeneous pattern structures as a mean to provide an object with descriptions in different spaces of data, to support both, its original pattern structure description and the relational scaling proposed in RCA.

The inspiration of this problem comes from a model known in information retrieval as "latent variable models" (LV-models), sometimes called "topic models" [11]. LV-models are a long used, cutting-edge and useful manner to index, cluster and retrieve documents [2]. They share the basic notion that the information in a document collection is "generated" by a reduced set of latent variables (LVs) hidden in data, i.e. terms in a given document are a manifestation of topics or LVs (e.g. in an article about "formal

C.V. Glodeanu, M. Kaytoue, and C. Sacarea (Eds.): ICFCA 2014, LNAI 8478, pp. 96–111, 2014.

concept analysis", the terms "formal context" and "concept lattice" are expected to be mentioned).

Latent variables, however, are abstractions. While they may represent topics, those topics lack a proper characterization, which makes difficult their interpretation. For example, in the case of latent semantic indexing (LSI) [2] (considered to be seminal work in topic models), LVs are represented by eigenvectors of a document-term matrix. Nevertheless, eigenvectors or convex regions in the eigenvector space (usually called "clusters") can be hardly recognizable as being, for instance, the topic of "formal concept analysis". Usually, we can try to manually recognize the documents and terms in a cluster to give it a "label", however this can be expensive and tedious. Moreover, LV-models do not allow the incorporation of external knowledge sources which could aid the "labelling" task.

Given the capabilities of FCA for classification and the extent/intent representation of concepts, LVs' characterization can be achieved by constructing a RCF containing a context of document descriptions in the latent variable space (a pattern structure), a formal context for terms' annotations from Wordnet[1] (e.g. a "lattice" *is a* "structure"[2]), and a relational context between documents and terms representing the binary relation document *contains* term. Accordingly, a key aspect of this work is to address the issue that relational scaling is not currently supported for pattern structures.

The main contributions of this work are the proposition of a coherent combination of pattern structures and RCA, the resulting description of heterogeneous pattern structures and a characterization technique for latent variables in a LV-model. The remainder of this paper is as follows. Section 2 provides the theoretical background of this work by describing the RCA process and the pattern structure framework. In Section 3 we describe the latent variable characterization problem in the context of the LSI technique and provide the problem statements. Section 4 describes our proposal for a pattern structures-RCA combination and defines the heterogeneous pattern structures framework. Finally, Section 5 answers both questions and discusses their implications, while providing the conclusions for this work.

## 2 Theoretical Framework

In the following, we define the basic notions which support our approach. The examples in this section are illustrative for RCA and pattern structures, respectively, however they do not represent our scenario which is actually introduced in the next section.

### 2.1 Relational Concept Analysis (RCA)

Hereafter, we briefly introduce the mechanism of RCA as detailed in [9,10]. Different from standard FCA, RCA considers the scenario where an object has not only attributes, but also relations with other objects which have attributes of their own. For example, consider a set of documents with authors as attributes (formal context $\mathcal{K}_1$ in Table 1a)

---

[1] http://wordnetweb.princeton.edu - Wordnet is an open lexical hierarchy available online.

[2] Hypernym of "lattice".

**Table 1.** Relational context family (RCF) - Table 1a: Formal context $\mathcal{K}_1$ of documents and their authors. Table 1b: Formal context $\mathcal{K}_2$ of terms and their Wordnet annotations. Table 1c: Relational context aw representing document "annotated with" term.

**(a)**

| | author1 | author2 | author3 | author4 |
|---|---|---|---|---|
| $g_1$ | x | x | | |
| $g_2$ | x | x | | |
| $g_3$ | x | x | | |
| $g_4$ | x | x | | x |
| $g_5$ | | x | | |
| $g_6$ | | | x | |
| $g_7$ | | x | x | |
| $g_8$ | | x | | |
| $g_9$ | x | | | x |

**(b)**

| | Person | Surgery | Illness | Artefact | Event | Activity |
|---|---|---|---|---|---|---|
| patient | x | | | | | |
| laparoscopy | | x | | | | x |
| scan | | | | x | | |
| user | x | | | | | |
| medicine | | | | | | x |
| response | | | | | x | |
| time | | | | | x | |
| MRI | | | | x | | |
| practice | | | | | | x |
| complication | | | x | | | |
| arthroscopy | | x | | | | x |
| infection | | | x | | | |

**(c)**

| | patient | laparoscopy | scan | user | medicine | response | time | MRI | practice | complication | arthroscopy | infection |
|---|---|---|---|---|---|---|---|---|---|---|---|---|
| $g_1$ | x | x | x | | | | | | x | | | |
| $g_2$ | | x | | x | x | x | x | | x | | | |
| $g_3$ | | x | | | x | x | | x | | | | |
| $g_4$ | x | | | | x | | | x | | | | |
| $g_5$ | | x | | | x | x | | | | | | |
| $g_6$ | | | | | | | x | | | x | | |
| $g_7$ | | | | | | | | | | x | x | |
| $g_8$ | | | | | | | | | | x | x | x |
| $g_9$ | | | | | | | | | | | x | x |

and a set of terms with entities extracted from Wordnet (formal context in Table 1b). Then, we can consider the relation "document *annotated with* term" (denoted as aw) which defines a *relational context* as the one shown in Table 1c. RCA defines a *relational context family* (RCF) as a set of contexts $\mathbf{K} = \{\mathcal{K}_1, \mathcal{K}_2\}$ and a set of binary relations $\mathbf{R} = \{r\}$. A relation $r \subseteq G_1 \times G_2$ connects two object sets, a *domain* $G_1$, $(\mathrm{dom}(r) = G_1)$ and a *range* $G_2$, $(\mathrm{ran}(r) = G_2)$. Moreover, a relation $r$ can be seen as a set-valued function $r : G_1 \rightarrow \wp(G_2)$ [9].

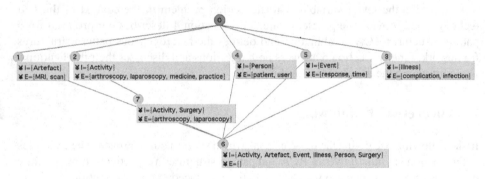

**Fig. 1.** Concept lattice of formal context $\mathcal{K}_2$ in Table 1b

For the current example, let $G_1$ be a set of documents and $G_2$ be a set of terms. Then the corresponding RCF is composed by contexts $\mathcal{K}_1 = (G_1, M_1, I_1)$ (with $M_1, I_1$ as shown in Table 1a), $\mathcal{K}_2 = (G_2, M_2, I_2)$ (with $M_2, I_2$ as shown in Table 1b) and the relational context aw in Table 1c.

RCA is based on a *relational scaling* mechanism that transforms a relation $r$ into a set of *relational attributes* that are added to complete the "initial context" describing the object set $G_1 = \mathrm{dom}(r)$. For each relation $r$, there is an *initial lattice* for each object set, i.e. $\mathcal{L}_1$ for $G_1$ and $\mathcal{L}_2$ for $G_2$.

The RCA mechanism starts from two initial lattices, $\mathcal{L}_1$ and $\mathcal{L}_2$, and builds a series of intermediate lattices by gradually completing the initial context $\mathcal{K}_1$ with new "relational attributes". Relational scaling follows the description logics (DL) semantics of role restrictions.

A relational attribute $\exists r : C$, C being a concept and $\exists$ the existential quantifier, is associated to an object $g \in G_1$ whenever $r(g) \cap \text{extent}(C) \neq \emptyset$ (other quantifiers are available, see [9]). The series of intermediate lattices converges toward a "fixpoint" or "final lattice" and the RCA mechanism is terminated. This is why there is one initial and one final lattice for each context of the considered RCF. For the running example, the lattice (in this case initial and final) in Figure 1 for the formal context $\mathcal{K}_2$ in Table 1b, along with the "relational context" in Table 1c, indicates the "relational attributes" that should be added to the formal context in Table 1a. For instance, using the existential quantifier, the relational attribute $\exists aw : C1$ (C1 is the concept with intent "Artefact" in Figure 1) should be added to all documents $g_i \in G_1$ in formal context $\mathcal{K}_1$ in Table 1a if $g_i$ contains terms "MRI" or "scan" in the relational context of Table 1c. Table 2 shows formal context $\mathcal{K}_1$ after the relational scaling process.

**Table 2.** Context $\mathcal{K}_1$ after relational scaling using existential quantifier

| | author1 | author2 | author3 | author4 | ∃aw : C1 | ∃aw : C2 | ∃aw : C3 | ∃aw : C4 | ∃aw : C5 | ∃aw : C6 | ∃aw : C7 |
|---|---|---|---|---|---|---|---|---|---|---|---|
| g1 | × | × | | | × | × | × | × | | | × |
| g2 | × | × | | | × | × | | | × | × | |
| g3 | × | × | | | × | × | | | × | | × |
| g4 | × | × | | | × | × | × | | × | | |
| g5 | | | × | | | | | | × | × | |
| g6 | | | × | | × | × | | | | | × |
| g7 | | × | × | | × | × | | | | | × |
| g8 | | | × | | × | × | | | | | × |
| g9 | × | | | | × | × | | | | | × |

**Table 3.** Many-valued formal context of term frequencies in each document

| | patient | laparoscopy | scan | user | medicine | response | time | MRI | practice | complication | arthroscopy | infection |
|---|---|---|---|---|---|---|---|---|---|---|---|---|
| g1 | 0.25 | 0.25 | 0.25 | 0 | 0 | 0 | 0 | 0 | 0 | 0.25 | 0 | 0 |
| g2 | 0 | 0 | 0.16 | 0.16 | 0.16 | 0.16 | 0.16 | 0 | 0.16 | 0 | 0 | 0 |
| g3 | 0 | 0.25 | 0 | 0.25 | 0.25 | 0 | 0 | 0.25 | 0 | 0 | 0 | 0 |
| g4 | 0.3 | 0 | 0 | 0 | 0.3 | 0 | 0 | 0.3 | 0 | 0 | 0 | 0 |
| g5 | 0 | 0 | 0 | 0.3 | 0 | 0.3 | 0.3 | 0 | 0 | 0 | 0 | 0 |
| g6 | 0 | 0 | 0 | 0 | 0 | 0 | 0 | 0 | 0.5 | 0 | 0.5 | 0 |
| g7 | 0 | 0 | 0 | 0 | 0 | 0 | 0 | 0 | 0 | 0.5 | 0.5 | 0 |
| g8 | 0 | 0 | 0 | 0 | 0 | 0 | 0 | 0 | 0 | 0.3 | 0.3 | 0.3 |
| g9 | 0 | 0 | 0 | 0 | 0 | 0 | 0 | 0 | 0 | 0 | 0.5 | 0.5 |

## 2.2  Pattern Structure Framework

Pattern structures model a FCA procedure when documents do not have attributes, but rather complex data descriptions such as numerical values, e.g. terms frequency values for each given document as shown in Table 3. In the following, we introduce the pattern structure framework firstly described in [3]. A pattern structure $(G_1, (D, \sqcap), \delta)$ is a generalization of a formal context where $G_1$ is a set of objects, $(D, \sqcap)$ is a semi-lattice of object descriptions and $\delta : G_1 \rightarrow D$ is a mapping associating a description to an object.

In the "interval pattern structures" setting (deeply discussed in [6]), an object descriptions $g \in G_1$ is a vector of intervals $d \in D$, $d = \langle [l_i, r_i] \rangle_{i \in \{1..|M|\}}$ with $l_i, r_i \in \mathbb{R}$ and $l_i \leq r_i$. For example, from Table 3 we have that the set of objects $G_1$ is composed by documents $g_1 - g_9$ (we use this notation for all documents between and including $g_1$ and $g_9$). The object description $\delta(g_1)$ is defined by the vector of intervals $\langle [0.25, 0.25], [0.25, 0.25], [0.25, 0.25], [0, 0], ..., [0.25, 0.25], [0, 0], [0, 0] \rangle$. An interval pattern defines a convex region within the given description space.

In $(D, \sqcap)$ the *similarity* operator $\sqcap$ applied to two object descriptions $d_1 = \langle [l_i^1, r_i^1] \rangle$ and $d_2 = \langle [l_i^2, r_i^2] \rangle$ with $i \in \{1..|M|\}$ and $d_1, d_2 \in D$, returns the convex hull described in Equation 1 while the subsumption order $\sqsubseteq$ between them is given by Equation 2.

$$d_1 \sqcap d_2 = \langle [\min(l_i^1, l_i^2), \max(r_i^1, r_i^2)] \rangle \tag{1}$$

$$d_1 \sqsubseteq d_2 \iff d_1 \sqcap d_2 = d_1 \tag{2}$$

A Galois connection between $\wp(G_1)$ (powerset of $G_1$) and $(D, \sqcap)$ for $A \subseteq G_1$ and $d \in D$ is defined as follows:

$$A^\square = \prod_{g \in A} \delta(g) \qquad\qquad d^\square = \{g \in G | d \sqsubseteq \delta(g)\} \tag{3, 4}$$

Where $A^\square$ represents the common description to all objects in $A$ while $d^\square$ represents the set of objects respecting the description $d$. A pair $(A, d)$ such as $A^\square = d$ and $d^\square = A$ is called an *interval pattern concept (ip-concept)* with extent $A$ and pattern intent $d$. Interval pattern concepts can be ordered in an interval pattern concept lattice (ip-concept lattice).

## 3   Inspiring Problem - Latent Semantic Indexing

### 3.1   Latent Variables Characterization Problem

As previously discussed, LV-models lack a proper characterization for the LVs found through its application. For instance, Latent Semantic Indexing (LSI) [2], a technique commonly used in information retrieval (IR) for indexation, clustering and dimension reduction purposes, is based on the idea that within a document-term matrix (as the one shown in Table 3) there is a set of hidden "latent variables" (LVs) that explain the data which constitutes the matrix. Consequently, LSI describes a technique to uncover these LVs through a "lower-rank approximation" of the original document-term matrix using linear algebra methods (specifically, singular value decomposition (SVD) [12]). Documents can later be described not as vectors of term frequencies, but as vectors of LV values in a reduced vectorial space. Latent variables are supposed to capture the "semantics" in the set of documents, nevertheless it is difficult to grasp this notion while documents are still described by vectors of numeric values. In the following, we provide a further description of the LSI process as described in [2].

### 3.2   Latent Semantic Indexing

Let us consider the values in the formal context in Table 3 as a matrix $A$ of dimensions $9 \times 12$. LSI works through the SVD of matrix $A$ and the consequent calculation of the reduced space of LVs as follows:

$$A_{(9 \times 12)} = U_{(9 \times 9)} \cdot \Sigma_{(9 \times 12)} \cdot V_{(12 \times 12)}^{T} \tag{5}$$

$$\tilde{A}_{(9 \times 12)} = U_{(9 \times k)} \cdot \Sigma_{(k \times k)} \cdot V_{(k \times 12)}^{T} \quad \text{(with } k \ll min(9, 12)) \tag{6}$$

$$A \sim \tilde{A} \tag{7}$$

$$\tilde{A} \cdot \tilde{A}^{T} = U_{(9 \times k)} \cdot \Sigma_{(k \times k)} \cdot V_{(k \times 12)}^{T} \cdot V_{(12 \times k)} \cdot \Sigma_{(k \times k)}^{T} \cdot U_{(k \times 9)}^{T} \tag{8}$$

$$\tilde{A} \cdot \tilde{A}^{T} = (U_{(9 \times k)} \cdot \Sigma_{(k \times k)}) \cdot (U_{(9 \times k)} \cdot \Sigma_{(k \times k)})^{T} \tag{9}$$

Where $(A)^{T}$ denotes the "transpose" of matrix $A$; $U, V$ are orthonormal matrices and $\Sigma$ is a diagonal matrix of "singular values". We have on one side the lower-rank approximation (Equation 7) to a matrix of rank $k$ which is ensured to be the best $k - rank$ matrix approximation by the Frobenius norm difference [12]. On the other hand, we have the dimensional reduction (Equation 9) using matrix $U_{(9 \times k)} \cdot \Sigma_{(k \times k)}$ as the space of documents in $k$ LVs. Table 4 shows this space for matrix $A$ with $k = 2$. Furthermore, Figure 2 presents a graphical representation of documents as points in a plane where we can appreciate the presence of 2 document groups, usually called "clusters". In this paper we use the notion of "cluster" as a convex region in the LV space. In fact, one of the main uses of LSI is to provide a more compact representation of documents so that clusters are easier to find in the space of LVs. Incidentally, an interval pattern in this space represents a cluster (rectangles in Figure 2).

### 3.3 Problem Statement

In Figure 2, while the clusters are easily distinguishable, it is not possible to say why they exist or what are their features. In order to characterize them we need to rely on their relations with terms. For example, we know that documents $g_6 - g_9$ share the term "arthroscopy". While this is not totally clear with documents $g_1 - g_5$ which do

Table 4. Documents in 2 LVs

|    | k1    | k2     |
|----|-------|--------|
| $g_1$ | 0.118 | -0.238 |
| $g_2$ | 0.046 | -0.271 |
| $g_3$ | 0.014 | -0.413 |
| $g_4$ | 0.014 | -0.368 |
| $g_5$ | 0.008 | -0.277 |
| $g_6$ | 0.519 | 0.002  |
| $g_7$ | 0.603 | -0.017 |
| $g_8$ | 0.469 | 0.02   |
| $g_9$ | 0.588 | 0.092  |

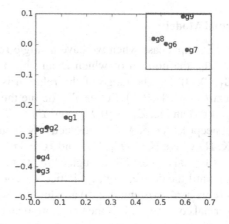

Fig. 2. Graphical representation of documents as points in a 2 dimensional LV space

not share a common term, we can see that documents $g_1 - g_4$ share the term "patient" and $g_2$, $g_3$ and $g_5$ share the term "user". Both terms are related through the annotation "People" extracted from Wordnet (see Table 1b) which lead us to think that LVs can represent differences in this concern.

One way to automatically make these characterizations is through the use of the RCA framework where we can model documents and terms as objects, LV values as document descriptions, and Wordnet annotations as term attributes, while the document-term relation is given by aw in Table 1c. Nevertheless, as explained in the previous section, LSI generates document descriptions in the form of vectors of LV values, while clusters in the LV-space are better represented by interval pattern structures.

The main problem tackled in this work is how to enable the application of RCA in these kinds of scenarios. We provide an adaptation of RCA which allows the relational scaling in pattern structures. We achieve this by the introduction of heterogeneous pattern structures described in the following section. A sub-goal of this work is to find out if domain knowledge can explain the existence and the "semantics" in LVs. We met this sub-goal by the characterization of LVs through the proposed combination of RCA and pattern structures. Given that LVs define a $k$-dimensional space ($k$ being the number of LVs) where documents are organized, we formulate the following questions: Is it possible for us to find sub-regions in the space of LV values related to domain knowledge elements such as Wordnet annotations? And if so, how can we characterize these sub-regions?.

## 4    Adapting RCA for Pattern Structures

In this section we firstly describe the formal model description in which pattern structures are considered into a RCF. We show that the adaptation of the relational scaling operators induces a new space of heterogeneous object descriptions which we support in the framework of heterogenous pattern structures. Following, we provide a full description of this novel pattern structures instance.

### 4.1    Formal Model

Consider the simple case when we have a single relation between two sets of objects $r \subseteq G_1 \times G_2$, the domain of which is an object set in a pattern structure such as $\mathcal{K}_1 = (G_1, (D, \sqcap), \delta)$. The range of the relation is an object set inside a binary formal context $\mathcal{K}_2 = (G_2, M, I)$. Let us also define the relation as the set-valued function $r : G_1 \to \wp(G_2)$ and let $\mathcal{L}_1 = \mathfrak{B}(\mathcal{K}_1)$ and $\mathcal{L}_2 = \mathfrak{B}(\mathcal{K}_2)$ be the pattern concept lattice and the concept lattice of $\mathcal{K}_1, \mathcal{K}_2$ respectively. Thus, we define the relational context family $(\mathbf{K}, \mathbf{R})$ where $\mathbf{K} = \{\mathcal{K}_1, \mathcal{K}_2\}$ and $\mathbf{R} = \{r\}$. The usual RCA procedure induces iterations of formal context $\mathcal{K}_1$ through a "relational scaling" task using $\mathcal{L}_2$ ("target lattice" of $r$), until the derived concept lattice $\mathcal{L}_1$ converges. For this reason, the scaling operators (universal, existential, etc.) are defined over a space of formal contexts into a space of formal contexts. This is the first complication in our model. Since in our setting $\mathcal{K}_1$ is a pattern structure and not a formal context, we cannot directly apply the scaling operators as defined in [9]. Thus, we move forward to redefine relational scaling operators which support pattern structures. To achieve this, let us define, for a relation $r$, a function that assigns a set of relational attributes to a given object in the pattern structure depending on the type of relational scaling applied (universal, existential, etc.).

**Definition 1.** Let $r \subseteq G_1 \times G_2$ be a relation between two object sets where $\mathcal{L}_2$ is its target lattice composed by formal concepts C. We define the potential set of all possible relational attributes $P_r$ scaled from relation $r$ as follows [3]:

$$P_r = \{r : C, \forall C \in \mathcal{L}_j\} \tag{10}$$

For $g \in G_1$, we also define two functions $\rho_r^{\exists}, \rho_r^{\forall\exists} : G_1 \to \wp(P_r)$ which assign a set of relational attributes to a given object using the 'existential quantifier operator ($\exists$)" and the "universal-existential quantifier operator ($\forall\exists$)" respectively.

$$\rho_r^{\exists}(g) = \{r : C \in P_r \mid r(g) \cap \text{extent}(C) \neq \emptyset\} \tag{11}$$
$$\rho_r^{\forall\exists}(g) = \{r : C \in P_r \mid r(g) \neq \emptyset, r(g) \subseteq \text{extent}(C)\} \tag{12}$$

Hereafter we refer to $\rho_r^{\exists}(g)$ or $\rho_r^{\forall\exists}(g)$ as the "relations of g".

**Example 1.** Let the following scenario be the running example for the remainder of this article. Consider a relational context family of two contexts $\mathbf{K} = \{\mathcal{K}_1, \mathcal{K}_2\}$ where $\mathcal{K}_1 = (G_1, (D, \sqcap), \delta)$ is the interval pattern structure of documents and their LV values shown in Table 4 and $\mathcal{K}_2$ is the formal context of terms and their Wordnet annotations shown in Table 1b. Consider as well the relation "document *annotated with* term" as shown in Table 1c such as $\mathbf{R} = \{r\}$. From the initial lattice shown in Figure 1 we can construct the set of relational attributes $P_r = \{aw : C_i\}$ where $i \in [0, 7]$ (i.e. each $C_i$ corresponds to one formal concept shown in the lattice[4]). Then, we have:

$$r(g_1) = \{\text{patient}, \text{laparoscopy}, \text{scan}, \text{complication}\}$$
$$\text{extent}(C1) = \{\text{MRI}, \text{scan}\}$$
$$r(g_1) \cap \text{extent}(C1) = \{\text{scan}\} \neq \emptyset \implies aw : C1 \in \rho_r^{\exists}(g_1)$$
$$\rho_r^{\exists}(g_1) = \{aw : C1, aw : C2, aw : C3, aw : C4, aw : C7\}$$

**Definition 2.** Let $(G_1, (D, \sqcap), \delta)$ be a pattern structure for a set of objects $G_1$ which are also associated with relational attributes in a set $P_r$ through $\rho_r^{\exists}$ or $\rho_r^{\forall\exists}$. We define the scaled pattern structure $(G_1, (H, \sqcap), \Delta)$ with mappings $\Delta^{\exists}, \Delta^{\forall\exists} : G_1 \to H$ as follows:

$$H = D \times \wp(P_r) \tag{13}$$
$$\Delta^{\exists}(g) = (\delta(g), \rho_r^{\exists}(g)) \tag{14}$$
$$\Delta^{\forall\exists}(g) = (\delta(g), \rho_r^{\forall\exists}(g)) \tag{15}$$

Where H contains heterogeneous descriptions of objects in $G_1$ combining both, a pattern $\delta(g) \in D$ and a set of relational attributes in $P_r$.

---

[3] Normally, the relational attributes $r : C$ have the operator $\exists$ or $\forall\exists$ attached as a prefix indicating the scaling operation applied. In this work, we omit the prefixes in favour of generality. Nevertheless, the scaling function will remain indicated at each step.

[4] aw stands for "annotated with". In the remainder of this article we will always work with the existential quantifier.

**Definition 3.** Let $r$ be a relation between two objects sets, then the existential scaling operator $(\text{sc}_r^{\exists})$ and the universal scaling operator $\text{sc}_r^{\forall\exists}$ for a pattern structure $\mathcal{K}_1$ are defined as:

$$\text{sc}_r^{\exists}(\mathcal{K}_1) = (G_1, (H, \sqcap), \Delta^{\exists}) \qquad \text{sc}_r^{\forall\exists}(\mathcal{K}_1) = (G_1, (H, \sqcap), \Delta^{\forall\exists}) \qquad (16, 17)$$

As shown in Definitions 2 and 3, in order to apply the relational scaling operation to a pattern structure, it is necessary to define a new different pattern structure in which we can consider the original object description $\delta(g)$ and its relational attributes $\rho_r^{\exists}(g)$ or $\rho_r^{\forall\exists}(g)$. This combination of descriptions or "heterogeneous descriptions" H is a Cartesian product between the set of object descriptions and the powerset of $P_r$ to which objects are mapped through $\Delta^{\exists} : G_1 \to H$. We denominate this new pattern structure instance "heterogeneous pattern structures". In the following, we provide a complete description of its characteristics and capabilities.

**Example 2.** Table 5 shows a representation of the heterogeneous pattern structure of documents with LVs and relational attributes, where we can find an object description such as:

$$\Delta^{\exists}(g_1) = (\delta(g_1), \rho_r^{\exists}(g_1))$$
$$\delta(g_1) = \langle [0.118, 0.118], [-0.238, -0.238] \rangle$$
$$\rho_r^{\exists}(g_1) = \{\text{aw} : \text{C1}, \text{aw} : \text{C2}, \text{aw} : \text{C3}, \text{aw} : \text{C4}, \text{aw} : \text{C7}\}$$

**Table 5.** Result of relational scaling in the example pattern structure represented in a hybrid formal context. We have removed the relational attribute $\text{aw} : \text{C0}$ usually assigned to every object.

| | D | | $P_r$ | | | | | | |
|---|---|---|---|---|---|---|---|---|---|
| | k1 | k2 | aw : C1 | aw : C2 | aw : C3 | aw : C4 | aw : C5 | aw : C6 | aw : C7 |
| $g_1$ | 0.118 | -0.238 | × | × | × | × | | | × |
| $g_2$ | 0.046 | -0.271 | × | × | | × | × | | |
| $g_3$ | 0.014 | -0.413 | × | × | | × | | | × |
| $g_4$ | 0.014 | -0.368 | × | × | | × | | | |
| $g_5$ | 0.008 | -0.277 | | | | × | × | | |
| $g_6$ | 0.519 | 0.002 | | × | × | | | | × |
| $g_7$ | 0.603 | -0.017 | | × | × | | | | × |
| $g_8$ | 0.469 | 0.02 | | × | × | | | | × |
| $g_9$ | 0.588 | 0.092 | | × | × | | | | × |

### 4.2   Heterogeneous Pattern Structures

**Definition 4.** Let $H = D \times \wp(P_r)$ be a set of heterogeneous object descriptions, where $h_1 = (d_1, B_1)$ and $h_2 = (d_2, B_2)$ are two heterogeneous object descriptions with $d_1$, $d_2 \in D$, $B_1, B_2 \subseteq P_r$ and $h_1, h_2 \in H$ (the elements d and B are referred to as the "components" of h). We define the "similarity operator" $\sqcap$ between $h_1$ and $h_2$ as:

$$h_1 \sqcap h_2 = (d_1 \sqcap d_2, B_1 \cap B_2) \qquad (18)$$

**Example 3.** The similarity operator applied to the object descriptions of $g_1$ and $g_2$ is:

$$\Delta^{\exists}(g_1) \sqcap \Delta^{\exists}(g_2) = (\delta(g_1) \sqcap \delta(g_2), \rho_r^{\exists}(g_1) \cap \rho_r^{\exists}(g_2))$$
$$\delta(g_1) \sqcap \delta(g_2) = \langle [0.046, 0.118], [-0.271, -0.238] \rangle$$
$$\rho_r^{\exists}(g_1) \cap \rho_r^{\exists}(g_2) = \{\text{aw} : \text{C1}, \text{aw} : \text{C2}, \text{aw} : \text{C4}\}$$

**Proposition 1.** $(H, \sqsubseteq)$ *with* $\sqcap$ *as described in Definition 4 is the direct product of the ordered sets* $(D, \sqsubseteq)$ *and* $(\wp(P_r), \subseteq)$ *and thus is an ordered set itself.*

*Proof.* In order to prove that $(H, \sqsubseteq)$ is the direct product of $(D, \sqsubseteq)$ and $(\wp(P_r), \subseteq)$, we show that $h_1 \sqsubseteq h_2 : \iff d_1 \sqsubseteq d_2$ and $B_1 \subseteq B_2$ (as described in [4]).

$$h_1 \sqsubseteq h_2 \iff h_1 \sqcap h_2 = h_1 \qquad \text{Equation 2} \qquad (19)$$
$$\iff (d_1 \sqcap d_2, B_1 \cap B_2) = (d_1, B_1) \qquad \text{Definition 4} \qquad (20)$$
$$\iff d_1 \sqcap d_2 = d_1 \text{ and } B_1 \cap B_2 = B_1 \qquad\qquad (21)$$
$$\iff d_1 \sqsubseteq d_2 \text{ and } B_1 \subseteq B_2 \qquad\qquad (22)$$

$$\square$$

Because of Proposition 1, we would like to know how the heterogeneous pattern concept lattice is related to the concept lattices of its components, namely the pattern concept lattice $(G_1, (D, \sqcap), \delta)$ and the concept lattice of the formal context of objects and their respective relational attributes $(G_1, P_r, I)$ where the incidence relation $I$ is defined in Equation 23. Regarding this, for the following definitions we introduce an alternative description for the standard FCA derivation operator $(\cdot)'$ in Equation 24 for a subset of objects $A \in G_1$ using the function $\rho_r^{\exists}$.

$$I = \bigcup_{g \in G_1} \{(g, m), \forall m \in \rho_r^{\exists}(g)\} \qquad\qquad A' = \bigcap_{g \in A} \rho_r^{\exists}(g) \qquad (23, 24)$$

**Definition 5.** The derivation operators $(\cdot)^{\circ}$ in $(G_1, (H, \sqcap), \Delta^{\exists})$ for an object set $A \in G_1$ and a heterogeneous element $h \in H$ are defined as:

$$A^{\circ} = \prod_{g \in A} \Delta^{\exists}(g) \qquad\qquad h^{\circ} = \{g \in G_1 \iff h \sqsubseteq \Delta^{\exists}(g)\} \quad (25, 26)$$

A heterogeneous pattern concept (hp-concept) is then defined as the pair $(A, h)$ where $h^{\circ} = A$ and $A^{\circ} = h$.

**Proposition 2.** *The derivation operator applied to a heterogeneous element* $h = (d, B)$ *is equal to the intersection of the derivation operator on its components:*

$$(d, B)^{\circ} = d^{\square} \cap B' \qquad\qquad (27)$$

**Table 6.** Table showing different object sets under different closures. $A_1$ is a proper extent of $(G_1, (H, \sqcap), \Delta^\exists)$ because its closed under $(\cdot)^\diamond$ while $A_2$ is not. $A_3$ and $A_4$ are examples of "pure hp-concepts". $A_5$ is an example of a "mixed hp-concept".

| Extent $A_i$ | $(A_i)^{\sqcup\sqcup}$ | $(A_i)''$ | $A^{\diamond\diamond} = A^{\sqcup\sqcup} \cap A''$ | $(A_i)^\diamond$ |
|---|---|---|---|---|
| $A_1 = \{g_1, g_3\}$ | $\{g_1 - g_4\}$ | $\{g_1, g_3\}$ | $\{g_1, g_3\}$ | - |
| $A_2 = \{g_5, g_9\}$ | $\{g_1, g_2, g_5, g_6, g_8, g_9\}$ | $G_1$ | $\{g_1, g_2, g_5, g_6, g_8, g_9\}$ | - |
| $A_3 = \{g_1, g_6 - g_9\}$ | $A_3$ | $A_3$ | $A_3$ | $(A_3^{\sqcup}, A_3')$ |
| $A_4 = \{g_6, g_7\}$ | $A_4$ | $A_3$ | $A_4$ | $(A_4^{\sqcup}, A_4')$ |
| $A_5 = \{g_1, g_3, g_7\}$ | $\{g_1 - g_4, g_7\}$ | $\{g_1, g_3, g_6 - g_9\}$ | $A_5$ | $(A_5^{\sqcup}, A_5')$ |

*Proof.* Let $g \in h^\diamond$, with $h = (d, B)$, by Equation 26 we have:

$$g \in h^\diamond \iff h \sqsubseteq \Delta^\exists(g) \iff d \sqsubseteq \delta(g) \text{ and } B \subseteq \rho_r^\exists(g) \qquad \text{Proposition 1}$$

The right side of last formula shows two conditions. Using Equation 4, we have that the first condition yields $d \sqsubseteq \delta(g) \iff g \in d^{\sqcup}$. As for the second condition, in Equation 23, we have that $(g, m) \in I, \forall m \in \rho_r^\exists(g)$. Then, $\forall m \in (B \subseteq \rho_r^\exists(g))$ we have that $(g, m) \in I$ and thus $g \in B'$. With this we have that:

$$g \in (d, B)^\diamond \iff g \in d^{\sqcup} \text{ and } g \in B'$$
$$(d, B)^\diamond = d^{\sqcup} \cap B' \qquad \square$$

**Proposition 3.** *The closure of a set of objects $A \in G_1$ (an extent) is equal to the intersection of its closures in each component.*

$$A^{\diamond\diamond} = A^{\sqcup\sqcup} \cap A'' \qquad (28)$$

*Proof.*

$$A^{\diamond\diamond} = \left( \bigsqcap_{g \in A} \Delta^\exists(g) \right)^\diamond = \left( \bigsqcap_{g \in A} \delta(g), \bigcap_{g \in A} \rho_r^\exists(g) \right)^\diamond = (A^{\sqcup}, A')^\diamond = A^{\sqcup\sqcup} \cap A'' \qquad \square$$

From Proposition 3, we can see three different conditions for a heterogeneous extent $A$, namely it can be closed in both of its components ($A^{\diamond\diamond} = A^{\sqcup\sqcup} = A''$), in only one (either $A^{\sqcup\sqcup} \subseteq A''$ or $A'' \subseteq A^{\sqcup\sqcup}$), or in none ($A^{\sqcup\sqcup} \not\subseteq A''$ or $A'' \not\subseteq A^{\sqcup\sqcup}$). Further in is this section, we provide a full description for these kinds of extents. Nevertheless, Proposition 3 provides us with two ways to calculate the set of heterogeneous pattern concepts. Firstly, Equation 28 is a canonical test which can be used in standard FCA algorithms such as AddIntent [13]. Secondly, we can calculate the complete set of extents from both, the formal context and the pattern structure separately and intersect them to calculate each possible heterogeneous extent.

**Example 4.** Consider the object set $A_1$ in Table 6. The closure in the fifth column shows that $A_1 = A_1^{\diamond\diamond}$ and thus it is a proper extent of $(G_1, (H, \sqcap), \Delta^\exists)$. This is not the case for $A_2$.

**Proposition 4.** *The closure of a heterogeneous description* $h \in H$ *is given by:*

$$h^{\diamond\diamond} = (h^{\diamond\square}, h^{\diamond\prime}) \tag{29}$$

Proposition 4 can be demonstrated analogously to Proposition 3. We are interested in Proposition 4 because it allows us to easily calculate the heterogeneous intents as we show next.

**Proposition 5.** *Let* $A_1$ *be an extent in* $(G_1, (D, \sqcap), \delta)$ *and* $A_2$ *be an extent in* $(G_1, P_r, I)$ *where* $A_1 \subseteq A_2$ *and for any other extent* $A$ *in* $(G_1, P_r, I)$ *we have* $A_1 \subseteq A \subseteq A_2 \iff A_2 = A$, *i.e.* $A_2$ *is the cover of* $A_1$. *Then for* $h = (A_1^\square, A_2')$, $h$ *is a heterogeneous intent and* $(A_1, h)$ *is a hp-concept.*

*Proof.* We show that $(A_1^\square, A_2')^{\diamond\diamond} = (A_1^\square, A_2')$

$$(A_1^\square, A_2')^{\diamond\diamond} = (A_1^{\square\square} \cap A_2'')^\diamond = (A_1 \cap A_2)^\diamond = (A_1)^\diamond$$

$$= \prod_{G \in A_1} \Delta^\exists(g) = \left( \prod_{G \in A_1} \delta(g), \bigcap_{G \in A_1} \rho_r^\exists(G) \right)$$

$$= (A_1^\square, A_1') = (A_1^\square, A_2')$$

The last step can be shown by the restrictions imposed to $A_1$ and $A_2$ as follows:
$$A_1 \subseteq A_2 \implies A_1 \subseteq A_1'' \subseteq A_2 \implies A_1'' = A_2 \implies A_1' = A_2' \qquad \square$$

Similarly, it can be shown that when $A_2 \subseteq A_1$, the hp-concept $(A_2, (A_2^\square, A_2'))$ exists. Proposition 5 shows that the extents in the pattern structure $(G_1, (D, \sqcap), \delta)$ and in $(G_1, P_r, I)$ will be present in the lattice of hp-concepts. Nevertheless, these do not cover the whole set of hp-concepts in $(G_1, (H, \sqcap), \Delta^\exists)$.

As previously discussed, the set of hp-concepts (denoted as $\mathfrak{B}((G_1, (H, \sqcap), \Delta^\exists))$) can be characterized as containing three types of extents, those that are closed under both components, those that are closed under one of its components and those that are an intersection of two different closed extents. We call these types "pure hp-concepts", "semi-pure hp-concepts" and "mixed hp-concepts" respectively.

**Definition 6.** Given a hp-concept $(A, h) \in (G_1, (H, \sqcap), \Delta^\exists)$ we say that:

$$(A, h) \text{ is "pure" iff } A^{\square\square} = A'' \tag{30}$$

$$(A, h) \text{ is "semi-pure" iff } A^{\square\square} \subseteq A'' \text{ or } A'' \subseteq A^{\square\square} \tag{31}$$

$$(A, h) \text{ is "mixed" iff } A^{\square\square} \cap A'' \neq \emptyset \text{ and } A^{\square\square} \not\subseteq A'' \text{ and } A'' \not\subseteq A^{\square\square} \tag{32}$$

**Example 5.** In Table 6, $A_3$ is a pure hp-concept extent since it is closed in both components. $A_4$ is a semi-pure hp-concept extent since it is closed in the pattern structure component but not in the relational attribute component. $A_5$ is a mixed hp-concept as it is closed in the hp-lattice but not in either of its components.

In order to obtain the whole set of hp-concepts, it is not sufficient to calculate the sets of pattern concepts and formal concepts from its respective components and match them using Proposition 5. Doing so only provides us with the set of pure and semi-pure hp-concepts, while the set of mixed hp-concepts will be missing. In the following, we describe our method to compute the whole set of hp-concepts.

**Table 7.** Scaled representation context for the running example. Patterns in D are represented by cardinals from 2 to 33 (number 1 was eliminated as it references the pattern concept ⊤).

| | D | | | | | | | | | | | | | | | | | | | | | | | | | | | | | | | | P_r | | | | | | |
|---|---|---|---|---|---|---|---|---|---|---|---|---|---|---|---|---|---|---|---|---|---|---|---|---|---|---|---|---|---|---|---|---|---|---|---|---|---|---|---|
| | 2 | 3 | 4 | 5 | 6 | 7 | 8 | 9 | 10 | 11 | 12 | 13 | 14 | 15 | 16 | 17 | 18 | 19 | 20 | 21 | 22 | 23 | 24 | 25 | 26 | 27 | 28 | 29 | 30 | 31 | 32 | 33 | aw:C0 | aw:C1 | aw:C2 | aw:C3 | aw:C4 | aw:C5 | aw:C7 |
| g1 | × | × | × | × | × | × | × | | | | | | | | | | | | | | | | | | | | | | | | | | × | × | × | × | × | | × |
| g2 | × | × | × | × | × | × | × | × | × | × | × | × | | | | | | | | | | | | | | | | | | | | | × | × | × | | × | × | |
| g3 | | × | × | | | | × | × | | | × | × | × | | | | | | | | | | | | | | | | | | | | × | × | × | | × | | × |
| g4 | | × | × | × | × | | | × | × | × | × | | | | | × | × | × | × | | | | | | | | | | | | | | × | × | × | | × | | |
| g5 | | | × | × | × | | | | × | × | × | | | × | | × | × | | | | | | | | | | | | | | | | × | | | | | × | × |
| g6 | | | | | | | | | | | | | | | | | | × | × | × | × | × | × | × | × | | | | | | | | × | | | × | × | | × |
| g7 | | | | | | | | | | | | | | | | | | | × | × | × | × | | | | × | × | | | | | | × | | | × | × | | × |
| g8 | | | | | | | | | | | | | | | | | | | | × | × | | | × | × | | | × | × | | | | × | | | × | × | | × |
| g9 | | | | | | | | | | | | | | | | | | | | × | × | | | × | × | | | × | | × | × | × | × | | | × | × | | × |

## 4.3 Calculating the hp-lattice

The heterogeneous pattern structure $(G_1, (H, \sqcap), \Delta^\exists)$ has been defined as a standard pattern structure and thus a standard algorithm to calculate pattern concept lattices can be used to obtain the hp-lattice. Some of these algorithms have been described and discussed in [6]. However, a much simpler manner to calculate the hp-lattice is through the use of a "scaled representation context".

A "representation context", as explained in [3], is a mechanism of complex data binarization. The pattern concepts of a pattern structure and the formal concepts of its derived representation context are in 1-1 correspondence and furthermore, their extents are the same [3,7]. In the particular case of a heterogeneous pattern structure as described in this work, we use the representation context of the pattern structure component which is later "relationally scaled" in terms of traditional RCA (see Section 2.1).

**Definition 7.** Let $(G_1, (H, \sqcap), \Delta^\exists)$ be a heterogeneous pattern structure with components $(G_1, (D, \sqcap), \delta)$ and $(G_1, P_r, I)$. The "scaled representation context" is defined as $(G_1, D \cup P_r, J)$ where the incidence relation is:

$$(g, x) \in J \iff x \sqsubseteq \delta(g) \text{ or } x \in \rho_r^\exists(g); \forall g \in G_1 \text{ and } x \in (D \cup P_r)$$

In other words, $(G_1, D \cup P_r, J)$ is the representation context of the pattern structure $(G_1, (D, \sqcap), \delta)$ plus the relational scaling of $(G_1, P_r, I)$. It can be shown that, in fact, this "scaled representation context" is isomorphic to the representation context of $(G_1, (H, \sqcap), \Delta^\exists)$. For the running example, we constructed the scaled representation context as depicted in Table 7. In this context, patterns and relational attributes are treated equally, hence the attribute set $D \cup P_r$. Patterns in D were filtered using a similarity threshold as described in [5], since the complete non-restricted pattern lattice contain a little more than 100 concepts. Incidentally, the filter by similarity applied to the calculation of D caused the hp-lattice derived from the context in Table 7 to contain only pure and semi-pure, i.e. their extents are either closed under $(\cdot)^\square$ or $(\cdot)'$ or both.

While there are some drawbacks w.r.t. the computational costs associated with the calculation of the formal concepts of the representation context, in this work we disregard them favouring the simplicity of the combined model.

# 5    Discussion and Conclusions

In Section 3.3 we proposed two questions that we discuss in the following.

**Is it possible for us to find sub-regions in the space of LV values related to domain knowledge elements?** Indeed, we can. A hp-concept describes exactly this in its intent as a relation of an interval pattern and a set of annotations in the Wordnet taxonomy. Moreover, these relations can be better described in the form of association rules [4]. Particularly, we are searching for those association rules with a premise in the space of latent variables and a consequence in the space of relational attributes, For example, we have the rule $6 \leftrightarrow aw : C4$ which means that the latent variable region in the interval pattern numbered 6 implies the Wordnet concept "People" as shown in Figure 3. While all kinds of association rules exist in the lattice of the scaled representation context, we are only interested in those related to our specific problem. Figure 3 presents a graphical representation for the association rules extracted on the running example. The map represents what can be called a "labelled hierarchical document clustering" [8] over the space of latent variables. In the map, the region marked as "Activity" is actually a union of two contiguous regions.

**How can we characterize the relations among sub-regions in the space of LV values and domain knowledge elements?** We have already described three types of hp-concepts, namely pure, semi-pure and mixed. In the following, we provide them with a characterization. Let us first introduce the Jaccard index [8] in terms of the hp-concept's extents and the extents of its components as follows ($| \cdot |$ represents set cardinality):

$$J(A^{\square\square}, A'') = \frac{|A^{\square\square} \cap A''|}{|A^{\square\square} \cup A''|} = \frac{|A^{\diamond\diamond}|}{|A^{\square\square} \cup A''|}$$

Pure hp-concepts are interesting since they represent strong coherent relations between clusters in different spaces. Moreover, for any given pure hp-concept $(A, h)$, the Jaccard index $J(A^{\square\square}, A'') = 1$. Consider for example, the pure hp-concept with extent $g_1 - g_5$ (region 6) which represents a "closed" region in the latent variable space related to the topic "People", i.e. outside this region, there are no documents related to "People". We can also relate "pure hp-concepts" as describing *necessary and sufficient conditions* of a defined concept in the description logics framework (DL) [1]. In this case, documents in region 6 have the necessary and sufficient condition of being labelled with the annotation "People".

A semi-pure hp-concept represents a directional coherence, i.e. either $A^{\square\square} \subseteq A''$ or $A'' \subseteq A^{\square\square}$. The Jaccard index is determined by $A^{\square\square}/A''$ in the first case or $A''/A^{\square\square}$ in the later. For example, the hp-concept with extent $g_6 - g_7$ (region 22) contains documents related to "Illness and "Surgery", but it does not contain all of them (i.e. $g_1$ is an exception). Thus, we can call a semi-pure hp-concept an "open" region in the latent variable space. In DL terms, semi-pure hp-concepts represent necessary conditions, i.e. region 22 have the necessary but not sufficient condition of being labelled with the annotation "Surgery". Mixed hp-concepts represent a weak coherence of clusters. In general, their Jaccard index will be lower than the index of semi-pure hp-concepts.

Finally, we can conclude that the technique presented in this paper is able to find useful relations among convex latent variable regions and domain knowledge which allows

**Table 8.** Table showing an imaginary mixed hp-concept

| $A_i$ | objects | $(A_i)^{\square\square}$ | $(A_i)''$ | $A^{\diamond\diamond}$ |
|---|---|---|---|---|
| $A_a$ | $\{g_1, g_3, g_4\}$ | $\{g_1, g_2, g_3, g_4\}$ | $\{g_1, g_3, g_4, g_5\}$ | $\{g_1, g_3, g_4\}$ |
| $A_a^{\diamond} = (\langle [0.014, 0.118][-0.413, -0.238]\rangle, \{aw : C4\})$ | | | | |

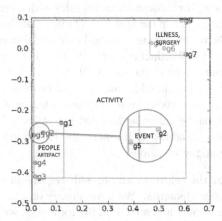

**Fig. 3.** Labelled document clusters using association rules from the hp-lattice with magnification on documents $g_2$ and $g_5$

giving a proper characterization to the latent variable space, and hence, the latent variables themselves. This is possible due to the simultaneous representation of documents in the latent variable vectorial space and the set of relational attributes as hp-concepts.

The implications of this work are multiple. In this work we have superficially described some connections with descriptions logics. Furthermore, the notion of mixed hp-concepts, left unexplored in this work, lead us to think that they may be useful for annotation and data correction purposes. Other application domains seem also to fit as heterogeneous pattern structures. For example, in image annotation, images are characterized as vectors of features which are then aligned with annotations in the Wordnet taxonomy.

# References

1. Baader, F., Calvanese, D., McGuinness, D.L., Nardi, D., Patel-Schneider, P.F. (eds.): The Description Logic Handbook: Theory, Implementation, and Applications. Cambridge University Press, New York (2003)
2. Deerwester, S., Dumais, S.T., Landauer, T.K., Furnas, G.W., Harshman, R.: Indexing by latent semantic analysis. Journal of the American Society for Information Science 41(6) (1990) 1097–4571
3. Ganter, B., Kuznetsov, S.O.: Pattern structures and their projections. In: Delugach, H.S., Stumme, G. (eds.) ICCS 2001. LNCS (LNAI), vol. 2120, pp. 129–142. Springer, Heidelberg (2001)

4. Ganter, B., Wille, R.: Formal Concept Analysis: Mathematical Foundations. Springer (December 1999)
5. Kaytoue, M., Assaghir, Z., Napoli, A., Kuznetsov, S.O.: Embedding tolerance relations in formal concept analysis. In: Proceedings of the 19th ACM International Conference on Information and Knowledge Management -CIKM 2010, p. 1689. ACM Press, New York (2010)
6. Kaytoue, M., Kuznetsov, S.O., Napoli, A.: Revisiting numerical pattern mining with formal concept analysis. In: Proceedings of the Twenty-Second International Joint Conference on Artificial Intelligence, vol. 2, pp. 1342–1347 (November 2011)
7. Kuznetsov, S.O.: Pattern Structures for Analyzing Complex Data. In: Sakai, H., Chakraborty, M.K., Hassanien, A.E., Ślęzak, D., Zhu, W. (eds.) RSFDGrC 2009. LNCS, vol. 5908, pp. 33–44. Springer, Heidelberg (2009)
8. Manning, C.D., Raghavan, P., Schütze, H.: Introduction to Information Retrieval (July 2008)
9. Rouane-Hacene, M., Huchard, M., Napoli, A., Valtchev, P.: A proposal for combining Formal Concept Analysis and description Logics for mining relational data. In: Kuznetsov, S.O., Schmidt, S. (eds.) ICFCA 2007. LNCS (LNAI), vol. 4390, pp. 51–65. Springer, Heidelberg (2007)
10. Rouane-Hacene, M., Huchard, M., Napoli, A., Valtchev, P.: Relational concept analysis: mining concept lattices from multi-relational data. Annals of Mathematics and Artificial Intelligence 67(1), 81–108 (2013)
11. Srivastava, A., Sahami, M.: Text Mining: Classification, Clustering, and Applications, 1st edn. Chapman & Hall/CRC (2009)
12. Trefethen, L., Bau, D.: Numerical Linear Algebra. Society for Industrial and Applied Mathematics, SIAM (1997)
13. van der Merwe, D., Obiedkov, S., Kourie, D.: AddIntent: A New Incremental Algorithm for Constructing Concept Lattices. In: Eklund, P. (ed.) ICFCA 2004. LNCS (LNAI), vol. 2961, pp. 372–385. Springer, Heidelberg (2004)

# RCA as a Data Transforming Method: A Comparison with Propositionalisation

Xavier Dolques[1], Kartick Chandra Mondal[2], Agnès Braud[2],
Marianne Huchard[3], and Florence Le Ber[1]

[1] ICube, University of Strasbourg/ENGEES, CNRS
{xavier.dolques,florence.leber}@engees.unistra.fr
[2] ICube, University of Strasbourg, CNRS
{mondal,agnes.braud}@unistra.fr
[3] LIRMM, University of Montpellier 2, CNRS
huchard@lirmm.fr

**Abstract.** This paper aims at comparing transformation-based approaches built to deal with relational data, and in particular two approaches which have emerged in two different communities: Relational Concept Analysis (RCA), based on an iterative use of the classical Formal Concept Analysis (FCA) approach, and Propositionalisation coming from the Inductive Logic Programming community. Both approaches work by transforming a complex problem into a simpler one, namely transforming a database consisting of several tables into a single table. For this purpose, a main table is chosen and new attributes capturing the information from the other tables are built and added to this table. We show the similarities between those transformations for what concerns the principles underlying them, the semantics of the built attributes and the result of a classification performed by FCA on the enriched table. This is illustrated on a simple dataset and we also present a synthetic comparison based on a larger dataset from the hydrological domain.

## 1 Introduction

In several applications, data present various characteristics (e.g. many-valued, temporal, spatial) which are not easy to take into account. Relational data in particular are generally transformed into a single table to be processed by data mining methods. In the field of Inductive Logic Programming, propositionalisation approaches (PA) aim at performing such transformations [1]. These approaches can be divided into database-oriented and logic-oriented such as the HiFi method [2]. HiFi allows to build features that are first-order logic conjunctions from related tables. In the field of Formal Context Analysis (FCA, [3]), relational information is addressed by Relational Concept Analysis (RCA, [4]). It has been designed to handle several formal contexts, corresponding to several categories of objects, and several relations between these objects, based on an iterative use of the classical Formal Concept Analysis algorithm. RCA classifies the objects of the different categories in lattices that are connected via relational

C.V. Glodeanu, M. Kaytoue, and C. Sacarea (Eds.): ICFCA 2014, LNAI 8478, pp. 112–127, 2014.

attributes. The analysis often focuses on a main category of objects, classified in a lattice which is the central point for analyzing data, while navigating towards the other, secondary lattices. Both methods enable us to turn the objects linked to a given object into special attributes, that are propositional features for PA or relational attributes for RCA.

In this paper we propose to compare the two methods, focusing on the semantics of the built attributes, in the context of acyclic data. FCA was used as a common classification method: it was applied on the propositional features obtained by the HiFi method from a given relational dataset and the resulting lattice was compared to the one obtained by the RCA method on the same relational dataset. We detail our comparison on a simple example about pizzas and their ingredients. Another comparison is also performed on a larger dataset from the hydrological domain. The lattices obtained appeared to be isomorphic and allowed to reveal the links between the propositional features in HiFi and the concept generators in RCA.

The paper is organized as follows. Section 2 describes a simple example that is used in Section 3 and 4 to introduce the principles of the RCA and Propositionalisation approaches. Section 5 details the results of the comparison performed both on the simple example and on the real dataset. Related work is described in Section 6. Section 7 concludes and draws some perspectives of this work.

## 2   A Motivating Example

The considered objects of our dataset (see Table 1) are people, pizzas, and ingredients. People are farmers described by their current production methodology (organic versus conventional). Pizzas are described by some typology of their shape (thin, thick, calzone). Ingredients are described by their category (fruit/vegetable, meat, fish, dairy). Two relations link these objects: People *prefer* some pizzas, pizzas *have* some ingredients.

A group of people (Juliet, Nancy and Alice) likes at least one pizza containing one dairy ingredient. A subgroup of this group (Nancy and Alice) corresponds to the conventional farmers and we deduce that in this dataset all conventional farmers like at least one pizza containing one dairy ingredient.

For extracting this kind of knowledge from the various relations, it is worth noting that several of them have to be crossed (here the relations **Prefers** and **HasIngredient**). Besides, the group Juliet, Nancy and Alice has initially no pizza in common and no common production methodology, because Juliet is an organic farmer, while Nancy and Alice are conventional farmers. Thus there is no direct reason for grouping these three people. The group Juliet, Nancy and Alice can be formed after two classification steps: (1) the recognition of pizzas Arctic, Lorraine, ThreeCheeses, and FourCheeses as belonging to the group $D$ of pizzas with at least one dairy ingredient; (2) the fact that Juliet, Nancy and Alice like at least one pizza from the $D$ group.

Such a kind of classification is the objective of the two approaches that we study in the following of this paper. This simple dataset can thus be used to exemplify the properties of these two approaches.

**Table 1.** The dataset

PEOPLE

| Name | ProdMethod |
|------|------------|
| Arthur | OrganicFarmer |
| John | OrganicFarmer |
| Alice | ConventionalFarmer |
| Juliet | OrganicFarmer |
| Nancy | ConventionalFarmer |

INGREDIENT

| IngName | Category |
|---------|----------|
| TomatoSauce | FruitVegetable |
| Cream | Dairy |
| Onion | FruitVegetable |
| Bacon | Meat |
| Salmon | Fish |
| SoyCream | FruitVegetable |
| Mozza | Dairy |
| GoatCheese | Dairy |
| Emmental | Dairy |
| FourmeAmbert | Dairy |
| EggPlant | FruitVegetable |
| Mushroom | FruitVegetable |

PIZZA

| PizzaName | Shape |
|-----------|-------|
| Forest | Thick |
| Occitane | Calzone |
| ThreeCheeses | Thin |
| FourCheeses | Thin |
| Lorraine | Thin |
| Arctic | Thick |

PREFERS

| Name | PizzaName |
|------|-----------|
| Arthur | Forest |
| John | Occitane |
| Alice | FourCheeses |
| | Lorraine |
| Juliet | ThreeCheeses |
| | Arctic |
| Nancy | Arctic |

HASINGREDIENT

| PizzaName | IngName |
|-----------|---------|
| Forest | SoyCream |
| | Mushroom |
| Occitane | TomatoSauce |
| | Onion |
| | EggPlant |
| ThreeCheeses | TomatoSauce |
| | Mozza |
| | GoatCheese |
| | Emmental |
| FourCheeses | TomatoSauce |
| | Cream |
| | Mozza |
| | GoatCheese |
| | Emmental |
| | FourmeAmbert |
| Lorraine | Cream Onion |
| | Bacon |
| | Mozza |
| Arctic | TomatoSauce |
| | Cream Salmon |
| | Mozza |

# 3  Relational Concept Analysis

In this part, the principles of relational concept analysis are presented based on the example described in Section 2. For more details about RCA, the reader is invited to read [5] which refines notations of [4].

The pizza dataset cannot be directly handled by RCA, it must first be transformed. Here, we choose to make a nominal scaling of the three tables PEOPLE, PIZZA and INGREDIENT to obtain three object-attribute contexts, respectively $\mathcal{K}_{People}$, $\mathcal{K}_{Pizza}$ and $\mathcal{K}_{Ingredient}$. For example, in $\mathcal{K}_{People}$ object-attribute context, objects ($G_{People}$) are people and attributes ($M_{People}$) are OrganicFarmer and ConventionalFarmer. $I_{People}$ contains a pair $(p, m)$ if and only if $p$ has the ProdMethod $m$ in PEOPLE table of the initial dataset, e.g., the pair ($Arthur$, $OrganicFarmer$) belongs to $I_{People}$. Tables PREFERS and HASINGREDIENT give rise to $r_{Prefers}$ and $r_{HasIngredient}$ object-object relations also using a nominal scaling, e.g. $r_{Prefers}$ contains ($Arthur, Forest$). Finally we obtain a set of contexts and a set of relations between these contexts: $\{\mathcal{K}_{People}, \mathcal{K}_{Pizza}, \mathcal{K}_{Ingredient}\}$, $\{r_{Prefers}, r_{HasIngredient}\}$. More generally, such a structure is called a Relational Context Family and defined as below.

**Definition 1 (Relational Context Family (RCF)).** *A Relational Context Family (denoted RCF) is a* (**K**, **R**) *pair where:*

- **K** $= \{\mathcal{K}_i\}_{i=1,...,n}$ *is a set of* $\mathcal{K}_i = (G_i, M_i, I_i)$ *formal contexts (object-attribute relations), where* $G_i$ *is the set of objects,* $M_i$ *is the set of attributes and* $I_i \subseteq G_i \times M_i$.
- **R** $= \{r_j\}_{j=1,...,m}$ *is a set of* $r_j$ *object-object relations where* $r_j \subseteq G_{i_1} \times G_{i_2}$ *for some* $i_1, i_2 \in \{1, ..., n\}$.

The principle of RCA consists in integrating object-object relations as new attributes (called *relational attributes*) in formal contexts. A naive approach would be to directly integrate relations as attributes of the form $(relation, targetobject)$, e.g. $(HasIngredient, Mushroom)$, an attribute that could be assigned to the *Forest* pizza. Such an approach would be able to discover the concept of pizzas with dairies. But it is limited to this one-step deduction and it cannot go beyond. The objective of RCA is to infer classifications based on the composition of several relations, e.g. RCA will be able to group people preferring pizzas having at least one dairy product among their ingredients. This is implemented in RCA via the transformation of the object-object relations into relations between objects of one category, and concepts formed on objects of another category. Such a transformation is made thanks to relational attributes and *scaling operators*. These relational attributes will have the form q r(C) where q is a *quantifier*, r is the relation and C is a concept. Theoretically, quantifiers can be chosen within the set $\mathbf{Q} = \{\forall, \exists, \forall\exists, \geq, \geq_q, \leq, \leq_q\}$. The most used quantifiers are:

- the *existential* quantifier ($\exists$) which encodes the fact that an object $o$ is in relation by $\exists r$ with a concept $C$ if $r(o)$ has a non-empty intersection with $Extent(C)$;
- the *strict universal* quantifier ($\forall\exists$) which encodes the fact that an object $o$ is in relation by $\forall\exists r$ with a concept $C$ if $r(o)$ is non-empty and included in the extent of $C$.

Let us now consider the concept lattices given in Fig. 1, built using any standard algorithm for FCA from the three formal contexts $\mathcal{K}_{People}$, $\mathcal{K}_{Pizza}$, and $\mathcal{K}_{Ingredient}$. In the following we examine the transformation of the pizza-ingredient relation $r_{HasIngredient}$ during its integration as new attributes for describing pizzas. In the lattice of ingredients, *Concept_Ingredient_5* represents the group of dairies. Besides, we observe that all pizzas, except *Forest* and *Occitane* pizzas, contain at least one ingredient which is a dairy. This is introduced as a relational attribute $\exists HasIngredient(Concept\_Ingredient\_5)$ shared by *Lorraine*, *Arctic*, *ThreeCheeses* and *FourCheeses* pizzas. Now, if we consider people, *Juliet*, *Nancy* and *Alice* prefer at least one pizza of this group, and they can be grouped into the concept of people that prefer at least one pizza that contains a dairy ingredient. Furthermore, to illustrate the universal scaling operator, let us have a look at *Concept_Ingredient_1*, grouping the fruits and vegetables. *Forest* and *Occitane* pizzas have all their ingredients in the extent of this concept. This is introduced as a new relational attribute $\forall\exists HasIngredient(Concept\_Ingredient\_1)$ which can be assigned to *Forest* and *Occitane* pizzas (highlighting the concept of vege pizzas). The pizzas that are preferred by Arthur and John are all in the group of vege pizzas, an indication to group these two people.

For defining the scaling operators, a generic function $\kappa$ is introduced and instantiated with (1) the existential and (2) the strict universal quantifiers:

$$\kappa : \mathbf{Q} \times \mathbf{R} \times \bigcup_{i=1,\ldots,n} 2^{G_i} \to \bigcup_{i=1,\ldots,n} 2^{G_i}$$
$$(1) \quad \exists \quad r \quad Extent(C) \quad \to \{o | r(o) \cap Extent(C) \neq \emptyset\}$$
$$(2) \quad \forall\exists \quad r \quad Extent(C) \quad \to \{o | r(o) \subseteq Extent(C) \text{ and } r(o) \neq \emptyset\}$$

A scaling operator can now be defined as follows.

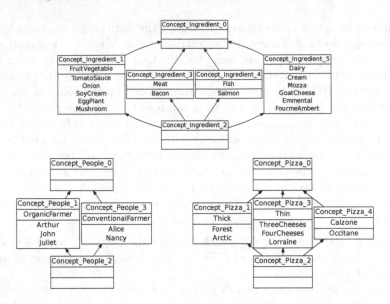

**Fig. 1.** Lattices for object-attribute relations Ingredient ($\mathcal{L}^0_{Ingredient}$), People ($\mathcal{L}^0_{People}$) and Pizza ($\mathcal{L}^0_{Pizza}$) (step 0 of RCA)

**Definition 2 (Scaling operator).** *Let $\mathcal{K} = (G, M, I)$ be a context, and $r$ a relation, where $G$ is the domain of $r$; let $G_{i_r}$ be the range of $r$, $\mathcal{K}_{i_r} = (G_{i_r}, M_{i_r}, I_{i_r})$ another context, and $\mathcal{L}_{i_r}$ a lattice built on $\mathcal{K}_{i_r}$; $q$ denotes a scaling quantifier. The scaling operator $\mathbb{S}_{(r,q),\mathcal{L}_{i_r}}$ over $\mathcal{K}$ yields the derived context $(G^+, M^+, I^+) = \mathbb{S}_{(r,q),\mathcal{L}_{i_r}}(\mathcal{K})$, where:*

- $G^+ = G$,
- $M^+ = \{'q\ r(c)'\ |\ c \in \mathcal{L}_{i_r}\}$,
- $I^+ = \bigcup_{c \in \mathcal{L}_{i_r}} \kappa(q, r, Extent(c)) \times \{'q\ r(c)'\}$.

The $r_{HasIngredient}$ transformed by the existential scaling, considering the lattice previously built for ingredients (see Fig. 1), is $\mathbb{S}_{(r_{HasIngredient}, \exists),\mathcal{L}^0_{Ingredients}}(\mathcal{K}_{Pizza})$. It is shown in Table 2 after the vertical triple bar. The original context $\mathcal{K}_{Pizza}$ can thus be extended with relational attributes representing the relation $r_{HasIngredient}$ between pizzas and ingredients.

Then, for each $\mathcal{K}$ context of **K**, the *apposition* of $\mathcal{K}$ (denoted by symbol '|') with the respective results of the scaling upon each $r_j$ of **R** with $G$ as domain $(1 \le j \le k)$, is used to build a new set of concepts (notations are taken from Def. 2). This apposition is the relational extension of the $\mathcal{K}$ context considering a scaling operator mapping $\rho$ and a set of lattices **L** which is a union of concept lattices including $\mathcal{L}_{i_{r_j}}$, $1 \le j \le k$:

$$\mathbb{E}_{\rho,\mathbf{L}}(\mathcal{K}) = \mathcal{K}\ |\ \mathbb{S}_{(r_1,\rho(r_1)),\mathcal{L}_{i_{r_1}}}(\mathcal{K})\ |\ \ldots\ |\ \mathbb{S}_{(r_k,\rho(r_k)),\mathcal{L}_{i_{r_k}}}(\mathcal{K})$$

Table 2 shows this result for $\mathcal{K}_{Pizza}$, when considering $\rho(r_{HasIngredient}) = \exists$ and the lattices of Fig. 1. If an additional relation connecting pizzas to another

**Table 2.** $\mathcal{K}_{Pizza}$ apposed to existential scaling of $r_{HasIngredient}$. CI stands for 'Concept_Ingredient'.

| | Thick | Thin | Calzone | ∃HasIngredient(CI_0) | ∃HasIngredient(CI_1) | ∃HasIngredient(CI_2) | ∃HasIngredient(CI_3) | ∃HasIngredient(CI_4) | ∃HasIngredient(CI_5) |
|---|---|---|---|---|---|---|---|---|---|
| Forest | × | | | × | × | | | | |
| Occitane | | × | | × | × | | | | |
| ThreeCheeses | × | | | × | × | | | | × |
| FourCheeses | × | | | × | × | | | | × |
| Lorraine | × | | | × | × | | | × | × |
| Arctic | × | | | × | × | | | × | × |

kind of objects, for example, *IsAppreciatedBy*, connecting pizzas to people had been present in the dataset, then the relational extension of $\mathcal{K}_{Pizza}$ would include the scaling upon *IsAppreciatedBy* too.

By extension, $\mathbb{E}^*_{\rho,\mathbf{L}}(\mathbf{K})$ denotes the relational extension of $\mathbf{K}$, which is composed of all the relational extensions of all $\mathcal{K}_i$ in $\mathbf{K}$ (and $\mathbf{L}$ is a union of concept lattices associated with all ranges of all relations).

$$\mathbb{E}^*_{\rho,\mathbf{L}}(\mathbf{K}) = \{\mathbb{E}_{\rho,\mathbf{L}}(\mathcal{K}_1), \dots, \mathbb{E}_{\rho,\mathbf{L}}(\mathcal{K}_n)\}$$

In our example, if we consider only the existential scaling and the lattices of Fig. 1, the relational extension of $\mathbf{K}$ would be composed of the relational extensions of $\mathcal{K}_{People}$, $\mathcal{K}_{Pizza}$ and $\mathcal{K}_{Ingredient}$. The relational extension of $\mathcal{K}_{Ingredient}$ is simply $\mathcal{K}_{Ingredient}$, because there is no outgoing relation. The relational extension of $\mathcal{K}_{Pizzas}$ has been shown in Table 2. The relational extension of $\mathcal{K}_{People}$ is $\mathcal{K}_{People}$ apposed to $\mathbb{S}_{(r_{Prefers},\exists),\mathcal{L}^0_{Pizza}}(\mathcal{K}_{People})$.

Now a whole construction process consists in building a finite sequence of contexts and concept lattices associated with $(\mathbf{K},\mathbf{R})$ and $\rho$. The last sequence is obtained when the fix point is reached. The first set of contexts (step 0) is $\mathbf{K}^0 = \mathbf{K}$. The contexts of step $p$ are used to build the associated concept lattices. The $\mathbf{L}_p$ set composed of the lattices at step $p$ is used to calculate the relational extension. The set of contexts at step $p+1$ is defined using the relational extension: $\mathbf{K}^{p+1} = \mathbb{E}^*_{\rho,\mathbf{L}_p}(\mathbf{K}_p)$.

For our example, the fix point is obtained after three steps. The lattice for ingredients is the same during all the process (see Fig. 1). The lattices for people and pizzas are shown in Fig. 2 and 3. In $\mathcal{L}^3_{Pizza}$ lattice, *Concept_Pizza_7* represents the group of pizzas which contain at least one ingredient which is a dairy. In $\mathcal{L}^3_{People}$ lattice, *Concept_People_12* represents the group of people which prefer at least one pizza which contains at least one dairy ingredient.

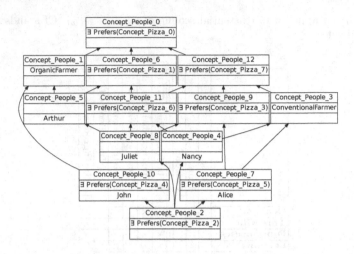

**Fig. 2.** Lattice of people ($\mathcal{L}^3_{People}$) (step 3 of RCA)

Figure 4 presents the three concepts involved in these groups of people, pizzas and ingredients respectively.

## 4    Propositionalisation: The HiFi Method

Propositionalisation has emerged within the field of Inductive Logic Programming (ILP) [6]. Initially ILP was concerned with learning logic programs, and ILP techniques have then been applied in relational data mining. In ILP, learning is performed directly in the first-order logic setting, so that the space to search is intractable when data are numerous. Propositionalisation [1] was proposed as a mean to reduce this complexity. The idea is to shift from a representation in first-order logic to an attribute-value one. This is usually done in two steps: (1) computation of new attributes, called features, for the attribute-value representation (2) computation of the extensions (the values in the resulting propositional table). For some techniques, the two steps are performed at the same time. It is then possible to apply one of the many efficient propositional systems on the propositional table. The logic-oriented approach HiFi [2] produces such a propositional table that can be then processed by FCA. Other logic-based approaches exist but we have chosen HiFi for its similarities with RCA.

A database can be seen as a couple $\mathcal{DB} = (\mathcal{R}, \mathcal{C})$, where $\mathcal{R}$ is a set of relations $r_i(a_{i_1}, ..., a_{i_n})$ and $\mathcal{C}$ is a set of reference constraints on some attributes of these relations $(c_i : a_{j_k} \to a_{l_m})$ (foreign keys). The database representation is directly transformed into first-order logic, each relation becoming a predicate.

In propositionalisation, a main relation, let say $r_1$, is chosen that corresponds to the description of the object of interest. The other relations are then called secondary. The aim of propositionalisation is to generate features that capture the relevant information from the secondary relations to enrich the description

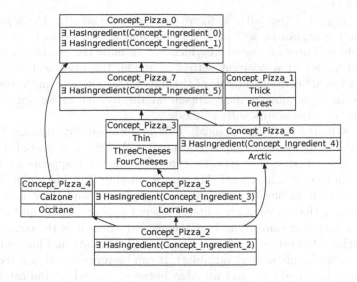

**Fig. 3.** Lattice of pizzas ($\mathcal{L}_{Pizza}^3$) (step 3 of RCA)

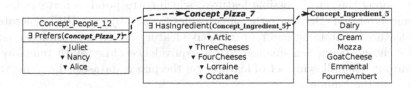

**Fig. 4.** Chained concepts for people (from $\mathcal{L}_{People}^3$), pizzas (from $\mathcal{L}_{Pizza}^3$) and ingredients (from $\mathcal{L}_{Ingredient}^3$) at step 3 of RCA (Objects in the extent that do not belong to the simplified extent are signaled by ▼)

of objects from the relation $r_1$. For example, if *People* is chosen as main table, *People* is the object of interest, that is the one on which we focus our study. *Pizza, Ingredient, HasIngredient* and *Prefers* are the secondary tables. Features will capture information on objects in those four tables that are linked to *People*. The two last relations will allow to work on relations between the different objects represented in the database. $\mathcal{C}$ gives the links between the relations.

HiFi produces features which are function-free first-order conjunctions. Those features are based on a template given by the user and belong to a specific class of features called hierarchical features. A template defines the literals that may appear in a feature, as well as some constraints on the arguments of a literal. Let $T$ be a template on the pizzas example:

$T = People(-Name), Prefers(+Name, -PizzaName), Pizza(+PizzaName, \#Shape), Pizza(+PizzaName, !Shape), HasIngredient(+PizzaName, -Ing Name), Ingredient(+IngName, \#Category).$

In this template, *Name, PizzaName, Shape, IngName* and *Category* act as types and indicate which arguments may share a variable. We can also notice modes: + (intput), - (output), # (constant) and ! (ignored). The input mode means that the argument will be a variable and it will be instantiated. The output mode indicates an argument which is a variable that receives an already instantiated value. At a position with a # mode, the argument should be a constant. A feature contains literals of the template, moreover any variable that appears as an input/output must appear in the feature as an output/input, except if the variable occurs with a ! mode.

Templates in HiFi are hierarchical. This is obtained by ensuring that: (1) every literal has at most one input argument, (2) there is a partial irreflexive order on types implying that type $t \prec$ type $t'$ whenever $t$ appears as an input and $t'$ as an output in some literal. The above template $T$ is hierarchical: we can check that any literal has at most one input argument and there is no pair of types $(t, t')$ such that there exists a literal where $t$ appears as an input argument and $t'$ as an output argument, and another literal where it is the contrary.

A hierarchical feature is based on a hierarchical template and has exactly one root (a literal with only output variables). It can be represented as a tree where each literal $l_i$ is a node $n_i$, and an edge between $n_i$ and $n_j$ indicates that a variable has an output occurrence in $l_i$ and an input occurrence in $l_j$.

HiFi avoids generating redundant features. Indeed, we can define equivalence classes among the set of possible features, which correspond to features having the same extension (they have the same values for all objects). HiFi generates a set of features containing one representative feature for each equivalence class, the one chosen being the smallest in the equivalence class. With template $T$, HiFi outputs the following set of features on the pizzas dataset (the _ notation comes from the ! mode):

$F_1$ : $People(A), Prefers(A, B), HasIngredient(B, C), Ingredient(C, Dairy)$
$F_2$ : $People(A), Prefers(A, B), HasIngredient(B, C), Ingredient(C, Fish)$
$F_3$ : $People(A), Prefers(A, B), HasIngredient(B, C), Ingredient(C, Meat)$
$F_4$ : $People(A), Prefers(A, B), Pizza(B, Calzone)$
$F_5$ : $People(A), Prefers(A, B), Pizza(B, Thick)$
$F_6$ : $People(A), Prefers(A, B), Pizza(B, Thick), Prefers(A, C), Pizza(C, Thin)$
$F_7$ : $People(A), Prefers(A, B), Pizza(B, Thin)$
$F_8$ : $People(A), Prefers(A, B), Pizza(B, \_)$

The corresponding propositional table is shown in Table 3. In this table, *ProdMethod* is a proper attribute of the object of interest *People* and $F_i$ are boolean features generated by HiFi to bring relational information from the secondary tables, and thus enrich the description of People. For example, $F_1$ is true for people who prefers at least one pizza with ingredients of the dairy category, it is false otherwise.

**Table 3.** Propositional table

| | ProdMethod | $F_1$ | $F_2$ | $F_3$ | $F_4$ | $F_5$ | $F_6$ | $F_7$ | $F_8$ |
|---|---|---|---|---|---|---|---|---|---|
| **Arthur** | OrganicFarmer | - | - | - | - | + | - | - | + |
| **John** | OrganicFarmer | - | - | - | + | - | - | - | + |
| **Alice** | ConventionalFarmer | + | - | + | - | - | - | + | + |
| **Juliet** | OrganicFarmer | + | + | - | - | + | + | + | + |
| **Nancy** | ConventionalFarmer | + | + | - | - | + | - | - | + |

## 5 Methods Comparison

### 5.1 Discussion on the Example

On the one hand, the scope of the propositionalisation approach extends to the building of features into a propositional table. On the other hand, the RCA approach goes one step further by building concept lattices from a relational extension. Figure 5 describes both approaches in parallel and highlights the comparison points. The left part of the figure stands for the data transformation part of the processes where relational tables are transformed into single propositional tables. The right part of the figure stands for a propositional algorithm, here FCA. To compare both approaches, we find relevant to consider:

- the *people* relational extension (together with the concept lattices) with the propositional table;
- the *people* concept lattice (together with the other concept lattices) with the concept lattice built from the propositional table.

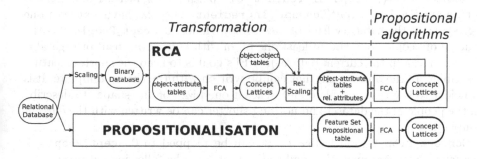

**Fig. 5.** RCA and propositionalisation processes described in parallel

The propositional table describes a binary relation in the same way as a formal context. Objects are the same in the context and in the propositional table and the attributes are the features found by the propositionalisation algorithm and the initial attributes (here *ProdMethod*). A pair $(o, a)$ is in the incidence relation if $a$ is an initial attribute and $o$ owns that initial attribute or if $a$ is a feature and $o$ is described by it. Thus, it is straightforward to build a concept lattice from a propositional table. The lattice from Fig. 6 has been built from Table 3.

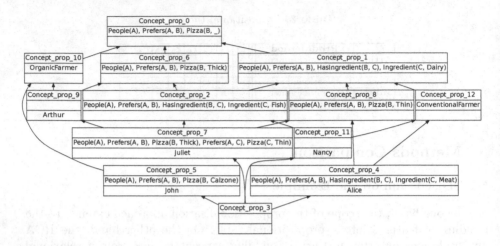

**Fig. 6.** The concept lattice $\mathcal{L}_{prop}$ of people described by features and proper attributes

This lattice structure is isomorphic to the one presented in Fig. 2 as they have the same set of concept extents. Thus it appears relevant to study the correspondences between concept intents as done below.

By considering concept extents, `Concept_prop_0` from lattice $\mathcal{L}_{prop}$ can be mapped to `Concept_People_0` from lattice $\mathcal{L}^3_{people}$. `Concept_prop_0` has for sole feature $People(A), Prefers(A, B), Pizza(B, \_)$ ("people preferring at least one pizza of any shape"). `Concept_People_0` has for sole relational attribute $\exists Prefers$(`Concept_Pizza_0`). `Concept_Pizza_0` has 2 relational attributes: $\exists HasIngredient$(`Concept_Ingredient_0`) ("pizza having at least one ingredient") and $\exists HasIngredient$(`Concept_Ingredient_1`) (pizza having at least one ingredient of the category fruit or vegetable). Hence, `Concept_People_0` is the concept of people preferring at least one pizza with at least one fruit or vegetable (i.e. any pizza in the current dataset). HiFi's goal is to keep the shortest feature describing all the objects and that can be written with the chosen template. It is sufficient to say that "people prefer at least one pizza of any shape" to describe all the people in the dataset and nothing shorter can be written with the current template.

`Concept_prop_8` from lattice $\mathcal{L}_{prop}$ can be mapped to `Concept_People_9`. `Concept_prop_8` groups *Alice* and *Juliet* that own the following features:

1. $People(A), Prefers(A, B), Pizza(B, Thin)$ which is in the proper intent
2. $People(A), Prefers(A, B), HasIngredient(B, C), Ingredient(C, Dairy)$
3. $People(A), Prefers(A, B), Pizza(B, \_)$

`Concept_People_9` also groups *Alice* and *Juliet* and owns the following relational attributes:

1. $\exists Prefers$(`Concept_Pizza_3`) where `Concept_Pizza_3` groups "thin pizzas". This attribute is in the proper intent of `Concept_People_9`

2. $\exists Prefers$(`Concept_Pizza_7`) where `Concept_Pizza_7` groups "pizzas which contain at least one dairy ingredient"
3. $\exists Prefers$(`Concept_Pizza_0`) where `Concept_Pizza_0` groups "all pizzas that contain at least one ingredient and at least one fruit or vegetable ingredient"

The mapping between `Concept_prop_8` and `Concept_people_9` relies on the mapping between the feature which is the proper intent of `Concept_prop_8` and the relational attribute that generates the construction of `Concept_people_9`.

`Concept_prop_7` describes *Juliet* and adds to the features inherited from `Concept_prop_8` the attribute *OrganicFarmer* and the following features:

1. $People(A), Prefers(A, B), Pizza(B, Thick), Prefers(A, C), Pizza(C, Thin)$
2. $People(A), Prefers(A, B), HasIngredient(B, C), Ingredient(C, Fish)$
3. $People(A), Prefers(A, B), Pizza(B, Thick)$

`Concept_people_8` owns the attribute *OrganicFarmer* and the following relational attributes:

1. $\exists Prefers$(`Concept_Pizza_3`)
2. $\exists Prefers$(`Concept_Pizza_7`)
3. $\exists Prefers$(`Concept_Pizza_0`)
4. $\exists Prefers$(`Concept_Pizza_6`)
5. $\exists Prefers$(`Concept_Pizza_1`)

The proper intent of `Concept_people_8` is empty. The minimal generators (i.e. the smallest by inclusion subsets of the intent which have the intent as image by the closure function) [7] of `Concept_people_8` are:

- $\{\exists Prefers(Concept\_Pizza\_6), OrganicFarmer\}$
- $\{\exists Prefers(Concept\_Pizza\_7), OrganicFarmer\}$
- $\{\exists Prefers(Concept\_Pizza\_3), OrganicFarmer\}$
- $\{\exists Prefers(Concept\_Pizza\_3), \exists Prefers(Concept\_Pizza\_6)\}$
- $\{\exists Prefers(Concept\_Pizza\_3), \exists Prefers(Concept\_Pizza\_1)\}$

If we discard the first three generators as they contain *OrganicFarmer* which is initially present in the main table for HiFi, we find 2 minimal generators. By replacing the references to other concepts by a generator of these concepts we obtain respectively $\{\exists Prefers(Thin), \exists Prefers(\exists HasIngredient(Fish))\}$ and $\{\exists Prefers(Thin), \exists Prefers(Thick)\}$. In `Concept_prop_7`, the feature of the proper intent is related to the second expression as it is the shortest one. Both `Concept_pizza_3` and `Concept_pizza_1` have a unique generator, respectively the attributes `thick` and `thin`.

The link between a concept $c_{RCA}$ from $\mathcal{L}^3_{people}$ and a concept $c_{prop}$ with same extents appears to reside in the link between a concept generator of $c_{RCA}$ and the feature from the proper intent of $c_{prop}$. The goal of both approaches can be seen as opposite. While HiFi will tend to provide the shortest description that can discriminate a concept from any other one, RCA will provide the most complete description of a concept.

## 5.2   Evaluation on a Real Dataset

We rely on a part of the Fresqueau database, representing data from Alsatian streams and water areas (North-East of France) [8]. The data are either issued from samples (e.g. biological data collected on stream sites), synthetic data (e.g. stream typology, land cover) or general information issued from the literature (e.g. information about the aquatic species living in the streams). More precisely in this paper we work with three many-valued tables. The first one describes 20 stream sites. The second table gives the level of population for 65 macro-invertebrates collected on these 20 sites. The third one describes the macro-invertebrates with 3 different life traits, i.e. their characteristics and functioning (maximal size, aquatic state and reproduction mode), each life trait being represented by several modalities (e.g. for the life trait maximal size there are 7 possible modalities going from less than 0.25cm to more than 8cm) and affinity values. The total number of the modalities for all life traits is 19.

This dataset has been processed by HiFi and RCA (with the ∃ scaling quantifier). HiFi template and RCA relational schema define the analysis framework. The following template is used for HiFi: *[Station(-s), presence(#abundance,+s, -macroInv), presence(#abundance, +s, !macroInv), affinity(#level,+macroInv, #modality), affinity(#level,+macroInv,!modality)]*. Accordingly, the relational schema for RCA has 3 formal contexts: *Station, MacroInv,* and *Modality* and 6 object-object relations: *abundance-1, abundance-2,* and *abundance-3* from *Station* to *MacroInv* and *affinity-1, affinity-2,* and *affinity-3* from *MacroInv* to *Modality*.

We found respectively 13460 features and 13461 concepts in the Station lattice. The extent of each feature is the extent of a concept. The additional concept is the bottom concept of the lattice, with an empty extent. So we verified that for each feature can be associated a concept and that the lattice obtained from the propositional table and the Station lattice are isomorphic.

## 6   Related Work

Data transformation is a main issue for all classification or automatic learning methods, when dealing with complex or numerous data. Scaling operators are used in FCA for transforming many-valued contexts into binary ones [3]. Such an approach was also used to analyze complex data about life traits of aquatic plants [9]. Statistical metrics can also be used for helping the transformation, e.g. the $\chi^2$ distance was used for selecting the best scaling operator upon a numerical context [10]. This last idea can be related to the metrics used to design decision trees. A comparison between decision trees and dichotomic lattices (i.e. lattices based on complemented contexts) has been presented in [11]. It was proven that the lattice contained all the trees built on the same context.

In [12], many-valued contexts are transformed into a family of formal contexts (under the guidance of a user objective) which is called the power context family (this notion has been introduced in [13]). It represents all the k-ary relations on the object set. From the concept lattices built on the formal contexts of the power context family, concept graphs are extracted which, in turn, are organized into a lattice. In [14], another approach for obtaining concept graphs is presented, that relies on temporal concept analysis, where the conceptual scales are used instead of the concept lattices of the k-ary relations. In these references, there is no use of different scaling operators and a single-step construction is done (comparatively to the iterative approach of RCA). In [12], graphs connecting objects are classified, while in RCA, objects are classified depending on their relations to other objects.

Relational data have been transformed into logical formulae within the framework of logical concept analysis [15]. Object contexts are combined with relational contexts and equipped with a combined logic. Relational attributes are defined as follows: $(\exists r.f)(x) =_{def} \exists x'.(r(x, x') \wedge f(x))$. The concepts' intents of the resulting lattice contain either classical attributes ($f$) or relational attributes ($\exists r.f$). Meta-relations are also built for navigating from a concept to another. Contrarily to RCA, no iteration is performed. In [16], authors propose a method for computing a basis of general concept inclusions in Description Logics $\mathcal{EL}_{gfp}$ where cyclic concept definition has close connections with RCA.

In [17], authors aim at redesigning a database schema. To this end, the database schema is encoded in a formal context and a kind of relational scaling is done in order to represent foreign keys. Here we do not work at the schema level, but at the object level, and the links between objects, rather than the relations between the tables are the focus of the transformation.

Boolean Factor Analysis is applied to multi-relational data in [18]. Their relational factors are tuples of boolean factors extracted independently from the various data tables. In this approach, several schemas of connection can be applied that are similar to the scaling operators of RCA (like existential or universal). Compared to RCA, the boolean factors (that are included in relational factors) are only a part of the formal concepts that could be built from the object-attribute tables, while in RCA all such formal concepts are initially considered. Besides, the process does not iterate.

The authors of [19] address the navigation of SPARQL query answers in concept lattices. They propose a transformation of an RDF graph to a formal context where relations are encoded as attributes. The concept lattice helps analyzing the query answers through their classification.

Reference [20] also considers objects connected by relations. It introduces a Galois connection (and the derived concept lattice) which associates a table (variables and the corresponding tuples) to a description that takes the form of a *windowed s-structure*. Such a windowed s-structure (designed to be a form of a query) is roughly a graph with edges labelled by the relations and with some nodes labelled by variables. There are some similarities between the windowed s-structures, the features and the relational attributes (when they are unfolded). In RCA, concepts correspond to tables with only one variable and finding the equivalent of the tables with more than one variable would rely on navigating on (potentially) several lattices and considering queries like in [21]. Besides, in [20] only existential queries are expressed and there is no iteration, thus no possibility to progressively find the concepts.

# 7  Conclusion

Several approaches exist in the literature to extract knowledge from relational data, using different data transformation methods. In this paper, we focus on two approaches, namely Relational Concept Analysis and Propositionalisation, which we compare on a small example and on a real dataset. We identify similarities in their objectives and between the features of the Propositionalisation approach and the generators in FCA approach. As future work we would like to evaluate the two approaches on other datasets to confirm the practical feasibility and the similar results, using different tunings including step number (for RCA), frequency or feature literal maximum number (for propositionalisation). We also plan to continue exploring the links between features and generators and in general the theoretical and practical advantages and limits of both approaches. In particular, we will study how other scaling operators used in the RCA framework (universal or involving cardinality restrictions) and cyclic schemas can be considered with the propositionalisation approach point of view. From this research, we expect to define a combined methodology that would improve the efficiency of knowledge extraction in relational data, for example by injecting HiFi results in RCA, or using relational attributes obtained at a given RCA step as information for HiFi.

**Acknowledgement.** This work was funded by ANR11_MONU14 Fresqueau. We acknowledge Corinne Grac (LIVE) for advices about the hydrological dataset.

# References

1. Lachiche, N.: Propositionalization. In: Sammut, C., Webb, G.L. (eds.) Encyclopedia of Machine Learning, pp. 812–817. Springer (2010)
2. Kuželka, O., Železný, F.: HiFi: Tractable Propositionalization through Hierarchical Feature Construction. In: Late Breaking Papers, the 18th Int. Conf. on Inductive Logic Programming, pp. 1–6 (2008)
3. Ganter, B., Wille, R.: Formal Concept Analysis: Mathematical Foundations. Springer (1999)
4. Rouane-Hacene, M., Huchard, M., Napoli, A., Valtchev, P.: Relational concept analysis: mining concept lattices from multi-relational data. Ann. Math. Artif. Intell. 67(1), 81–108 (2013)
5. Rouane-Hacene, M., Huchard, M., Napoli, A., Valtchev, P.: Soundness and completeness of relational concept analysis. In: Cellier, P., Distel, F., Ganter, B. (eds.) ICFCA 2013. LNCS, vol. 7880, pp. 228–243. Springer, Heidelberg (2013)
6. Muggleton, S., Raedt, L.D.: Inductive logic programming: Theory and methods. Journal of Logic Programming 19(20), 629–679 (1994)
7. Agrawal, R., Imielinski, T., Swami, A.N.: Mining association rules between sets of items in large databases. In: Proceedings of the 1993 ACM SIGMOD Int. Conference on Management of Data, pp. 207–216 (1993)

8. Grac, C., Le Ber, F., Braud, A., Trémolières, M., Bertaux, A., Herrmann, A., Manné, S., Lafont, M.: Programme de recherche-développement Indices – rapport scienfique final. Contrat pluriannuel 1463 de l'Agence de l'Eau Rhin-Meuse, LHYGES – LSIIT – ONEMA – CEMAGREF (2011)
9. Bertaux, A., Le Ber, F., Braud, A., Trémolières, M.: Identifying ecological traits: A concrete FCA-based approach. In: Ferré, S., Rudolph, S. (eds.) ICFCA 2009. LNCS (LNAI), vol. 5548, pp. 224–236. Springer, Heidelberg (2009)
10. Hereth, J., Stumme, G., Wille, R., Wille, U.: Conceptual knowledge discovery and data analysis. In: Ganter, B., Mineau, G.W. (eds.) ICCS 2000. LNCS (LNAI), vol. 1867, pp. 421–437. Springer, Heidelberg (2000)
11. Guillas, S., Bertet, K., Ogier, J.M., Girard, N.: Some links between decision tree and dichotomic lattice. In: 8th Int. Conf. on Concept Lattices and Applications, CLA 2008, Olomouc, Czech Republic, pp. 193–205 (2008)
12. Prediger, S., Wille, R.: The Lattice of Concept Graphs of a Relationally Scaled Context. In: Tepfenhart, W.M. (ed.) ICCS 1999. LNCS, vol. 1640, pp. 401–414. Springer, Heidelberg (1999)
13. Wille, R.: Conceptual Graphs and Formal Concept Analysis. In: Delugach, H.S., Keeler, M.A., Searle, L., Lukose, D., Sowa, J.F. (eds.) ICCS 1997. LNCS, vol. 1257, pp. 290–303. Springer, Heidelberg (1997)
14. Wolff, K.E.: Relational Scaling in Relational Semantic Systems. In: Rudolph, S., Dau, F., Kuznetsov, S.O. (eds.) ICCS 2009. LNCS, vol. 5662, pp. 307–320. Springer, Heidelberg (2009)
15. Ferré, S., Ridoux, O., Sigonneau, B.: Arbitrary Relations in Formal Concept Analysis and Logical Information Systems. In: Dau, F., Mugnier, M.-L., Stumme, G. (eds.) ICCS 2005. LNCS (LNAI), vol. 3596, pp. 166–180. Springer, Heidelberg (2005)
16. Baader, F., Distel, F.: A finite basis for the set of $\mathcal{EL}$-implications holding in a finite model. In: Medina, R., Obiedkov, S. (eds.) ICFCA 2008. LNCS (LNAI), vol. 4933, pp. 46–61. Springer, Heidelberg (2008)
17. Stanley, R., Astudillo, H., Codocedo, V., Napoli, A.: A Conceptual-KDD Approach and its Application to Cultural Heritage. In: 10th Int. Conf. on Concept Lattices and Their Applications, CLA. CEUR Workshop Proceedings, vol. 1062, pp. 163–174 (2013)
18. Krmelova, M., Trnecka, M.: Boolean Factor Analysis of Multi-Relational Data. In: 10th Int. Conf. on Concept Lattices and Their Applications, CLA 2013, La Rochelle, France. CEUR Workshop Proceedings, vol. 1062, pp. 187–198 (2013)
19. Chekol, M.W., Napoli, A.: An FCA Framework for Knowledge Discovery in SPARQL Query Answers. In: Int. Semantic Web Conference (Posters & Demos), ISWC 2013, Sydney, Australia. CEUR Workshop Proceedings, vol. 1035, pp. 197–200 (2013)
20. Kötters, J.: Concept Lattices of a Relational Structure. In: Pfeiffer, H.D., Ignatov, D.I., Poelmans, J., Gadiraju, N. (eds.) ICCS 2013. LNCS, vol. 7735, pp. 301–310. Springer, Heidelberg (2013)
21. Azmeh, Z., Huchard, M., Napoli, A., Hacene, M.R., Valtchev, P.: Querying relational concept lattices. In: 8th Int. Conf. on Concept Lattices and Their Applications, Nancy, France. CEUR Workshop Proceedings, vol. 959, pp. 377–392 (2011)

# Ordinal Factor Analysis of Graded Data

Cynthia Vera Glodeanu[1] and Jan Konecny[2,*]

[1] Technische Universität Dresden, 01062 Dresden, Germany
[2] Department of Computer Science, Palacky University, Olomouc, Czech Republic
`Cynthia-Vera.Glodeanu@tu-dresden.de`,
`jan.konecny@upol.cz`

**Abstract.** In the last few years, concept factor analysis has been an object of study in the FCA community. Its main idea is to use formal concepts as factors to explain the data in a more concise way. We study factorisation of graded tabular data by means of well-structured families of concepts which have an ordinal character. This method enables us to obtain a smaller number of items which explain the data while they still have a clear and comprehensible meaning. We illustrate the method and its applicability on a sports data set.

**Keywords:** Factor Analysis, Formal Concept Analysis, Fuzzy data, Ordinal factor.

## 1 Introduction

We present a generalisation of the ordinal factors introduced in [1] to the fuzzy setting. Unlike the factorisation of graded data by means of **L**-concepts studied so far in the literature [2], we propose the usage of well-structured families of concepts with an ordinal character as factors for this task. The present method naturally yields fewer factors while the factors still have a clear and comprehensible meaning.

The paper is structured as follows. In Section 2 we recall some basic notions from fuzzy sets, fuzzy logic and Formal Fuzzy Concept Analysis needed in the sequel. Subsection 2.2 contains as well our running example. In Section 3 we recall the factorisation of graded data by means of **L**-concepts. The main work starts in Section 4 where we introduce and study ordinal factors for graded data. The results are illustrated on a factorisation of a sports data set. Concluding remarks and future work are presented in the last section.

## 2 Preliminaries

In Subsection 2.1 we briefly recall some notions from fuzzy sets and fuzzy logic needed in the sequel. Subsection 2.2 contains a short introduction to Formal Fuzzy Concept Analysis and our running example.

---

\* Supported by the ESF project No. CZ.1.07/2.3.00/20.0059, the project is cofinanced by the European Social Fund and the state budget of the Czech Republic.

C.V. Glodeanu, M. Kaytoue, and C. Sacarea (Eds.): ICFCA 2014, LNAI 8478, pp. 128–140, 2014.

## 2.1   L-Sets

In this section we present some basics about fuzzy sets and fuzzy logic. The interested reader may find more details for instance in [3,4].

The underlying ideas of fuzzy sets and fuzzy logics were born in 1965, when Zadeh published [5]. There, he noted that the descriptions used by humans are neither black nor white and that there is a gradual transition from black to white. He pointed out that classical mathematics is not able to grasp these unsharp notions. Contradicting the principle of bivalence, Zadeh stated that there are different cases of belonging to a fuzzy set besides "fully belonging" and "fully not belonging". Hence, being a member of a fuzzy set is a graded matter.

Thus, instead of having just "yes" and "no", or 1 and 0, we have a potentially infinite set of truth values. This set is denoted by $L$ and one usually takes for it the real unit interval $[0,1]$ with its natural ordering, where 0 denotes (full) falsity and 1 (full) truth. Now we are looking for operations on $L$ which model the logical connectives. Since fuzzy theory is a generalisation of classical mathematics, these operations should coincide with the classical ones if we restrict them to the truth values 0 and 1, i.e., $L = \{0,1\}$. The algebraic structures that satisfy the desired properties (see [3]) are named *residuated lattices*.

Fuzzy theory was successfully used in both theoretical and real-world applications, extensive references can be found, for instance, in [6].

An algebra $\mathbf{L} := (L, \wedge, \vee, \otimes, \rightarrow, 0, 1)$ is a **complete residuated lattice** if:

1. $(L, \wedge, \vee, 0, 1)$ is a complete lattice;
2. $(L, \otimes, 1)$ is a commutative monoid;
3. the adjointness property, i.e., $a \otimes b \leq c \Leftrightarrow a \leq b \rightarrow c$ holds for all $a, b, c \in L$.

Elements of $L$ are called **truth degrees**, $\otimes$ and $\rightarrow$ are (truth functions of) "fuzzy conjunction" and "fuzzy implication" and are called **multiplication** and **residuum**, respectively.

A common choice of $\mathbf{L}$ has $L = [0,1]$, $\wedge$ and $\vee$ as minimum and maximum, and $\otimes$ and $\rightarrow$ as one of the three most important pairs of adjoint operations on $[0,1]$:

Łukasiewicz: $a \otimes b := \max(0, a + b - 1)$   and   $a \rightarrow b := \min(1, 1 - a + b)$,

Gödel:   $a \otimes b := \min(a, b)$   and   $a \rightarrow b := \begin{cases} 1, & a \leq b, \\ b, & \text{otherwise,} \end{cases}$

Product:   $a \otimes b := a \cdot b$   and   $a \rightarrow b := \begin{cases} 1, & a \leq b, \\ b/a, & \text{otherwise.} \end{cases}$

An **L-set** $A$ on a set $U$ is a mapping $A : U \rightarrow L$. In an L-set $A$, $A(u)$ is interpreted as "the degree to which $u$ belongs to $A$". We denote by $u \in A$ the fact that $A(u) = 1$. If $U = \{u_1, \ldots, u_n\}$, then $A$ can be denoted by $A = \{^{a_1}/u_1, \ldots, ^{a_n}/u_n\}$ meaning that $A(u_i)$ equals $a_i$ for each $i \in \{1, \ldots, n\}$.

Let $\mathbf{L}^U$ denote the collection of all L-sets on $U$. The operations on L-sets are defined component-wise. For instance, the binary intersection of L-sets $A, B \in$

$\mathbf{L}^U$ is the **L**-set $A \cap B$ in $U$ given by $(A \cap B)(u) = A(u) \wedge B(u)$ for each $u \in U$, etc.
The **L-subsethood degree** of two **L**-sets $A, B \in \mathbf{L}^U$ is defined as $S(A, B) :=$
$\bigwedge_{u \in U}(A(u) \to B(u))$. Thus, $S(A, B)$ represents the degree to which $A$ is a subset
of $B$. In particular, we write $A \subseteq B$ if and only if $S(A, B) = 1$.

For an **L**-set $A \in \mathbf{L}^U$ and a truth value $a \in L$, the **shift** of $A$ by $a$ is an **L**-set
$a \to A \in \mathbf{L}^U$ given by $(a \to A)(u) := a \to A(u)$ for all $u \in U$.

A **binary L-relation** $R$ between the sets $X$ and $Y$ is an **L**-set $R : X \times Y \to L$.

For binary **L**-relations $R \in \mathbf{L}^{X \times F}, S \in \mathbf{L}^{F \times Y}$ define composition $R \circ S \in$
$\mathbf{L}^{X \times Y}$ as follows

$$(R \circ S)(x, y) = \bigvee_{f \in F} R(x, f) \otimes S(f, y) \quad \text{for each } x \in X, y \in Y.$$

## 2.2   Formal Fuzzy Concept Analysis

There are various approaches to Formal Fuzzy Concept Analysis. A survey can
be found in [7]. The first works connecting Formal Concept Analysis and Fuzzy
theory are [8,9]. In the following we give a brief introduction to Formal Fuzzy
Concept Analysis [9,10,4].

A triple $(G, M, I)$ is called an **L-context** if $I : G \times M \to L$ is a binary **L**-
relation between the sets $G$ and $M$ and $L$ is the support set of some residuated
lattice. Elements from $G$ and $M$ are called **objects** and **attributes**, respectively.

The **L**-relation $I$ assigns to each $g \in G$ and each $m \in M$ the truth degree
$I(g, m) \in L$ to which the object $g$ has the attribute $m$. The verbal meaning of
$I(g, m) = l$ is "the object $g$ has attribute $m$ with the truth degree $l$".

Small **L**-contexts can be represented by tables, such as the one in Figure 1.
The rows of the table are named after the objects and the columns after the
attributes. A value $l$ in row $g$ and column $m$ means $I(g, m) = l$.

Given $(G, M, I)$ the **derivation operators** $(-)^{\uparrow} : \mathbf{L}^G \to \mathbf{L}^M$ and $(-)^{\downarrow} :$
$\mathbf{L}^M \to \mathbf{L}^G$ for **L**-sets $A \in \mathbf{L}^G$ and $B \in \mathbf{L}^M$ are defined by

$$A^{\uparrow}(m) := \bigwedge_{g \in G} (A(g) \to I(g, m)), \tag{1}$$

$$B^{\downarrow}(g) := \bigwedge_{m \in M} (B(m) \to I(g, m)) \tag{2}$$

where $g \in G$ and $m \in M$. Then, $A^{\uparrow}(m)$ is the truth degree of the statement
"$m$ is shared by all objects from $A$", and $B^{\downarrow}(g)$ is the truth degree of "$g$ has all
attributes from $B$".

To distinguish between the derivation operators in different **L**-contexts, we
sometimes use the **L**-relations of the **L**-contexts instead of $(-)^{\uparrow}$ and $(-)^{\downarrow}$.

An **(L)-concept** of $(G, M, I)$ is a tuple $(A, B)$ with $A \in \mathbf{L}^G, B \in \mathbf{L}^M$ such
that $A^{\uparrow} = B$ and $B^{\downarrow} = A$. Then, $A$ is called the **extent** and $B$ the **intent**
of $(A, B)$. We denote the set of all **L**-concepts of a given context $(G, M, I)$ by
$\mathfrak{B}(G, M, I)$. Further, $(A, B)$ is called an **(L)-preconcept** of $(G, M, I)$ if $A \subseteq B^{\downarrow}$
and $B \subseteq A^{\uparrow}$.

Let $(A_1, B_1)$ and $(A_2, B_2)$ be two **L**-concepts of $(G, M, I)$. The **L**-concept $(A_1, B_1)$ is called a **subconcept** of $(A_2, B_2)$, written $(A_1, B_1) \leq (A_2, B_2)$, if and only if $A_1 \subseteq A_2$ (or, equivalently, $B_1 \supseteq B_2$). Then, we call $(A_2, B_2)$ a **superconcept** of $(A_1, B_1)$. The set of all **L**-concepts of $(G, M, I)$ ordered by this concept order is called the **L-concept lattice** and is denoted by $\mathfrak{B}(G, M, I) := (\mathfrak{B}(G, M, I), \leq)$. That this name is not misleading is shown by the *Main Theorem on Concept Lattices* [10,4] which proves that every concept lattice is a complete lattice. For a stronger version of this theorem, including completely lattice **L**-ordered sets, see [4].

*Example 1.* The **L**-context in Figure 1 will serve as our running example. The data has been taken from [2] and contains the performances of the top 5 athletes in the 2004 Olympic Decathlon games.

|            | 10   | lj   | sp   | hj   | 40   | hu   | di   | pv   | ja   | 15   |
|------------|------|------|------|------|------|------|------|------|------|------|
| S: Sebrle  | 0.50 | 1.00 | 1.00 | 1.00 | 0.75 | 1.00 | 0.75 | 0.75 | 1.00 | 0.75 |
| C: Clay    | 1.00 | 1.00 | 0.75 | 0.75 | 0.50 | 1.00 | 1.00 | 0.50 | 1.00 | 0.50 |
| K: Karpov  | 1.00 | 1.00 | 1.00 | 0.75 | 1.00 | 1.00 | 1.00 | 0.25 | 0.25 | 0.75 |
| M: Macey   | 0.50 | 0.50 | 0.75 | 1.00 | 0.75 | 0.75 | 0.75 | 0.25 | 0.50 | 1.00 |
| W: Warners | 0.75 | 0.75 | 0.50 | 0.50 | 0.75 | 1.00 | 0.25 | 0.50 | 0.25 | 0.75 |

**Fig. 1.** Scores of top 5 athletes in the 2004 Olympic Decathlon scaled into 5-element chain. Data taken from [2]. The abbreviations of the attributes have the following meaning: *10* – 100 meters sprint race; *lj* – long jump; *sp* – shot put; *hj* – high jump; *40* – 400 meters sprint race; *hu* – 110 meters hurdles; *di* – discus throw; *pv* – pole vault; *ja* – javelin throw; *15* – 1500 meters run.

Using the Łukasiewicz adjoint pair, we obtain 129 **L**-concepts. For instance

$$(\{S, {}^{.75}/C, {}^{.25}/K, {}^{.5}/M, {}^{.25}/W\}, \{{}^{.5}/10, lj, sp, hj, {}^{.75}/40, hu, {}^{.75}/di, {}^{.75}/pv, ja, {}^{.75}/15\})$$

is an **L**-concept. Looking at its intent, we see that it contains the attributes long jump, shot put, high jump, 110 meters hurdles and javelin with degree 1. Thus, this concept can be interpreted as the ability to apply very high force in a very short term, as explosiveness; . From the extent we see that Sebrle is an "explosive athlete", Clay fits this description strongly, Macey is partially explosive, whereas Karpov and Warners are not so explosive athletes.

The initial variant of Formal Concept Analysis [11] was developed for discrete data. Roughly speaking, by using $\mathbf{L} = \{0, 1\}$ one obtains the crisp setting from the so-far introduced notions. **Double-scaling** [10] is a procedure that transforms an **L**-context into a crisp formal context. The method works as follows: Let $(G, M, I)$ be an **L**-context and define for an **L**-set $A \in \mathbf{L}^G$ the crisp set $A^\square$ by

$$A^\square := \{(g, \nu) \mid g \in G, \nu \in L, \nu \leq A(g)\}.$$

Hence, $A^\square \subseteq G^\square := G \times L$. For the **L**-relation $I$ between $G$ and $M$ define a crisp incidence relation $I^\square$ between $G^\square$ and $M^\square$ given by

$$(g, \nu) \, I^\square \, (m, \lambda) :\Longleftrightarrow \nu \otimes \lambda \leq I(g, m),$$

where $\otimes$ is the multiplication in the residuated lattice **L**. We have the following important result:

**Theorem 1 ([10]).** *Let $(G, M, I)$ be an **L**-context and $(G^\square, M^\square, I^\square)$ the corresponding double-scaled context. Then, $\mathfrak{B}(G, M, I) \cong \mathfrak{B}(G^\square, M^\square, I^\square)$.*

## 3   Conceptual Factorisation

The factorisation of **L**-contexts was introduced in [2]. In accordance with the rest of this paper we deviate from the authors' notations.

**Definition 1.** *A **factorisation** of an **L**-context $(G, M, I)$ consists of two **L**-contexts $(G, F, I_{GF})$ and $(F, M, I_{FM})$ such that*

$$I(g, m) = \bigvee_{f \in F} I_{GF}(g, f) \otimes I_{FM}(f, m) \quad \text{for all } g \in G, m \in M.$$

*The set $F$ is called the **(L-)factor set**, its elements the **(L-)factors**, and $(G, F, I_{GF})$ and $(F, M, I_{FM})$ are said to be the **first** and **second factorisation contexts**. We write*

$$(G, M, I) = (G, F, I_{GF}) \cdot (F, M, I_{FM})$$

*to indicate a factorisation.*

We may associate to each factorisation a **factorising family** $\{(A_f, B_f) \mid f \in F\}$ given by the **L**-sets $A_f \in \mathbf{L}^G$ and $B_f \in \mathbf{L}^M$ defined as $A_f(g) := I_{GF}(g, f)$ and $B_f(m) := I_{FM}(f, m)$ for all $g \in G$ and for all $m \in M$. $\{(A_f, B_f) \mid f \in F\}$ is a factorising family of $(G, M, I)$ if and only if

$$I(g, m) = (\bigcup_{f \in F} A_f \circ B_f)(g, m) := \bigvee_{f \in F} A_f(g) \otimes B_f(m) \tag{3}$$

for each $g \in G, m \in M$.

Expressed differently, $\{(A_f, B_f) \mid f \in F\}$ is a factorising family of $(G, M, I)$ if and only if

$$I(g, m) = (\bigcup_{f \in F} f^{I_{GF}} \circ f^{I_{FM}})(g, m) := \bigvee_{f \in F} f^{I_{GF}}(g) \otimes f^{I_{FM}}(m)$$

for each $g \in G, m \in M$.

These factorising families correspond precisely to those families of **L**-preconcepts of $(G, M, I)$ that cover the **L**-relation $I$. By enlarging these preconcepts we obtain a factorising family of **L**-concepts. Note however that this enlargement is not unique. The advantage is thus that we are searching in a smaller set for a covering of the **L**-relation without increasing the number of factors.

*Remark 1.* In the following we will only work with factorisations that are conceptual, i.e., each $(A_f, B_f)$ of a factorising family of $(G, M, I)$ is also an **L**-concept of $(G, M, I)$.

From the definition of the factorisation contexts it is straightforward to see that the relationship between objects and attributes from $(G, M, I)$ is explained by the factors of $F$. Indeed, object $g$ has attribute $m$ if and only if there is a factor $f$ which applies to $g$ and for which $m$ is one of its manifestations. As we are dealing with **L**-sets the notions "applies to" and "is a manifestation of" have truth values. Thus, for a factor $f$ there is a degree $A_f(g)$ to which $f$ applies to $g$ and a degree $B_f(m)$ to which $m$ is a manifestation of $f$. To obtain the degree to which "$f$ applies to $g$ and $m$ is a manifestation of $f$", we have to compute $A_f(g) \otimes B_f(m)$.

It was shown in [2] that using **L**-concepts in the factorisation of **L**-contexts yields the smallest possible number of factors. It follows trivially from the crisp case [12] that finding an optimal factorisation is NP-hard. In the light of this fact, [2] provides us with greedy approximation algorithms.

*Example 2.* In [13] the authors performed a conceptual factorisation on the decathlon data from Figure 1. The factorisation contexts are displayed in Figure 2. One can immediately see, that we have a data reduction. Instead of using 10 attributes to describe the athletes we only need 6. Further, the factors have also a verbal meaning that can be deduced from their intents [13]. For instance, factor $f_2$ is the **L**-concept from the previous example and stands for explosiveness, factor $f_1$ can be interpreted as the ability to run fast for short distances.

The first three factors are the most important ones as they cover 91 % of the **L**-relation of the **L**-context (i.e. $I_{GE} \circ I_{EM}\ (g, m) = I(g, m)$ in 91 % of the pairs $(g, m) \in G \times M$). To cover the remaining 9 % we need the last three factors as well. Further, by using only the first factor we can cover 56 % of the **L**-relation, and the first and the second cover 82 %. Note that only exact matchings were counted; for example $I_{GE} \circ I_{EM}\ (g, m) = 0.5$; $I(g, m) = 0.75$ is not counted toward the coverage.

|   | $f_1$ | $f_2$ | $f_3$ | $f_4$ | $f_5$ | $f_6$ |
|---|---|---|---|---|---|---|
| S | 0.50 | 1.00 | 0.75 | 1.00 | 0.75 | 0.75 |
| C | 1.00 | 0.75 | 0.50 | 0.75 | 0.50 | 1.00 |
| K | 1.00 | 0.25 | 0.75 | 0.75 | 1.00 | 0.25 |
| M | 0.50 | 0.50 | 1.00 | 0.75 | 0.75 | 0.50 |
| W | 0.75 | 0.25 | 0.50 | 1.00 | 0.25 | 0.25 |

|   | 10 | lj | sp | hj | 40 | hu | di | pv | ja | 15 |
|---|---|---|---|---|---|---|---|---|---|---|
| $f_1$ | 1.00 | 1.00 | 0.75 | 0.75 | 0.50 | 1.00 | 0.50 | 0.25 | 0.25 | 0.50 |
| $f_2$ | 0.50 | 1.00 | 1.00 | 1.00 | 0.75 | 1.00 | 0.75 | 0.75 | 1.00 | 0.75 |
| $f_3$ | 0.50 | 0.50 | 0.75 | 1.00 | 0.75 | 0.75 | 0.75 | 0.25 | 0.50 | 1.00 |
| $f_4$ | 0.50 | 0.75 | 0.50 | 0.50 | 0.75 | 1.00 | 0.25 | 0.50 | .025 | 0.75 |
| $f_5$ | 0.75 | 0.75 | 1.00 | 0.75 | 1.00 | 1.00 | 1.00 | 0.25 | 0.25 | 0.75 |
| $f_6$ | 0.75 | 1.00 | 0.75 | 0.75 | 0.50 | 1.00 | 1.00 | 0.50 | 1.00 | 0.50 |

**Fig. 2.** Factorisation contexts of an **L**-conceptual factorisation of the **L**-context from Figure 1

In a conceptual factorisation of $(G, M, I)$ the second factorisation context is determined by the first. Indeed, we get from $B_f = A_f^I$ that

$$I_{FM}(f, m) = l \iff l = A_f^I(m) = (f^{I_{GF}})^I(m).$$

*Remark 2.* It is possible to give necessary and sufficient conditions for **L**-contexts $(G, F, I_{GF})$ and $(F, M, I_{FM})$ such that they are the conceptual factorisation contexts of $(G, M, I)$. In the crisp setting this is done via a linkage between each factorisation context to the complementary of the other. However, in the fuzzy setting the law of double negation does not hold in general.

In the fuzzy setting, we can use isotone (non-dual) derivation operators [14] to make such characterisation. However, the presentation of this result goes beyond the scope of this paper.

## 4    Ordinal Factors

As we have seen in the previous section, the set $F$ of factors may be large. However even a large factorisation may be of avail provided the factors can be divided into conceptually meaningful subsets. An instance of such a structure is an ordinal factor, which represents a chain of conceptual factors.

**Proposition 1.** *Let* $(G, F, I_{GF})$ *and* $(F, M, I_{FM})$ *be conceptual factorisation contexts of an* **L***-context* $(G, M, I)$ *and let* $E \subseteq F$ *with* $I_{GE} := I_{GF} |_{G \times E} \in \mathbf{L}^{G \times E}$ *and* $I_{EM} := I_{FM} |_{E \times M} \in \mathbf{L}^{E \times M}$. *Then,* $(G, E, I_{GE})$ *and* $(E, M, I_{EM})$ *also are conceptual factorisation contexts.*

*Proof.* Let

$$(G, M, I_E) := (G, E, I_{GE}) \circ (E, M, I_{EM})$$

and $e \in E$. Since $E \subseteq F$ it follows that $(e^{I_{GF}}, e^{I_{FM}})$ is an **L**-concept of $(G, M, I)$ and, since $I_E \subseteq I$, also of $(G, M, I_E)$. Thus, $(G, E, I_{GE})$ and $(E, M, I_{EM})$ are conceptual factorisation contexts.

**Definition 2.** *Let* $(G, M, I)$ *be an* **L***-context. We call* $I$ *a* **row-staircase L-relation** *if there is a linear order* $\leqslant_G$ *on* $G$ *defined by*

$$g_1 \leqslant_G g_2 :\iff I(g_1, m) \leq I(g_2, m) \text{ for all } m \in M.$$

*Similarly, we call* $I$ *a* **column-staircase L-relation** *if there is a linear order* $\leqslant_M$ *on* $M$ *given by* $m_1 \leqslant_M m_2 :\iff I(g, m_1) \leq I(g, m_2)$ *for all* $g \in G$.

*Remark 3.*  1. Note that $I \in \mathbf{L}^{G \times M}$ is a row-staircase **L**-relation if and only if $g_1^I \subseteq g_2^I \subseteq \cdots$ and $I$ is a column-staircase **L**-relation if and only if $m_1^I \subseteq m_2^I \subseteq \cdots$.

2. Unlike the crisp case, in the fuzzy setting the property row-staircase of a **L**-relation does not imply column-staircase and vice versa. Consider therefore the **L**-relation $I \in \mathbf{L}^{G \times M}$ given in the left table in Figure 3. Evidently, $I$ is a row-staircase **L**-relation since $g_2 \leqslant_G g_1$, i.e., $g_2^I \subseteq g_1^I$. However, $I$ is not a column-staircase **L**-relation since neither $m_1^I \subseteq m_2^I$ nor $m_2^I \subseteq m_1^I$ holds. Observe that the incidence relation of the double-scaled context, the table on the right in Figure 3, is neither column-staircase nor row-staircase. Thus these notions cannot be directly derived from the crisp case.

|     | $m_1$ | $m_2$ |
|-----|-------|-------|
| $g_1$ | 1   | 0.5   |
| $g_2$ | 0   | 0.5   |

|            | $(m_1,1)$ | $(m_2,1)$ | $(m_1,0.5)$ | $(m_2,0.5)$ |
|------------|-----------|-----------|-------------|-------------|
| $(g_1,1)$   | 1         | 0         | 1           | 1           |
| $(g_2,1)$   | 0         | 0         | 0           | 1           |
| $(g_1,0.5)$ | 1         | 1         | 1           | 1           |
| $(g_2,0.5)$ | 0         | 1         | 1           | 1           |

**Fig. 3.** An **L**-context and its double-scaled context with the Łukasiewicz logic

**Proposition 2.** *Let $(G, M, I)$ be an **L**-context and let $(G, F, I_{GF})$, $(F, M, I_{FM})$ be its factorisation contexts. If $(G, F, I_{GF})$ is a column-staircase relation, then $(F, M, I_{FM})$ is a row-staircase relation and vice versa.*

*Proof.* Let $f_1, f_2 \in F$ with $f_1 \leqslant_F f_2$ in $(G, F, I_{GF})$. By Remark 3 we have that $A_{f_1} = f_1^{I_{GF}} \subseteq f_2^{I_{GF}} = A_{f_2}$ and thus $B_{f_1} = A_{f_1}^I \supseteq A_{f_2}^I = B_{f_2}$. Hence, we obtain $f_1^{I_{FM}} \supseteq f_2^{I_{FM}}$, i.e., $f_2 \leqslant_F f_1$ in $(F, M, I_{FM})$. The converse is similar.

**Definition 3.** *Let $(G, F, I_{GF})$ be the first factorising context of $(G, M, I)$ and $E \subseteq F$ as in Proposition 1. We call $(G, E, I_{GE})$ an **ordinal factor**, if it is a column-staircase relation. We say that $(G, M, I)$ has an **ordinal factorisation**, if its first factorising context can be written as an apposition of ordinal factors.*

Proposition 2 has a number of evident consequences that we sum up in the following corollary:

**Corollary 1.** *1. An **L**-context is an ordinal factor of $(G, M, I)$ iff its attribute extents are a linearly ordered family of concept extents of $(G, M, I)$.*

*2. For an ordinal factorisation there must be a partition $\{F_d \mid d \in D\}$ of the set $F$ of factors such that within each class the attribute order of $(G, F, I_{GF})$ is linear.*

*3. If $(G, M, I)$ has an ordinal factorisation, then $(F, M, I_{FM})$ can be written as the subposition of ordinal factors.*

*Remark 4.* The fuzzy ordinal factorisation can be considered to be a generalisation of both **L**-conceptual factorisation [2] and crisp ordinal factorisation [1]. That is because fuzzy concepts can be considered to be one-element chains and because the defined notions become identical to those in [1] when $\{0, 1\}$ is used as the structure of truth-degrees.

*Example 3.* We have seen in Example 2 that the **L**-context from Figure 1 can be factorised using 6 **L**-conceptual factors. Now, in Figure 4 there is another **L**-conceptual factorisation of the same context but with 10 **L**-conceptual factors. These however can be grouped into 3 ordinal factors.

|   | $f_1^1$ | $f_2^1$ | $f_3^1$ | $f_4^1$ | $f_5^1$ | $f_1^2$ | $f_2^2$ | $f_3^2$ | $f_1^3$ | $f_2^3$ |
|---|---|---|---|---|---|---|---|---|---|---|
| S | 1 | .5 | .5 | .5 | .5 | 1 | .75 | .75 | 1 | 1 |
| C | 1 | 1 | 1 | .75 | .5 | .75 | .5 | .5 | 1 | .75 |
| K | 1 | 1 | 1 | 1 | 1 | 1 | .75 | .25 | .25 | .25 |
| M | .75 | .5 | .5 | .5 | .5 | 1 | 1 | .25 | .5 | .5 |
| W | 1 | .75 | .25 | .25 | .25 | 1 | .5 | .5 | .25 | .25 |

|   | 10 | lj | sp | hj | 40 | hu | di | pv | ja | 15 |
|---|---|---|---|---|---|---|---|---|---|---|
| $f_1^1$ | .5 | .75 | .5 | .5 | .5 | 1 | .25 | .25 | .25 | .5 |
| $f_2^1$ | 1 | 1 | .75 | .75 | .5 | 1 | .5 | .25 | .25 | .5 |
| $f_3^1$ | 1 | 1 | .75 | .75 | .5 | 1 | 1 | .25 | .25 | .5 |
| $f_4^1$ | 1 | 1 | 1 | .75 | .75 | 1 | 1 | .25 | .25 | .75 |
| $f_5^1$ | 1 | 1 | 1 | .75 | 1 | 1 | 1 | .25 | .25 | .75 |
| $f_1^2$ | .5 | .5 | .5 | .5 | .75 | .75 | .25 | .25 | .25 | .75 |
| $f_2^2$ | .5 | .5 | .75 | 1 | .75 | .75 | .75 | .25 | .5 | 1 |
| $f_3^2$ | .75 | 1 | 1 | 1 | 1 | 1 | .75 | 1 | .75 | 1 |
| $f_1^3$ | .5 | 1 | .75 | .75 | .5 | 1 | .75 | .5 | 1 | .5 |
| $f_2^3$ | .5 | 1 | 1 | 1 | .75 | 1 | .75 | .75 | 1 | .75 |

**Fig. 4.** Ordinal factorisation of the **L**-context from Figure 1 using 3 ordinal factors

First let us turn our attention to the percentage of covering by the factors of the **L**-relation of the context. Afterwards, we will discuss their interpretation.

By using only the first ordinal factor we can cover 60% of the **L**-relation of the context and by using the first two ordinal factors 88% of the **L**-relation is covered. Three ordinal factors are sufficient for 100%. In Figure 6 the **L**-relations induced by the three ordinal factors are displayed.

Also in the case of ordinal factorisations, we have a verbal meaning of the factors. The first factor $f^1$ corresponds to the ability to run. Its most specific attributes (that appear in $(f_5^1)^{I_{FM}}$) in degree 1 are 100 and 400 meters sprint race, 110 meters hurdles, but also long jump, shut put and discus throw while 1500 meters run is present with degree 0.75. Clearly, Karpov has the best ability to run, since each step of the factor applies to him with degree 1, i.e., $(f_1^1)^{I_{FM}}(S) = (f_2^1)^{I_{FM}}(S) = \cdots = (f_5^1)^{I_{FM}}(S) = 1$. The second athlete with the best ability to run is Clay, the third one is Sebrle tightly followed by Macey.

Factor $f^2$ can be interpreted as endurance. Among its most specific attributes are 1500 meters run and 400 meters sprint race. Macey has the first two steps of the factor with degree one, however the most specific attributes only apply with 0.25 to him. These attributes apply to Sebrle with degree 0.75.

Factor $f^3$ corresponds to explosiveness, since its last step is factor $f_2$ from the conceptual factorisation from Figure 2.

In Factor Analysis it is popular to have a graphical representation of the factors. Our attempt to represent the ordinal factors from Figure 4 is shown in Figure 5. From there one can easily read how high the athletes load on the different factors, i.e., what is the truth value to which each step of the factor

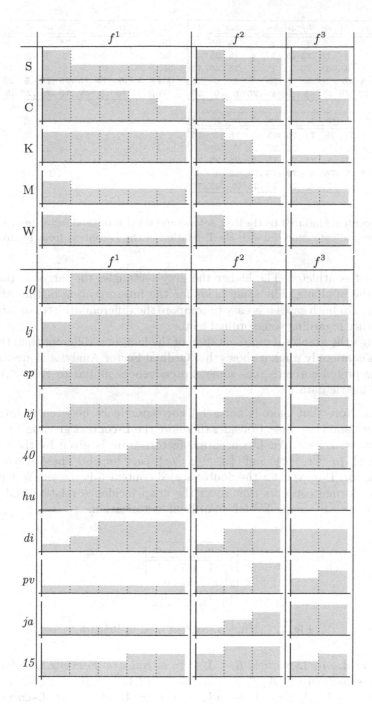

**Fig. 5.** Graphical representation of the ordinal factors from Figure 4. For explanations see Example 3.

| $f^1$ | 10 | lj | sp | hj | 40 | hu | di | pv | ja | 15 |
|---|---|---|---|---|---|---|---|---|---|---|
| S | .5 | 1 | .5 | .75 | .5 | 1 | .5 | .25 | .25 | .5 |
| C | 1 | 1 | .75 | .75 | .5 | 1 | 1 | .25 | .25 | .5 |
| K | 1 | 1 | 1 | .75 | 1 | 1 | 1 | .25 | .25 | .75 |
| M | .5 | .5 | .5 | .25 | .5 | .75 | .5 | 0 | 0 | .25 |
| W | .75 | .75 | .5 | .5 | .5 | 1 | .25 | 0 | 0 | .25 |

| $f^2$ | 10 | lj | sp | hj | 40 | hu | di | pv | ja | 15 |
|---|---|---|---|---|---|---|---|---|---|---|
| S | .5 | .75 | .5 | .5 | .75 | .75 | .25 | .5 | .25 | .75 |
| C | .25 | .5 | .5 | .5 | .5 | .5 | .25 | .25 | .25 | .5 |
| K | .25 | .5 | .5 | .5 | .5 | .5 | .25 | .25 | .25 | .75 |
| M | .5 | .5 | .75 | .75 | .75 | .75 | .5 | .25 | .5 | .1 |
| W | .5 | .75 | .5 | .5 | .75 | .75 | .25 | .5 | .25 | .75 |

| $f^3$ | 10 | lj | sp | hj | 40 | hu | di | pv | ja | 15 |
|---|---|---|---|---|---|---|---|---|---|---|
| S | .5 | .75 | .75 | .75 | .75 | .75 | .5 | .75 | .5 | .75 |
| C | 0 | 0 | .25 | .5 | .25 | .25 | .25 | 0 | 0 | .25 |
| K | .5 | .5 | .5 | .75 | .75 | .75 | .5 | .25 | .25 | .75 |
| M | .5 | .5 | .75 | 1 | .75 | .75 | .75 | .25 | .5 | 1 |
| W | .5 | .5 | .5 | .5 | .75 | .75 | .25 | .5 | .25 | .75 |

**Fig. 6.** L-contexts induced by the first, second and third L-ordinal factor from Figure 4. Bold values indicate exact fit with the L-relation of the L-context from Figure 1.

applies to the athletes. The higher the grey rectangle, the larger is the truth value of the applying. The same holds for the interpretation of the attributes and factors. In both cases it is easy to compare the different athletes or attributes with another regarding some ordinal factor.

Although the graphical representation of the factors is different than the visualisations commonly used, it shows that Ordinal Factor Analysis of graded data, when interpreted correctly, has some expressiveness similar to Factor Analysis based on metric data.

*Remark 5.* Note that there is not a 1-1 correspondence between factorisations in the fuzzy and crisp case. Consider therefore the L-context given in Fig. 7 with the Łukasiewicz logic. An L-conceptual factorisation is given by the concepts $(\{x, {}^{.75}/y, z\}, \{a, {}^{.5}/b, {}^{.5}/c\})$ and $(\{x, {}^{.75}/y, {}^{.5}/z\}, \{a, {}^{.75}/b, c\})$. These also form one ordinal factor. However, for the double-scaled context 5 is the smallest possible number of formal concepts that cover the crisp incidence relation and 2 crisp ordinal factors [1] are needed for an ordinal factorisation.

|  | a | b | c |
|---|---|---|---|
| $x$ | 1 | 0.75 | 1 |
| $y$ | 0.75 | 0.5 | 0.75 |
| $z$ | 1 | 0.5 | 0.5 |

**Fig. 7.** Example L-context used in Remark 5

**Definition 4.** *An* L-relation $R \in \mathbf{L}^{G \times M}$ *is called a* **Ferrers** L-relation *iff there are subsets* $A_1 \subset A_2 \subset A_3 \ldots \in \mathbf{L}^G$ *and* $\mathbf{L}^M \ni B_1 \supset B_2 \supset B_3 \supset \ldots$ *such that* $R = \bigcup_i A_i \circ B_i$. $R$ *is called a Ferrers* L-relation *of* L-concepts *of* $(G, M, I)$ *iff there are* L-concepts $(A_1, B_1) \leq (A_2, B_2) \leq (A_3, B_3) \leq \ldots$ *such that* $R = \bigcup_i A_i \circ B_i$.

|        | 10 | lj | sp | hj | 40 | hu | di | pv | ja | 15 |
|--------|----|----|----|----|----|----|----|----|----|----|
| Sebrle | 1  | 1  | 3  | 3  | 2  | 1  | 3  | 2  | 3  | 2  |
| Clay   | 1  | 1  | 1  | 1  | 1  | 1  | 1  | 3  | 3  | 1  |
| Karpov | 1  | 1  | 1  | 1  | 1  | 1  | 1  | 1  | 1  | 1  |
| Macey  | 1  | 1  | 2  | 2  | 2  | 1  | 2  | 2  | 2  | 2  |
| Warners| 1  | 1  | 1  | 1  | 2  | 1  | 1  | 2  | 2  | 2  |

**Fig. 8.** Covering of the **L**-relation of $(G, M, I)$ by three Ferrers **L**-relations, where the numbers correspond with the three different **L**-relations. Note that these **L**-relations are not disjoint. "1" means that the context cell is covered by the first Ferrers **L**-relation; "2" means that the context cell is covered by second Ferrers **L**-relation and not by the first; "3" means that the context cell is covered by the third Ferrers **L**-relation and not by the first two.

*Remark 6.* Note that unlike the crisp case, the concept lattice $\mathfrak{B}(G, M, R)$ of a Ferrers **L**-relation $R \in \mathbf{L}^{G \times M}$ is not a chain. However, a different result about Ferrers relation still holds in a fuzzy setting, as shown in Proposition 3.

**Proposition 3.** *Any Ferrers* **L**-*relation* $R \subseteq I$ *is contained in a Ferrers* **L**-*relation of concepts of* $(G, M, I)$.

*Proof.* Similarly to the discrete case. If $A_i \circ B_i \subseteq I$, then $A_i \circ B_i \subseteq A_i^{\uparrow\downarrow} \circ A_i^{\uparrow}$. Thus, if $R = \bigcup_i A_i \circ B_i \subseteq I$, then $R \subseteq \overline{R} := \bigcup_i A_i^{\uparrow\downarrow} \circ A_i^{\uparrow} \subseteq I$, and $\overline{R}$ is a Ferrers **L**-relation of concepts.

**Definition 5.** *The* **width** *of a factorising family* $\mathcal{F}$ *of* **L**-*concepts is the largest number of pairwise incomparable elements of* $\mathcal{F}$. *The* **ordinal factorisation width** *of* $(G, M, I)$ *is the smallest width of a factorising family of* **L**-*concepts.*

**Theorem 2.** *The following are equivalent:*

1. $(G, M, I)$ *has ordinal factorisation width* $\leq n$.
2. $(G, M, I)$ *has an ordinal factorisation with* $\leq n$ *ordinal factors.*
3. $I$ *can be written as a union of* $\leq n$ *Ferrers* **L**-*relations.*

*Proof.* (1) $\Rightarrow$ (2): The ordinal factorisation width of $(G, M, I)$ is $\leq n$ if and only if there is a factorising family $\mathcal{F} \subseteq \mathfrak{B}(G, M, I)$ that as an ordered subset of the concept lattice has width $\leq n$. By Dilworth's Theorem we have that $\mathcal{F}$ can be covered by $\leq n$ chains, i.e., linear ordered families of concepts. Thus, the extents of each such family induces a column-staircase relation and therefore an ordinal factor.

(2) $\Rightarrow$ (3): Each ordinal factor is a column-staircase relation, i.e., the attribute extents form a chain which can be written as a Ferrers **L**-relation.

(3) $\Rightarrow$ (1): If $I$ is the union of $\leq n$ Ferrers **L**-relations, then it can also be written as a union of $\leq n$ Ferrers **L**-relations of concepts as shown in Proposition 3. Each Ferrers **L**-relation of **L**-concepts contains a chain of **L**-concepts of $(G, M, I)$ and hence these concepts form a factorising family of width $\leq n$.

*Example 4.* Consider the **L**-context from Figure 1. Its **L**-relation can indeed be covered by three Ferrers **L**-relations, as shown in Figure 8. So the ordinal width of the **L**-context from Figure 1 equals three. This is not surprising since an ordinal factorisation with three ordinal factors was given in Figure 4.

## 5    Conclusion

We presented a generalisation of ordinal factorisations [1] to the fuzzy setting. The main idea is to use chains of concepts in the factorisation of graded data instead of concepts alone, as it was done so far in the literature [2].

It turned out that the generalisation was not as smooth as expected. However, this rises many interesting questions for future research. First of all we will investigate the connection between crisp and fuzzy ordinal factorisations and the possible linkage between the ordinal dimension and order dimension. Further, we plan to apply the method on various real world data sets and compare the results with the ones of well-established methods.

## References

1. Ganter, B., Glodeanu, C.V.: Ordinal Factor Analysis. In: Domenach, F., Ignatov, D.I., Poelmans, J. (eds.) ICFCA 2012. LNCS, vol. 7278, pp. 128–139. Springer, Heidelberg (2012)
2. Bělohlávek, R., Vychodil, V.: Factor analysis of incidence data via novel decomposition of matrices. In: Ferré, S., Rudolph, S. (eds.) ICFCA 2009. LNCS, vol. 5548, pp. 83–97. Springer, Heidelberg (2009)
3. Hájek, P.: The Metamathematics of Fuzzy Logic. Kluwer (1998)
4. Bělohlávek, R.: Fuzzy Relational Systems: Foundations and Principles. In: Systems Science and Engineering. Kluwer Academic/Plenum Press (2002)
5. Zadeh, L.: Fuzzy sets. Information and Control 8, 338–353 (1965)
6. Klir, G., Yuan, B.: Fuzzy sets and fuzzy logic - theory and applications. Prentice Hall P T R, Upper Saddle River (1995)
7. Bělohlávek, R., Vychodil, V.: What is a fuzzy concept lattice? In: Concept Lattices and Their Applications, Olomouc, vol. 162, pp. 34–45. CEUR-WS.org (2005)
8. Burusco, A., Fuentes-Gonzáles, R.: The study of the L-fuzzy concept lattice ma. Mathware and Soft Computing 1(3), 209–218 (1994)
9. Umbreit, S.: Formale Begriffsanalyse mit unscharfen Begriffen. PhD thesis, Martin-Luther-Universitaet Halle-Wittenberg (1994)
10. Pollandt, S.: Fuzzy Begriffe. Springer, Heidelberg (1997)
11. Ganter, B., Wille, R.: Formal Concept Analysis: Mathematical Foundations. Springer, Heidelberg (1999)
12. Bělohlávek, R., Vychodil, V.: Formal concepts as optimal factors in boolean factor analysis: Implications and experiments. In: Eklund, P.W., Diatta, J., Liquiere, M. (eds.) CLA. CEUR Workshop Proceedings, vol. 331. CEUR-WS.org (2007)
13. Bělohlávek, R., Krmelova, M.: Factor analysis of sports data via decomposition of matrices with grades. In: Priss, U., Szathmary, L. (eds.) Concept Lattices and Their Applications. CLA Conference Series, pp. 293–316, Universidad de Malaga (Dept. Matemática Aplicada), Spain (2012)
14. Georgescu, G., Popescu, A.: Non-dual fuzzy connections. Archive for Mathematical Logic 43(8), 1009–1039 (2004)

# On Knowledge Spaces and Item Testing

Immanuel Albrecht and Hermann Körndle

Institute of Educational and Developmental Psychology,
Technische Universität Dresden, Dresden, Germany
{immanuel.albrecht,hermann.koerndle}@tu-dresden.de

**Abstract.** First, we briefly introduce some of the fundamental notions
of knowledge space theory and how they relate to formal concept analysis.
Knowledge space theory has a probabilistic extension which allows it to
be utilized in order to assess knowledge states by looking at responses to
a variety of test items, which are designed to demand performing different
sets of cognitive operations. Second, we introduce an easy extension to
lambda calculus in order to incorporate extra-logical operations. Further
we define a weight function on term reductions, which is to be used
as a model to calculate item response probabilities for test items after
task analysis. We use the new model in order to review the probabilistic
extension of knowledge space theory.

**Keywords:** knowledge spaces, lambda calculus, item testing.

## 1 Knowledge Spaces

### 1.1 Introduction to Knowledge Space Theory

The notion of knowledge spaces was first introduced by Doignon and Falmange
in 1985 as follows:

> « The information regarding a particular field of knowledge is concep-
> tualized as a large, specified set of questions (or problems). The *knowl-*
> *edge state* of an individual with respect to that domain is formalized as
> the subset of all the questions that this individual is capable of solving.
> A particularly appealing postulate on the family of all possible knowl-
> edge states is that it is closed under arbitrary unions. A family of sets
> satisfying this condition is called a *knowledge space.* » [2].

Furthermore, there is a tightly connected notion of *learning spaces*, which are
knowledge spaces with two additional properties: First, one can achieve every
higher level of knowledge relative to one's knowledge state by learning one item at
a time in a finite number of steps, and second, having a higher level of knowledge
does not prevent learning another item, which could have been learned from a
subordinate knowledge state. For the formal definition, see [4] 2.2.1.

Also, there is a probabilistic theory of knowledge spaces, which allows for
assessment of a subject's knowledge state by evaluating realistic test item responses, which are prone to measurement errors – so called careless errors and
lucky guesses.

C.V. Glodeanu, M. Kaytoue, and C. Sacarea (Eds.): ICFCA 2014, LNAI 8478, pp. 141–156, 2014.

## 1.2   Definitions and Relation to Formal Concept Analysis

**Definition 1.** *(See [4], 2.1.2 and 2.2.2) A pair $(Q, \mathcal{K})$ is called a* knowledge structure, *whenever $Q$ is a set, and $\mathcal{K}$ is a family of subsets of $Q$, i.e. $\mathcal{K} \subseteq 2^Q$, such that $\emptyset \in \mathcal{K}$ and $Q \in \mathcal{K}$. Furthermore such $(Q, \mathcal{K})$ is called a* knowledge space, *if $\mathcal{K}$ is also closed under $\bigcup$, i.e. if for all families with $\mathcal{F} \subseteq \mathcal{K}$, $\bigcup \mathcal{F} \in \mathcal{K}$.*

Clearly, if $(Q, \mathcal{K})$ is a knowledge structure, then $\mathcal{K}$ together with the set inclusion form a complete lattice: Let $\mathcal{F} \subseteq \mathcal{K}$, then define the lattice join by $\bigvee \mathcal{F} := \bigcup \mathcal{F}$, which yields an element of $\mathcal{K}$ because $\mathcal{K}$ is closed under arbitrary unions. In order to calculate the lattice meet, we may use the canonical equivalence

$$\bigwedge \mathcal{F} = \bigvee \{K \in \mathcal{K} \mid \forall F \in \mathcal{F} : K \subseteq F\}$$

The most obvious interpretation of knowledge spaces in terms of formal contexts would be $(\mathcal{K}, \mathcal{K}, \subseteq)$, but it hides some of the structure of $(Q, \mathcal{K})$ in the sense that you have to look at the elements of the objects (or attributes) in order to reconstruct $Q$. – Yet this operation is outside the scope of formal concept analysis.

Another obvious link between knowledge structures and formal contexts is a *knowledge context*[1] $(A, Q, I)$ where an individual $a \in A$ incides with an item $q \in Q$ if the individual $a$ is *not* capable of solving the item $q$ [3, p.161].

For knowledge spaces $(Q, \mathcal{K})$, it is possible to look at the context $(\mathcal{K}, Q, \not\ni)$: "The intents of the concepts induced by this knowledge context are the complements of the states in $\mathcal{K}$ with respect to $Q$." [3, p.163]

Last, there is another obvious link that might seem odd at first, but resembles some nice connection between features of knowledge space theory and features of formal concept analysis.

**Definition 2.** *(See [4], 3.7.1) Let $(Q, \mathcal{K})$ be a knowledge structure. The* precedence relation *$\precsim$ wrt. $(Q, \mathcal{K})$ is defined as the binary relation on $Q$ where for $q, r \in Q$*

$$q \precsim r :\Longleftrightarrow \forall K \in \mathcal{K} : r \in K \Rightarrow q \in K$$

The precedence relation offers another way to think about the knowledge space $(Q, \mathcal{K})$ in terms of formal concept analysis. The defining equivalence for $q \precsim r$ may be read as attribute implication: if the concept related to $q$ is a subconcept of the concept related to $r$, clearly all attributes of the latter are supposed attributes of the former, too. This yields the formal context $(Q, \mathcal{K}, \epsilon)$ where

$$\epsilon := \{(q, K) \in Q \times \mathcal{K} \mid q \in K\}$$

Let $(Q, \mathcal{K})$ be some knowledge space. Then for $q \in Q$, wrt. $(Q, \mathcal{K}, \epsilon)$, we get

$$\{q\}' = \mathcal{K}_q := \{K \in \mathcal{K} \mid q \in K\}$$

---

[1] The term *knowledge context* is used for various notions that involve formal concept analysis and knowledge spaces, though.

Thus for $q, r \in Q$, if $\{q\}' \supseteq \{r\}'$, then $\mathcal{K}_r \subseteq \mathcal{K}_q$, which means that for all $K \in \mathcal{K}$, $r \in K \Rightarrow q \in K$, and so

$$\{r\}' \supseteq \{q\}' \Leftrightarrow r \precsim q$$

and it is easy to see that the single item extents for two indiscernible elements $q_1, q_2 \in Q$ – i.e. for all $K \in \mathcal{K}$, $q_1 \in K \Leftrightarrow q_2 \in K$ – are the same.

Furthermore, for $X \in \mathcal{K}$, since $X \subseteq Q$, we see that the intent of a knowledge state viewed as a set of objects is the principal filter of that state wrt. $(\mathcal{K}, \subseteq)$:

$$X' = \{K \in \mathcal{K} \mid \forall r \in X : r \in K\} = \{K \in \mathcal{K} \mid X \subseteq K\} = \uparrow_{(\mathcal{K}, \subseteq)} X$$

And we see further, that $X \in \mathcal{K}$ is a concept extent, i.e. $X = X''$:

$$X'' = \{q \in Q \mid \forall K \in X' : q \in K\} = \bigcap \uparrow_{(\mathcal{K}, \subseteq)} X = X$$

So we see that knowledge states $X \in \mathcal{K}$ correspond to concepts $(X, X')$ of $\mathcal{B}(Q, \mathcal{K}, \epsilon)$.

Now, consider any extent $X \subseteq Q$, i.e. $X'' = X$. In this case, we know by analogous arguments, that for $K \in X'$, also $\uparrow_{(\mathcal{K}, \subseteq)} K \subseteq X'$, but we cannot infer that $X'$ is a principal filter wrt. $(\mathcal{K}, \subseteq)$, as this short example demonstrates:

*Example 1.* Let $Q = \{a, b, c\}$ and $\mathcal{K} = \{\emptyset, \{a\}, \{c\}, \{a, b\}, \{a, c\}, \{b, c\}, Q\}$. Since $\mathcal{K}$ is closed under union, $(Q, \mathcal{K})$ is a knowledge space.

$$\{b\}'' = \{\{a, b\}, \{b, c\}, Q\}' = \{b\} \notin \mathcal{K}$$

We see that $\{b\}'$ is not a principal filter wrt. $(\mathcal{K}, \subseteq)$, because the meet of $\mathcal{K}$ viewed as complete lattice is incompatible with the set meet that is involved in the attribute derivation operator of $(Q, \mathcal{K}, \epsilon)$.

This situation usually arises, when performing a task requires the ability to perform at least one of several distinct subtasks: For instance, if $a$ means that a student knows how to draw a circle with a pen using the left hand, and $c$ means that a student knows how to draw a circle using the right hand, then $b$ could be the ability to write the letter "o" in cursive. In this case, you cannot say that $b$ requires $a$, or that $b$ requires $c$; and thus there is an abstract concept $(\{b\}, \{b\}')$ in the concept lattice $\mathcal{B}(Q, \mathcal{K}, \epsilon)$, which does not correspond to a measurable knowledge state with regard to the test items.

To sum it up, we may view $\mathcal{K}$ as a complete sub join-semi-lattice of $\mathcal{B}(Q, \mathcal{K}, \epsilon)$ with the possibility that in some cases $\mathcal{K}$ may or may not be a complete sub lattice of $\mathcal{B}(Q, \mathcal{K}, \epsilon)$.

## 1.3   Probabilistic Extension

First, we want introduce the general framework which is needed to establish probabilistic methods for knowledge space theory.

**Definition 3.** *(See [4], 11.1.2) A triple $(Q, \mathcal{K}, p)$ is called* probabilistic knowledge structure, *if $(Q, \mathcal{K})$ is a partial knowledge structure, where $Q$ and $\mathcal{K}$ are finite, and if $p \colon \mathcal{K} \to [0, 1]$ is a probability distribution on $\mathcal{K}$, i.e. $\sum_{K \in \mathcal{K}} p(K) = 1$.*

**Definition 4.** *(See [4], 11.1.2) A quadruple $(Q, \mathcal{K}, p, r)$ is called* basic probabilistic model, *if $(Q, \mathcal{K}, p)$ is a probabilistic knowledge structure, and if $r$ is a
map $r: 2^Q \times \mathcal{K} \to [0, 1]$ – called the* response function *– such that for all $K \in \mathcal{K}$,
$r(\cdot, K)$ is a probability distribution on $2^Q$, i.e. $\sum_{R \subseteq Q} r(R, K) = 1$.*

**Local Independence** The measurement of a test subject's knowledge state is
most easy, if the probability that the test item response does not correctly reflect
the subjects knowledge state was only dependent on the item $q \in Q$ and whether
the subjects state $K$ contains $q$ or not.

**Definition 5.** *(See [4], 11.1.2) Let $(Q, \mathcal{K}, p, r)$ be a basic probabilistic model.
This model satisfies* local independence, *if there are vectors $\beta, \eta \in \mathbb{R}^Q$, such that
for all $R \subseteq Q$,*

$$
r(R, K) = \left( \prod_{R \not\ni q \in K} \beta(q) \right) \cdot \left( \prod_{R \ni q \in K} (1 - \beta(q)) \right) \cdot \left( \prod_{R \ni q \notin K} \eta(q) \right) \cdot \left( \prod_{R \not\ni q \notin K} (1 - \eta(q)) \right)
$$

Clearly, local independence means that careless error probability $\beta(q)$ wrt. to a
test item $q$ is the same even for two subjects with huge differences between their
respective knowledge states, which may not be appropriate in all situations, yet
it dramatically reduces the amount of parameters involved and seems reasonable
if both error probabilities are always small. The fact that the lucky guess probability $\eta(q)$ is the same for these two subjects is even less of a problem, since
guessing probabilities usually can be reduced to negligibility by appropriate test
item design ([4], Remark 11.1.3 (b)). With local independence, one can easily
employ a standard $\chi^2$ test on the ratio of the maximum likelihood estimations
in order to verify, whether the negligibility of guessing is achieved. If this is the
case, the number of free parameters to deal with effectively halves in turn.

**Learning Spaces**

**Definition 6.** *Let $(Q, \mathcal{K})$ be a knowledge space. $(Q, \mathcal{K})$ is called a* learning space,
*if for all $K, L \in \mathcal{K}$ with $K \subseteq L$, there are $n \in \mathbb{N}$, and $q_1, \ldots, q_n \in Q$, such
that for all $i \in \{1, \ldots, n\}$ there is a state $K \cup \{q_1, \ldots, q_i\} \in \mathcal{K}$; and such that
$K \cup \{q_1, \ldots, q_n\} = L$. In this case we know that $Q$ and $\mathcal{K}$ are finite.*

The basic principle of the probabilistic extensions of learning in knowledge structures is stated as follows:

> « The probability that, at the time of the test, a subject is in a state
> $K$ of the structure is expressed as the probability that this subject *(i)*
> has successively mastered all the items in the state $K$, and *(ii)* has failed
> to master any item immediately accessible from $K$. » [4, p.198]

This principle may be employed for assessing the knowledge state of some test
subject by looking at the subject's responses to a series of test items. The details

of such a procedure are given in *Knowledge Spaces: Applications in Education* [3, pp.140-145] and another procedure involving Markov chains is given in *Learning Spaces* [4]. For the sake of brevity we will give only a quick informal description of the *assessment algorithm* found in [3]:

The assessment algorithm is based on a stochastic process defined by a sequence of random probability distributions $L_n$ of the subject being in a certain state $K \in \mathcal{K}$, a sequence of random variables $Q_n$ that designate which test item $q \in Q$ is asked in the corresponding trial, and a sequence of random variables $R_n$ that yield 1, if the subject's response in the trial was correct and 0 otherwise. The process starts with some initial distribution $L_1$, the probability of a certain test item being asked in a trial depends on the history of the previous trials and the probability distribution of that trial. The probability of the correct response in a trial depends on the question $q \in Q$ asked, the history of the previous trials and the knowledge state distribution for that trial. The correct response probability is deemed to be $1 - \beta_q$ if the test item is mastered in the subject's latent state. This fact may be used to construct the assessed knowledge state of the student, as the process "$L_n(K_0)$ almost surely converges to 1" for the latent state $K_0$ of the subject [3, p.143].

## 2   λ-μ-Calculus

Knowledge and learning space theory defines knowledge states such that they are determined by the ability of a test subject to solve a test item. Therefore we need to investigate how subjects solve test items, or how solution candidates for test items may be constructed. This investigation benefits from a formal apparatus of constructions and operations that is as general as possible. In *Item Construction and Psychometric Models* [9] Tatsuoka argues that according to Glaser [5] achievement tests must reflect the underlying cognitive processes of problem solving, dynamic changes in strategies, and the structure of knowledge and cognitive skills:

« The correct response to the item is determined by whether all cognitive tasks involved in the item can be answered correctly. Therefore the hypothesis would be that if any of the tasks would be wrong, then there would be a high probability that the final answer would also be wrong. » [9, p.108]

Tatsuoka suggests further, that properties and relations among microlevel and invisible tasks should be explored and predicted [9], which "involves a painstaking and detailed task analysis in which goals, subgoals, and various solution paths are identified in a procedural network (or a flow chart). This process of uncovering all possible combinations of subtasks at the microlevel is essential for making a tutoring system perform the role of the master teacher [...]" [9].

« Identifying subcomponents of tasks in a given problem-solving domain and abstracting their attributes is still an art. It is necessary that

the process can be made automatic and objective. However, we here assume [...] that any task in the domain can be expressed by a combination
of cognitively relevant prime subcomponents.» [9, p.110]

We conclude that the formal apparatus for problem solving must have some way
to express extra-logical operations – which may be interpreted as the cognitively
relevant prime subcomponents of tasks in a given domain – and it must have
some way to express composed tasks and combinations of subtasks in great
generality. Our choice of the following extra-logically extended typed lambda
calculus as the formalization framework for test item analysis is motived as
follows: First, typed lambda calculus typically comes with an algorithm that
allows to check whether a given (untyped) lambda term is typable or not, which
may be seen as a very rough plausibility test of solution path candidates. Second,
types gracefully govern the input and output domains of prime and compound
operations. Third, complex task solutions do not have a strict order in which
subtasks have to be carried out, this corresponds to different reductions starting
from the same term. And last, a formal notion of extra-logical reductions may be
interpreted multifariously, for instance as invoking a random process that leads
to either a correct or an incorrect solution, as a skill requirement for a given
solution path, or as a step-by-step instruction, guideline, hint, etc.

## 2.1  $\lambda$-Calculus

First, we want to fix some definitions regarding the lambda calculus in order
to have something precise to refer to, but at the same time we want to point
out that our extension of the lambda calculus is quite natural and does not
depend on a specific way of formalization. As a general framework for different
possible lambda calculi, we use Pure Type Systems in a presentation found in
Kamareddine, Laan, and Nederpelt [7], which originates from Berardi [1] and
Terlouw [10].

**Definition 7.** *(See [7], 4.16) Let* $\mathbb{V}$ *and* $\mathbb{C}$ *be disjoint sets, that do not contain
any of these symbols: "$\lambda$", "$\Pi$", "(", ")", ".", and ":". The set of terms wrt.*
$\mathbb{V}$ *and* $\mathbb{C}$ *is denoted by* $\mathcal{T}(\mathbb{V}, \mathbb{C})$. *It is defined to be the smallest subset of the
support set* $|\langle \mathbb{V} \cup \mathbb{C} \cup \{\lambda, \Pi, (,), ., : \} \rangle|$ *of the free monoid[2] generated by* $\mathbb{V} \cup \mathbb{C} \cup$
$\{\lambda, \Pi, (,), ., : \}$, *such that* $\mathbb{V} \cup \mathbb{C} \subseteq \mathcal{T}(\mathbb{V}, \mathbb{C})$ *and for all* $A, B \in \mathcal{T}(\mathbb{V}, \mathbb{C})$ *and all*
$x \in \mathbb{V}$:

$$((A)(B)) \in \mathcal{T}(\mathbb{V}, \mathbb{C}), \ (\lambda v : A.B) \in \mathcal{T}(\mathbb{V}, \mathbb{C}), \ and \ (\Pi v : A.B) \in \mathcal{T}(\mathbb{V}, \mathbb{C})$$

*As a notational convention, we may drop parentheses, if they can be restored by
successively adding parentheses, where the "("-parentheses are added at the left-
most possible positions, and ")"-parentheses are added at the right-most possible
positions for* $\lambda$ *and* $\Pi$ *terms, and at the left-most possible position for application*

---

[2] The operation of the free monoid is denoted by juxtaposition, and the neutral element
is denoted by $\varepsilon$.

*terms: For instance, we may write* $\lambda x : A.\lambda y : B.C$ *to denote the term* $(\lambda x : A.(\lambda y : B.C))$, *and we may write* $xyz$ *to denote* $((((x)(y)))(z))$, *where* $A, B, C \in \mathcal{T}(\mathbb{V}, \mathbb{C})$ *and* $x, y, z \in \mathbb{V}$.

**Definition 8.** *Let* $\mathbb{V}$ *and* $\mathbb{C}$ *be sets that satisfy the conditions in Def. 7, and let* $A, X \in \mathcal{T}(\mathbb{V}, \mathbb{C})$ *and* $x \in \mathbb{V}$. *The* substitution *of* $x$ *in* $A$ *by* $X$ *is denoted by* $A[x := X]$. *A formal definition can be found in [6] 1A7 on page 3.*[3] *Since the concept of variable substitution is quite natural to any mathematics, here we give only an informal definition: in* $A$, *we replace every occurrence of* $x$ *with the word* $X$, *unless it is part of* $B$ *of a term sub word*[4] *of the form* $(\lambda x : A.B)$ *or* $(\Pi x : A.B)$, *i.e. "$\lambda$" and "$\Pi$" bind variables right of ".". For instance*

$$((x)((\lambda x : x.x)))[x := ((y)(z))] = (((((y)(z)))((\lambda x : ((y)(z)).x)))$$

**Definition 9.** *(See [7], 4.13) Let* $\mathbb{V}$ *and* $\mathbb{C}$ *be sets that satisfy the conditions in Def. 7, and let* $\rightarrow \subseteq \mathcal{T}(\mathbb{V}, \mathbb{C}) \times \mathcal{T}(\mathbb{V}, \mathbb{C})$ *a binary relation on terms.* $\rightarrow$ *is called* compatibility, *if for all* $A, A', B \in \mathcal{T}(\mathbb{V}, \mathbb{C})$ *with* $A \rightarrow A'$, *also the following holds:* $(A)B \rightarrow (A')B$, $(B)A \rightarrow (B)A'$, $\lambda x : A.B \rightarrow \lambda x : A'.B$, $\lambda x : B.A \rightarrow \lambda x : B.A'$, $\Pi x : A.B \rightarrow \Pi x : A'.B$, *and* $\Pi x : B.A \rightarrow \Pi x : B.A'$.

**Definition 10.** *(See [7], 4.13) Let* $\mathbb{V}$ *and* $\mathbb{C}$ *be sets that satisfy the conditions in Def. 7. The* $\beta$-reduction *relation wrt.* $(\mathbb{V}, \mathbb{C})$ *is the smallest compatibility on* $\mathcal{T}(\mathbb{V}, \mathbb{C})$, *denoted by* $\rightarrow_\beta$, *such that* $\rhd_\beta \subseteq \rightarrow_\beta$; *where* $\rhd_\beta \subseteq \mathcal{T}(\mathbb{V}, \mathbb{C}) \times \mathcal{T}(\mathbb{V}, \mathbb{C})$ *such that for all* $A, B, C \in \mathcal{T}(\mathbb{V}, \mathbb{C})$ *and* $x \in \mathbb{V}$

$$(((\lambda x : A.B))(C)) \quad \rhd_\beta \quad B[x := C]$$

*The reflexive and transitive closure of* $\rightarrow_\beta$ *is denoted by* $\twoheadrightarrow_\beta$, *the reflexive, transitive and symmetric closure of* $\rightarrow_\beta$ *is denoted by* $=_\beta$.

*Remark 1.* The notion of $\beta$-reduction (see Definition 11) is closely related to the notion of $\alpha$-conversion, which is a compatibility and equivalence relation on $\mathcal{T}(\mathbb{V}, \mathbb{C})$, such that

$$(\lambda x : A.B) \equiv_\alpha (\lambda y : A.B[x := y]) \text{ and } (\Pi x : A.B) \equiv_\alpha (\Pi y : A.B[x := y])$$

if the variable $y$ is not free in the left-hand term. This means that you may rename bound variables, unless you would capture a free variable with the new name. You may read the rest of this paper either by thinking of terms as terms or as $\equiv_\alpha$-equivalence classes. This is a standard feature of lambda calculus.

**Definition 11.** *Let* $\mathbb{V}$ *and* $\mathbb{C}$ *be sets that satisfy the conditions in Def. 7, let* $\rhd_x$ *be a binary relation on* $\mathcal{T}(\mathbb{V}, \mathbb{C})$ *and* $\rightarrow_x$ *be the smallest compatibility with* $\rhd_x \subseteq \rightarrow_x$; *further let* $\underline{n} = \{1, \ldots, n\} \subseteq \mathbb{N}$. *A map* $r : \underline{n} \rightarrow \mathcal{T}(\mathbb{V}, \mathbb{C})$ *is a finite*

---

[3] Or see [7] 4.12, but beware of missing $x \not\equiv y$ and typos in 4.8 (swap $A_1$ and $A_2$ on one side).

[4] A term sub word of a term $A$ is a term $B$, such that there are elements of the free monoid $C, D \in |\langle \mathbb{V} \cup \mathbb{C} \cup \{\lambda, \Pi, (,), ., : \}\rangle|$ with $CBD = A$.

$\triangleright_x$-reduction, *if for all* $i \in \underline{n-1} = \underline{n} \backslash \{n\}$, $r(i) \to_x r(i+1)$; *and if there are maps*

$$\text{pre}, \text{post}: \underline{n-1} \to |\langle \mathbb{V} \cup \mathbb{C} \cup \{\lambda, \Pi, (,), ., : \} \rangle|, \quad \text{redex}: \underline{n-1} \to \mathcal{T}(\mathbb{V}, \mathbb{C})$$

*such that for all* $i \in \underline{n-1}$ :

$$\text{pre}(i)\text{redex}(i)\text{post}(i) = r(i) \quad and$$

$$\exists t_i \in \mathcal{T}(\mathbb{V}, \mathbb{C}) : \text{redex}(i) \triangleright_x t_i \text{ such that } \text{pre}(i)t_i\text{post}(i) = r(i+1)$$

*The set of all finite* $\triangleright_x$*-reductions is denoted by* $\nabla_x$, *its subset of all finite* $\triangleright_x$*-reductions with* $r(1) = t$ *for* $t \in \mathcal{T}(\mathbb{V}, \mathbb{C})$ *is denoted by* $\nabla_x^t$.

**Definition 12.** *(See [7], 4.18) A tuple* $(\mathbb{V}, \mathbb{C}, \mathbf{S}, \mathbf{A}, \mathbf{R})$ *is called* pure type system specification, *if* $\mathbb{V}$ *and* $\mathbb{C}$ *are sets that satisfy the conditions in Def. 7; and if* $\mathbf{S} \subseteq \mathbb{C}$, $\mathbf{A} \subseteq \mathbf{S} \times \mathbf{S}$, *and* $\mathbf{R} \subseteq \mathbf{S} \times \mathbf{S} \times \mathbf{S}$. *In this context, we call* $\mathbf{S}$ *the set of* sorts, $\mathbf{A}$ *the set of* axioms, *and* $\mathbf{R}$ *the set of* $\Pi$*-formation rules.*

These definitions are sufficient for our purposes. For a complete introduction to lambda calculus and pure type systems, we refer you to [7], sections 4a through 4c.

## 2.2   $\mu$-Extension

In this section, we want to define a generic extension of lambda calculus that allows to deal with extra-logical operations. Since these operations are not part of the lambda calculus, we need to specify a set of symbols which are regarded as new constants from the point of some given lambda calculus, and we need to specify a set of new derivation rules for typed terms that govern the correct formal types of these symbols. Part of this work can be done by altering the pure type system specification of the underlying lambda calculus, part of this work has to be done by hand.

**Definition 13.** *The* $\mu$*-extension alphabet is defined to be the set* $\mathbb{M}$ *that contains the distinct symbols* "[", "]", ";", "$\mu_i$", "$\nu_i$", *and* "$m_i$" *for all* $i \in \mathbb{N}$,[5] *i.e.*

$$\mathbb{M} = \{[,;,]\} \cup \{\mu_i, \nu_i, m_i \mid i \in \mathbb{N}\}$$

*The set of* $\nu$*-constants* $\mathcal{M}_\nu$ *is defined to be the smallest subset of* $|\langle \mathbb{M} \rangle|$, *that has the following properties: For all* $i \in \mathbb{N}$, $m_i \in \mathcal{M}_\nu$; *and for all* $i, k \in \mathbb{N}$, *if* $x_1, \ldots, x_k \in \mathcal{M}_\nu$, *then* $\nu_i[x_1; \ldots; x_k] \in \mathcal{M}_\nu$.
*The set of* $\mu$*-constants* $\mathcal{M}$ *is defined as* $\mathcal{M} := \{\mu_i \mid i \in \mathbb{N}\} \cup \mathcal{M}_\nu$.

**Definition 14.** *A pair* $(\mathbf{S}, S_0, M, a, p, v)$ *is called* $\mu$*-specification, if* $S_0 \in \mathbf{S}$, *and if* $M$, $a$, $p$, *and* $v$ *are maps, such that* $M: \mathbb{N} \to \mathbf{S}$, $a: \mathbb{N} \to \mathbb{N}$, $p: \mathbb{N} \to \mathbf{S}^{(\mathbb{N})}$, $v: \mathbb{N} \to \mathbf{S}$, *and such that for all* $i \in \mathbb{N}$, $p(i) \in \mathbf{S}^{a(i)}$. *Here,* $M(i)$ *specifies the type of the constant* $m_i$; $a(i)$ *specifies the arity of the symbol* $\mu_i$, *whereas* $p(i)$ *specifies the parameter types of* $\mu_i$ *and* $v(i)$ *specifies the value type of* $\mu_i$.

---

[5] Of course you could use a finite subset of $\mathbb{N}$ as well.

For instance, each of the symbols $\mu_i$ may encode one of the operations listed in Table 1 in [8] with appropriate $a(i)$ and $p(i)$.

**Definition 15.** *Let* $(\mathbb{V}, \mathbb{C}, \mathbf{S}, \mathbf{A}, \mathbf{R})$ *be a pure type system specification, such that* $\mathcal{M} \cap \mathcal{T}(\mathbb{V}, \mathbb{C}) = \emptyset$; *and let* $(\mathbf{S}', S_0, M, a, p, v)$ *be a* $\mu$-*specification, such that* $\mathbf{S}' \subseteq \mathbb{C}$. *The* $\mu$-*extension of* $(\mathbb{V}, \mathbb{C}, \mathbf{S}, \mathbf{A}, \mathbf{R})$ *wrt.* $(\mathbf{S}', S_0, M, a, p, v)$ *is defined to be the tuple* $(\mathbb{V}, \mathbb{C}_\mu, \mathbf{S}_\mu, \mathbf{A}_\mu, \mathbf{R}, S_0, a, p, v)$, *where* $\mathbb{C}_\mu := \mathcal{M} \cup \mathbb{C}$,[6] $\mathbf{S}_\mu := \mathbf{S} \cup \mathbf{S}'$ *and*

$$\mathbf{A}_\mu := \mathbf{A} \cup \{(m_i, M(i)), (M(i), S_0), (v(i), S_0) \mid i \in \mathbb{N}\}$$

*A tuple* $(\mathbb{V}, \mathbb{C}_\mu, \mathbf{S}_\mu, \mathbf{A}_\mu, \mathbf{R}, S_0, a, p, v)$ *that is a* $\mu$-*extension of a pure type system specification wrt. some* $\mu$-*specification is called* $\lambda$-$\mu$-specification *from here on.*

For technical reasons, we cannot express the axioms for the correct function type of $\mu_i$ ($i \in \mathbb{N}$) by elements of the set $\mathbf{A}_\mu$, since $\mu_i$ may have a compound type[7] that is represented by a term from the set $\mathcal{T}(\mathbb{V}, \mathbb{C}_\mu) \backslash \mathbb{C}_\mu$. Furthermore, $\nu_i$ involves a different derivation rule that takes care of the input and output types of the $\mu_i$ operation. Therefore we need to define three and a half additional derivation rule schemes for the $\lambda$-$\mu$-calculus.

**Definition 16.** *Let* $(\mathbb{V}, \mathbb{C}_\mu, \mathbf{S}_\mu, \mathbf{A}_\mu, \mathbf{R}, S_0, a, p, v)$ *be a* $\lambda$-$\mu$-*specification.*
*The* derivation rules of the corresponding $\lambda$-$\mu$-calculus *are the derivation rules of the pure type system corresponding to* $(\mathbb{V}, \mathbb{C}_\mu, \mathbf{S}_\mu, \mathbf{A}_\mu, \mathbf{R})$ *plus the following additional rules for all* $i \in \mathbb{N}$, $x \in \mathbb{V}$, $x_1, \ldots, x_{a(i)} \in \mathcal{M}_\nu$:

**(axiom$\mu_i$)**      $\langle \rangle \quad \vdash \quad \mu_i \quad : \quad \Pi x : p(i)_1. \ldots. \Pi x : p(i)_{a(i)}.v(i)$

**(axiom$\mu_i$')**     $\langle \rangle \quad \vdash \quad \Pi x : p(i)_1. \ldots. \Pi x : p(i)_{a(i)}.v(i) \quad : \quad S_0$

**(appl$\mu_i$)**      $$\frac{\langle \rangle \vdash x_1 : p(i)_1 \quad \ldots \quad \langle \rangle \vdash x_{a(i)} : p(i)_{a(i)}}{\langle \rangle \quad \vdash \quad \mu_i x_1 \ldots x_{a(i)} \quad : \quad v(i)}$$

**(appl$\nu_i$)**      $$\frac{\langle \rangle \vdash x_1 : p(i)_1 \quad \ldots \quad \langle \rangle \vdash x_{a(i)} : p(i)_{a(i)}}{\langle \rangle \quad \vdash \quad \nu_i[x_1; \ldots; x_{a(i)}] \quad : \quad v(i)}$$

*A term* $t \in \mathcal{T}(\mathbb{V}, \mathbb{C}_\mu)$ *that may be derived using the above derivation rules is called* typable *wrt. the* $\lambda$-$\mu$-*specification.*

*Remark 2.* The derivation rule (**appl$\mu_i$**) is unnecessary from a purist point of view, since it may be expressed using $a(i)$ subsequent application rules (**appl**). Yet this does not treat $\mu_i$ as function with multiple parameters which are applied at once, but as a Curry-transformed version that maps its single parameter to another single parameter function, to which then the next parameter is applied

---

[6] If $\mathbb{C}$ was finite or countably infinite, $\mathbb{C}_\mu$ is also countably infinite; thus we do not break any countable infinity assumptions on $\mathbb{V}$ and $\mathbb{C}$, as made on p.112 in [7].
[7] This is the case, whenever $a(i) \neq 0$.

and so on. Since prime subcomponents of tasks should be indivisible into subtasks, the partial applications of formal operations appear to be purely logic. In order to reflect the intuition of prime formal operations with multiple parameters, we introduce a rule scheme that allows application of all parameters of formal operations in one step.

For instance, a $\mu$-reduction steps may correspond to performance of operations as given by a line in Table 4 [8].

**Definition 17.** *Let* $(\mathbb{V}, \mathbb{C}_\mu, \mathbf{S}_\mu, \mathbf{A}_\mu, \mathbf{R}, S_0, a, p, v)$ *be a $\lambda$-$\mu$-specification. The $\mu$-reduction relation wrt. that $\lambda$-$\mu$-specification is the smallest compatibility on* $\mathcal{T}(\mathbb{V}, \mathbb{C}_\mu)$ *– denoted by* $\to_\mu$ *– with* $\triangleright_\mu \subseteq \to_\mu$, *where* $\triangleright_\mu \subseteq \mathcal{T}(\mathbb{V}, \mathbb{C}_\mu) \times \mathcal{T}(\mathbb{V}, \mathbb{C}_\mu)$ *such that for all $i \in \mathbb{N}$, $x_1, \ldots, x_{a(i)} \in \mathcal{M}_\nu$,*

$$\underbrace{(\ldots(\mu_i)(x_1))\underbrace{\ldots}_{\substack{")(x_i))" \; for \\ 1 < i < a(i)}})(x_{a(i)}))}_{2 \cdot a(i) \times \text{"("}} \quad \triangleright_\mu \quad \nu_i[x_1; \ldots; x_{a(i)}]$$

This defines the notion of finite $\triangleright_\mu$-reductions (see Definition 11), that reduce the formal operations $\mu_i$ ($i \in \mathbb{N}$), which are applied to the extra-logical value constants $x_1, \ldots, x_{a(i)} \in \mathcal{M}_\nu$, to their canonical result $\nu_i[x_1; \ldots; x_{a(i)}] \in \mathcal{M}_\nu$.

We now have to check that this reduction works in a way, such that if a term $T \in \mathcal{T}(\mathbb{V}, \mathbb{C}_\mu)$ is well-typed wrt. the $\lambda$-$\mu$-specification, then also all terms $R \in \mathcal{T}(\mathbb{V}, \mathbb{C}_\mu)$ with $T \to_\mu R$ are well-typed wrt. that $\lambda$-$\mu$-specification. Since $\mu_i$ has the type $\Pi x : p(i)_1. \ldots. \Pi x : p(i)_{a(i)}.v(i)$, and if $x_i, \ldots, x_{a(i)}$ are terms such that $x_j$ has the type $p(i)_j$ for $j \in \{1, \ldots, a(i)\}$, the term $\mu_i x_1 \ldots x_{a(i)}$ is well-typed and has the type $v(i)$. This means, that if we have a derivation for the type of some term $T \in \mathcal{T}(\mathbb{V}, \mathbb{C}_\mu)$, i.e. if $T$ is well-typed, we can obtain a derivation for every term $R \in \mathcal{T}(\mathbb{V}, \mathbb{C}_\mu)$ with $T \to_\mu R$ by replacing the appropriate $\mu_i$-subterms with the appropriate $\nu_i[\ldots]$-subterms, and then using the derivation rule (**appl**$\nu_i$) as a replacement for (**appl**$\mu_i$)[8].

## 2.3    Stateful $\mu$-Actions

In this section, we give the common abstraction that will help interpreting finite $\triangleright_\mu$-reductions as operations performed one after another.

**Definition 18.** *Let* $(\mathbb{V}, \mathbb{C}_\mu, \mathbf{S}_\mu, \mathbf{A}_\mu, \mathbf{R}, S_0, a, p, v)$ *be a $\lambda$-$\mu$-specification. A triple* $(X, A, \mathbb{A})$ *is called stateful $\mu$-action, if $X$ is a set – called the set of* subject states; *$A$ is a set – called the set of* auxiliary states, *and if*

$$\mathbb{A} \colon X \times \mathbb{N} \to (X \times A)^{A^{(\mathbb{N})} \times \mathcal{M}_\nu}$$

*is a map, such that for all $i \in \mathbb{N}$ and $x \in X$; the map $\mathbb{A}(x, i)$ – called the $\mu_i$-action for $x$ – has the following signature:*

$$\mathbb{A}(x, i) \colon A^{a(i)} \times \mathcal{M}_\nu \to X \times A$$

---

[8] or the $a(i)$ usages of (**appl**).

**Definition 19.** *Let* $(\mathbb{V}, \mathbb{C}_\mu, \mathbf{S}_\mu, \mathbf{A}_\mu, \mathbf{R}, S_0, a, p, v)$ *be a* $\lambda$-$\mu$-*specification, and* $\mathcal{A} = (X, A, \mathbb{A})$ *be a stateful* $\mu$-*action. Furthermore, let* $n \in \mathbb{N}$, $r \in \nabla_\mu$ *be a finite* $\triangleright_\mu$-*reduction,*

$$\text{redex: } \underline{n-1} \to \mathcal{T}(\mathbb{V}, \mathbb{C}_\mu)$$

*and*

$$\text{pre: } \underline{n-1} \to |\langle \mathbb{V} \cup \mathbb{C}_\mu \cup \{\lambda, \Pi, (,), ., : \}\rangle|$$

*be the corresponding functions as in Definition 11. The triple* $(r, \text{redex}, \text{pre})$ *is a* solution strategy, *if the last term is a constant symbol, i.e.* $r(n) \in \mathcal{M}_\nu$; *and if the first term* $r(1)$ *is typable wrt. the* $\lambda$-$\mu$-*specification and consists only of symbols from* $\{\mu_i, m_i \mid i \in \mathbb{N}\} \cup \{(,)\}$, *i.e.*

$$r(1) \in |\langle \{\mu_i, m_i \mid i \in \mathbb{N}\} \cup \{(,)\}\rangle|$$

*Since* redex *and* pre *are canonical for* $r$, *we may denote the solution strategy just by* $r$. *The set of all solution strategies is denoted by* $\Diamond_\mu$.

**Definition 20.** *Let* $(\mathbb{V}, \mathbb{C}_\mu, \mathbf{S}_\mu, \mathbf{A}_\mu, \mathbf{R}, S_0, a, p, v)$ *be a* $\lambda$-$\mu$-*specification, and* $\mathcal{A} = (X, A, \mathbb{A})$ *be a stateful* $\mu$-*action and* $(r, \text{redex}, \text{pre})$ *be a solution strategy. Then the* performance map *of* $r$ *wrt.* $\mathcal{A}$ *and the* $\lambda$-$\mu$-*specification*

$$r^{\mathcal{A}} : X \times A^{\mathbb{N}} \to X \times A,$$

$$(x, (\alpha_i)_{i \in \mathbb{N}}) \mapsto \left(r_X^{\mathcal{A}}(x, (\alpha_i)_{i \in \mathbb{N}}), r_A^{\mathcal{A}}(x, (\alpha_i)_{i \in \mathbb{N}})\right)$$

*is defined by the sequence of* $\triangleright_\mu$-*reduction steps of* $r$:

*Let* $x \in X$ *and* $(\alpha_i)_{i \in \mathbb{N}} \in A^{\mathbb{N}}$, *and let* $i \in \{1, \dots, n\}$ *be the running index for the rest of this definition. We define the map*

$$\bar{r}_i : \{1, \dots, k_i\} \to \mathbb{V} \cup \mathbb{C}_\mu \cup \{\lambda, \Pi, (,), ., : \}$$

*to be the map such that* $k_i \in \mathbb{N}$ *and* $\bar{r}_i(1)\bar{r}_i(2)\dots\bar{r}_i(k_i) = r(i)$ *wrt. the freely generated monoid* $\langle \mathbb{V} \cup \mathbb{C}_\mu \cup \{\lambda, \Pi, (,), ., : \}\rangle$, *i.e.* $\bar{r}_i$ *is the symbol-at-index map of the term* $r(i)$ *viewed as a word. Further let* $l_i \in \mathbb{N}$ *be the length of the word* $\text{pre}(i)$, *i.e. the number such that there are* $\sigma_1, \dots, \sigma_{l_i} \in \mathbb{V} \cup \mathbb{C}_\mu \cup \{\lambda, \Pi, (,), ., : \}$ *with* $\text{pre}(i) = \sigma_1 \sigma_2 \dots \sigma_{l_i}$. *We define the auxiliary maps* $\bar{X} : \underline{n} \to X$ *and, for* $i \in \{1, \dots, n\}$, $A_i : \{1, \dots, k_i\} \to A$: *We set* $\bar{X}(1) = x$, *and*

$$A_1(j) = \begin{cases} \alpha_h & \text{if } \bar{r}_i(j) = m_h \text{ for } h \in \mathbb{N} \\ \alpha_0 & \text{else} \end{cases}$$

*and for* $g \in \underline{n-1}$:

$$A_{g+1}(j) = \begin{cases} A_g(j) & \text{if } j \le l_g \\ \pi_A(\mathbb{A}(\bar{X}(g), f_g)((A_g(e_{g,1}), \dots, A_g(e_{g,a(i)})), \bar{r}_{g+1}(l_g + 1))) & \text{if } j = l_g + 1 \\ A_g(j + k_g - 1) & \text{if } l_g + 1 < j \end{cases}$$

*and*

$$\bar{X}(g+1) = \pi_X(\mathbb{A}(\bar{X}(g), f_g)((A_g(e_{g,1}), \dots, A_g(e_{g,a(i)})), \bar{r}_{g+1}(l_g + 1)))$$

*where $e_{g,d_g} = 5 \cdot d_g + 2 \cdot a(f_g) - 1$ for $d_g \in \{1, \ldots, a(f_g)\}$, and where $f_g \in \mathbb{N}$ such that $\bar{r}_g(l_g + 1) = \mu_{f_g}$, whereas $\pi_X$ and $\pi_A$ denote the respective coordinate projections of $X \times A$.*

*Finally, we set*

$$r^{\mathcal{A}}(x, (\alpha_i)_{i \in \mathbb{N}}) = (\bar{X}(n), A_n(1))$$

It is obvious that the above definition requires explanation: Given is a finite $\triangleright_\mu$-reduction $r$ that ends in some result $r(n) \in \mathcal{M}_\nu$ of extra-logical operations, which is a single constant symbol term in $\mathcal{T}(\mathbb{V}, \mathbb{C}_\mu)$. Furthermore, the given reduction starts with a well-typed term $r(1)$ that consists only of $\mu_i$ and $m_i$ symbols, and the symbols for their respective applications.

We are interested in the state transition corresponding to the operation sequence of $r$, which is a simultaneous transition of an input subject state and a vector of auxiliary states associated with the extra-logical constants $m_i$ into an output subject state and a single output auxiliary state associated with the result $r(n)$. The reduction $r$ induces an order in which different actions are carried out, and the map $\bar{X}$ represents the state of the subject between the actions. Furthermore the maps $A_g$ represent the auxiliary states of the intermediate results associated with the symbols from $\mathcal{M}_\nu$.[9]

Here, the subject state may be a representation of the current knowledge, skills, short and long term memory, motivation and fatigue of a subject, whereas the auxiliary state may measure the correctness of the intermediate results, partial response times and the amount of effort that was put into solving the subtask. Or the subject state may be a distribution of knowledge states and the auxiliary state may be a distribution of the correctness of the intermediate results; or – slightly abusing the original idea – the subject state may keep track of the required skills, whereas the auxiliary state may keep track of the required effort.

Last, we want to point out, that when given a $\lambda$-$\mu$-term that contains no $\nu_i[\ldots]$ symbols, and if that term has a finite $\triangleright_{\beta,\mu} = \triangleright_\beta \cup \triangleright_\mu$-reduction $r$ with $r(n) \in \mathcal{M}_\nu$, we may postpone all $\triangleright_\mu$-reduction steps to the end, since neither $\mu$-redexes nor $\mu$-reducts have any '$\lambda$' symbol, which is part of the $\beta$-redex – roughly speaking: a $\triangleright_\mu$-reduction step cannot create new or remove old work for the $\triangleright_\beta$-reduction. This means, that if we have some typable term $t \in \mathcal{T}(\mathbb{V}, \mathbb{C}_\mu)$, we may calculate its $\beta$-normal form before invoking the extra-logical $\mu$-machinery.

## 2.4   Solution Probabilities for Test Items Formalized by $\lambda$-$\mu$-Specifications

**Definition 21.** *Let $(\mathbb{V}, \mathbb{C}_\mu, \mathbf{S}_\mu, \mathbf{A}_\mu, \mathbf{R}, S_0, a, p, v)$ be a $\lambda$-$\mu$-specification. A pair $(m_q, S_q)$ is called* test item *wrt. the $\lambda$-$\mu$-specification, if $m_q$ is a $\mu$-constant symbol $m_i$, i.e. $m_q \in \{m_i \mid i \in \mathbb{N}\} \subseteq \mathcal{M}_\nu$, and if $S_q$ is the type of the solution, i.e. $S_q \in \mathbf{S}_\mu$.*

*The* correct type *corresponding to $(m_q, S_q)$ is the type of functions from $M(i)$ to $S_q$, i.e. $\Pi x : M(i).S_q$ where $i \in \mathbb{N}$ such that $m_q = m_i$.*

---

[9] Partial maps $A_g$ fit the situation better, but require more technical details.

**Definition 22.** *Let* $(\mathbb{V}, \mathbb{C}_\mu, \mathbf{S}_\mu, \mathbf{A}_\mu, \mathbf{R}, S_0, a, p, v)$ *be a* $\lambda$-$\mu$-*specification, and let* $t \in \mathcal{T}(\mathbb{V}, \mathbb{C}_\mu)$ *be a term. We call* $t$ *a solution candidate, if* $t$ *is typable wrt. the* $\lambda$-$\mu$-*specification. In this case,* $t_\beta$ *denotes the* $\triangleright_\beta$-*normal form of* $t$, *which then exists (see [7], Theorem 4.40: Strong Normalization Theorem for ECC).*

**Definition 23.** *Let* $(\mathbb{V}, \mathbb{C}_\mu, \mathbf{S}_\mu, \mathbf{A}_\mu, \mathbf{R}, S_0, a, p, v)$ *be a* $\lambda$-$\mu$-*specification, let* $t$ *be a solution candidate, and* $q = (m_i, S_q)$ *be a test item wrt. the* $\lambda$-$\mu$-*specification, where* $i \in \mathbb{N}$ *and* $m_i \in \mathcal{M}$. *Then* $t$ *is a* solution procedure *for* $q$, *if* $t$ *does not contain any symbols from* $\mathcal{M}_v$ *and has the correct type corresponding to* $q$, *i.e. if there is a valid derivation tree with the root judgement*

$$\langle\rangle \;\vdash\; t \;:\; \Pi x : M(i).S_q$$

*We denote the set of all solution procedures for* $q$ *by* $\Xi_q$.

For instance, every method listed in Table 1 [8] corresponds to such a solution procedure.

*Remark 3.* Clearly, if $q = (m_i, S_q)$ and $r = (m_j, S_r)$ are test items such that $S_q = S_r$ and $M(i) = M(j)$, then every solution procedure for $q$ is a solution procedure for $r$ and vice versa.

Now consider a test item $q = (m_q, S_q)$ that is given to some test subject. There are two cases: *(i)* the subject does not find a solution procedure for $q$, or *(ii)* we may view the solution procedure as a discrete random variable $\xi$ which may take values from $\Xi_q$. In the first case, the probability of giving a correct response equals the probability of a correct guess. In the second case, we may determine the probability of a correct response under the condition, that $\xi = t$ for $t \in \Xi_q$ by utilizing a stateful $\mu$-action:

**Definition 24.** *Let* $(\mathbb{V}, \mathbb{C}_\mu, \mathbf{S}_\mu, \mathbf{A}_\mu, \mathbf{R}, S_0, a, p, v)$ *be a* $\lambda$-$\mu$-*specification, Q is a set, such that all its elements* $q \in Q$ *are test items wrt. the* $\lambda$-$\mu$-*specification. A tuple* $(x_0, \alpha, \gamma, \beta, \eta, \bar{\eta}, \equiv_v, \delta)$ *is called* formal test subject *wrt. Q, if*

1. $\alpha = (\alpha_i)_{i \in \mathbb{N}} \in [0,1]^{\mathbb{N}}$ – *the probabilities for correctly understanding* $m_i$,
2. $\gamma = (\gamma_q)_{q \in Q} \in [0,1]^Q$ – *the prob. for discovery of a correct solution procedure,*
3. $\bar{\eta} = (\bar{\eta}_q)_{q \in Q} \in [0,1]^Q$ – *the prob. for guessing if no procedures was discovered,*
4. $\beta = (\beta_i)_{i \in \mathbb{N}} \in [0,1]^{\mathbb{N}}$ – *the probabilities for failing to perform* $\mu_i$,
5. $\eta = (\eta_i)_{i \in \mathbb{N}} \in [0,1]^{\mathbb{N}}$ – *the prob. for guessing the correct results for* $\mu_i$,
6. $\equiv_v \subseteq \mathcal{M}_v \times \mathcal{M}_v$ *is an equivalence relation identifying results obtained in different ways[10],*
7. $\delta = (\delta_x)_{x \in \mathcal{M}_v / \equiv_v} \in [0,1]^{\mathcal{M}_v / \equiv_v}$ – *the prob. for keeping the result* $x$ *in memory,*
8. $x_0 : \mathcal{M}_v / \equiv_v \to [0,1] \times [0,1]$ – $x_0(x) = (p_1, p_2)$ *means that* $x$ *is taken from memory with prob.* $p_1$, *and* $p_2$ *is the prob. that the retrieved result is correct.*

---

[10] A nice property for $\equiv_v$ would be that it only identifies results with that have the same type, but this is not necessary from a formal point of view.

**Definition 25.** *Let* $(\mathbb{V}, \mathbb{C}_\mu, \mathbf{S}_\mu, \mathbf{A}_\mu, \mathbf{R}, S_0, a, p, v)$ *be a* $\lambda$-$\mu$-*specification, and* $s = (x_0, \alpha, \gamma, \beta, \eta, \bar{\eta}, \equiv_\nu, \delta)$ *be a formal test subject. The* stateful $\mu$-*action associated with* $s$ *is defined to be the triple* $\mathcal{A}_s = (X_s, [0, 1], \mathbb{A}_s)$ *where*

$$X_s = ([0, 1] \times [0, 1])^{\mathcal{M}_\nu / \equiv_\nu}$$

*and where*

$$\mathbb{A}_s \colon X_s \times \mathbb{N} \to (X_s \times [0, 1])^{[0,1]^{(\mathbb{N})} \times \mathcal{M}_\nu}$$

*such that for all* $x \in X_s$ *and* $i \in \mathbb{N}$;

$$\mathbb{A}_s(x, i) \colon [0, 1]^{a(i)} \times \mathcal{M}_\nu \to X_s \times [0, 1],$$
$$((p_1, \dots, p_{a(i)}), r) \mapsto (y, p_c + p_g + p_m)$$

*Where:*

$$p_c = \left( \prod_{i=0}^{a(i)} p_i \right) \cdot (1 - \beta_i) \cdot \left( 1 - \pi_1 \left( x \left( [r]_{\equiv_\nu} \right) \right) \right)$$

$$p_g = \eta_i \cdot \beta_i \cdot \left( 1 - \pi_1 \left( x \left( [r]_{\equiv_\nu} \right) \right) \right)$$

$$p_m = \pi_1 \left( x \left( [r]_{\equiv_\nu} \right) \right) \cdot \pi_2 \left( x \left( [r]_{\equiv_\nu} \right) \right)$$

*and*

$$y \colon \mathcal{M}_\nu / \equiv_\nu \to [0, 1] \times [0, 1],$$
$$t \mapsto \begin{cases} x(t) & \text{if } t \neq [r]_{\equiv_\nu} \\ (\pi_1(x(t)) + (1 - \pi_1(x(t))) \cdot \delta_i, & \\ \qquad p_c + p_g + p_m) & \text{if } t = [r]_{\equiv_\nu} \end{cases}$$

*Here,* $\pi_1$ *and* $\pi_2$ *denote the respective coordinate projections of* $[0, 1] \times [0, 1]$; *and* $[r]_{\equiv_\nu}$ *denotes the equivalence class of* $r$ *wrt.* $\equiv_\nu$.

The above definitions interact in the following way: We assume, that we have a $\mu$-specification $(\mathbf{S}', S_0, M, a, p, v)$ which models the domain of knowledge we are investigating. Each $S \in \mathbf{S}'$ stands for a certain way of purposeful information representation. The symbols $m_i \in \mathcal{M}_\nu$ stand for some information represented in the way of $M(i)$. The symbols $\mu_i$ stand for operations that process $a(i)$ pieces of information represented in the ways of $p(i)_1, \dots p(i)_{a(i)}$ to some piece of information represented in the way of $v(i)$.

A test item is then formalized as some given problem represented by the symbol $m_q$ and a task objective $S_q \in \mathbf{S}'$, where we view the task as re-representing the given information in a certain way that elucidates the answer. In order to solve that item, a test subject has to find a series of $\mu_i$ operations that turn $M_q$-representations into $S_q$-representations, where $m_q : M_q$. In general, the subject will have to perform general logic operations in order to create a solution strategy, such as using the given information $m_q$ in different ways to perform different operations – and this is where $\lambda$-calculus is needed.

Consider that you want solve the question $1+5 =?$, and that your operation at hand is 'add two single digit decimal numbers'. Then you would have to extract two different pieces of information: the left and right operands, and your solution candidate for that kind of tasks could be

$$\lambda x : S_{\text{easyAddition}} \cdot \mu_1 \left( \mu_2\, x \right) \left( \mu_3\, x \right) \quad : \quad \Pi x : S_{\text{easyAddition}} \cdot S_{\text{integerNumber}}$$

where $\mu_1$ represents the addition operation, and where $\mu_2$ and $\mu_3$ represent the operand extraction. Please note that there is no need for logic decisions based on the results of the $\mu_i$ operations, since the correct decisions may be encoded by $S'$: carrying information about two easy numbers together with the purpose 'addition' means inhabiting the type $S_{\text{easyAddition}}$. This way, it is possible to decide whether a solution strategy may work or not based on the type of the representing term alone.

After choosing a solution strategy, the test subject has to perform the operations accordingly. The various sources of errors and lucky guesses – which may be due to actual operation or due to remembering correct or wrong (intermediate) results – give rise to a formal test subject, which corresponds to the test subject and the particular time of testing.

The stateful $\mu$-action that is associated with the formal test subject then acts in the following way: The operations are carried out according to the finite $\mu$-reduction in a probabilistic manner, such that if the result is available from memory, then that result is used, otherwise the result is derived from the input, and in case that the input is correct, there is a probability of introducing a new error, and a probability of guessing the correct result. Then the memory is updated with the new derived result.

We sketch a patch of the assessment algorithm [3, pp.140-145], which may lead to a reduction in free parameters for knowledge domains, if there are less operations than test items – which may be achieved by appropriate design. We give a definition that replaces the response rule axiom [R]:

**Definition 26.** *Let $Q$ be a set of test items wrt. a $\lambda$-$\mu$-specification, $Q \ni q = (m_q, S_q)$, $s = (x_0, \alpha, \gamma, \beta, \eta, \bar{\eta}, \equiv_\nu, \delta)$ a formal test subject wrt. $Q$; let $R$ be a random variable with outcomes $\{0,1\}$, and $\xi$ be a random variable with outcomes $\Xi_q \cup \{\bot\}$.*

*The pair $(R, \xi)$ is called* formal response *of the subject $s$ to the test item $q$, if the following equations are satisfied for all $t \in \Xi_q$: $\mathbb{P}(\xi = \bot) = \gamma_q$ and $\mathbb{P}(R = 1 | \xi = \bot) = \bar{\eta}_q$ and for $\Upsilon := \{(r, \text{redex}, \text{pre}) \in \Diamond_\mu \mid r \in \nabla_\mu^{(t\, m_q)_\beta}\}$;*

$$\mathbb{P}(R = 1 | \xi = t) = \frac{1}{\#\Upsilon} \sum_{r \in \Upsilon} r^{\mathcal{A}_s}(x_0, \alpha)$$

*where $(t\, m_q)_\beta$ denotes the $\triangleright_\beta$-normal form of the application term $(t\, m_q)$, and $\mathcal{A}_s$ the stateful $\mu$-action associated with $s$.*

# 3   Discussion

Although the assessment algorithm of knowledge space theory may be utilized to uncover the latent knowledge state of a test subject, it is merely blind to any learning process involved during the trials. The careless error and lucky guess parameters do not allow to investigate relations between the underlying cognitive processes of different test items. We want to investigate those relations, and thus we cannot assume that the process of solving a test item does not involve any state changes – as it is implausible that the same cognitive operation with the same input parameters is repeated within the short timespan of a few solution steps.

The model we introduced implies that expertise available to the test subject may grow steadily just by solving test items, which means that the careless error probabilities for items that belong to basic knowledge states should in fact sink as the subject reaches a higher knowledge state – the probability of careless errors when doing single digit additions should be less for students in senior class compared to when they newly learned it.

In future work, the described models may be adapted to assess a concept associated with some measurement, where the different trials would be the objects of the context and the interdependencies between attributes would be modeled using $\lambda$-$\mu$-calculus, leading to a stochastic assessment procedure.

# References

1. Berardi, S.: Towards a mathematical analysis of the coquand-huet calculus of constructions and the other systems in barendregt's cube. Tech. rep., Dept. of Computer Science, Carnegie-Mellon University and Dipartimento Matematico, Universita di Torino (1988)
2. Doignon, J.P., Falmagne, J.C.: Spaces for the assessment of knowledge. International Journal of Man-Machine Studies 23(2), 175–196 (1985)
3. Falmagne, J., Doble, C., Albert, D., Eppstein, D., Hu, X.: Knowledge Spaces: Applications in Education. Springer-Verlag New York Incorporated (2013)
4. Falmagne, J.C., Doignon, J.P.: Learning Spaces. Springer (2010)
5. Glaser, R.: The integration of instruction and testing. In: Freeman, E. (ed.) The Redesign of Testing in the 21st Century: Proceedings of the 1985 ETS Invitational Conference, pp. 45–58. Educational Testing Service, Princeton (1985)
6. Hindley, J.R.: Basic simple type theory. Cambridge University Press (1997)
7. Kamareddine, F., Laan, T., Nederpelt, R.: A Modern Perspective on Type Theory: From its Origins until Today. Applied logic series. Kluwer Academic Publishers (2006)
8. Korossy, K.: Modeling Knowledge as Competence and Performance. In: Albert, D., Lukas, J. (eds.) Knowledge Spaces: Theories, Empirical Research, and Applications, Mahwah, NJ, pp. 103–132 (1999)
9. Tatsuoka, K.K.: Item Construction and Psychometric Models Appropriate for Constructed Responses. In: Bennett, R., Ward, W. (eds.) Construction Versus Choice in Cognitive Measurement: Issues in Constructed Response, Performance Testing, and Portfolio Assessment, ch. 6, pp. 107–133. Routledge, New York (2009)
10. Terlouw, J.: Een nadere bewijstheoretische analyse von gstt's. Tech. rep., Department of Computer Science, University of Nijmegen (1989)

# Scalable Estimates of Concept Stability

Aleksey Buzmakov[1,2], Sergei O. Kuznetsov[2], and Amedeo Napoli[1]

[1] LORIA (CNRS – Inria NGE – U. de Lorraine), Vandœuvre-lès-Nancy, France
[2] National Research University Higher School of Economics, Moscow, Russia
aleksey.buzmakov@inria.fr, amedeo.napoli@loria.fr, skuznetsov@hse.ru

**Abstract.** Data mining aims at finding interesting patterns from
datasets, where "interesting" means reflecting intrinsic dependencies in
the domain of interest rather than just in the dataset. Concept stability is a popular relevancy measure in FCA. Experimental results of this
paper show that high stability of a concept for a context derived from
the general population suggests that concepts with the same intent in
other samples drawn from the population have also high stability. A
new estimate of stability is introduced and studied. It is experimentally
shown that the introduced estimate gives a better approximation than
the Monte Carlo approach introduced earlier.

**Keywords:** formal concept analysis, stability, pattern selection, experiment.

## 1 Introduction

Given a dataset, data mining methods may reveal a huge number of patterns,
so filtering patterns w.r.t. some relevancy measures can be necessary. The question of how much a pattern is interesting arises in many areas of data mining,
including those that employ tools of Formal Concept Analysis (FCA). FCA is a
mathematical formalism having many applications in data analysis [1]. It aims
at computing concepts and their lattices from a formal context, a triple $(G, M, I)$
where $G$ is a set of objects (experiments or elements of a dataset), $M$ is a set
of attributes used to build the description of every object, and $I \subseteq G \times M$ is
a relation between objects and attributes. The number of concepts for a given
context can be exponential w.r.t. the size of the context, and thus, a special
procedure for selecting the most relevant concepts is needed. Two options can
be distinguished. The first one is to introduce background knowledge into the
procedure for computing concepts [2–6]. Background knowledge allows one to
sort concepts which are likely to be useful for the current goal. In this case,
although the number of concepts can be significantly reduced, the size of the
lattice can still be huge. The second option is to rank concepts in the lattice
using a relevance measure.

The authors of [7] provide several measures for ranking concepts that stem
from human behavior. Stability is another measure for ranking concepts, introduced in [8] and later revised in [9–11]. Several other methods are considered

C.V. Glodeanu, M. Kaytoue, and C. Sacarea (Eds.): ICFCA 2014, LNAI 8478, pp. 157–172, 2014.

**Table 1.** A toy formal context

|     | $m_1$ | $m_2$ | $m_3$ | $m_4$ | $m_5$ | $m_6$ |
|-----|-------|-------|-------|-------|-------|-------|
| $g_1$ | x |   |   |   |   | x |
| $g_2$ |   | x |   |   |   | x |
| $g_3$ |   |   | x |   |   | x |
| $g_4$ |   |   |   | x |   | x |
| $g_5$ |   |   |   |   | x |   |

**Fig. 1.** Concept Lattice for Table 1 with corresponding stability indexes

in [12], where it is shown that stability is more reliable for artificially noised data. Although there is a number of methods for ranking concepts, there is neither a reliable comparison nor a deep research on relevancy of the selection methods mentioned above. In this work we focus on the stability measure and its estimates. The intuition behind stability is the probability of preserving the concept intent when some objects of the context are removed. In this paper we study the behavior of stability computed in several datasets coming from the same general population. It is done by spliting given datasets into two disjoint subsets called reference and test datasets. The stability behaviour is shown to be similar in reference and test datasets independently of the general population.

Since computing stability is #P-complete [8, 9] one needs to use estimates or approximations in order to compute stability over large lattices. Correspondingly, in the second part of our paper we introduce estimates of stability. It is shown empirically that their performance is better then the performance of the known Monte Carlo approximation [13].

The rest of the paper is organized as follows. Section 2 introduces the formal definition of stability, its estimate and Monte Carlo approximation and discusses their relation. In Section 3 experiments on relevancy of stability are explained and discussed. Then Section 4 validates the introduced estimate.

## 2    Stability of a Formal Concept

### 2.1    The Definition of Stability

Stability is a relevancy measure of a formal concept introduced in [8] and later revised in [9–11].

---

**Function** FindStabilityLimits
    **Data:** A context $\mathbb{K} = (G, M, I)$, A concept $C$.
    **Result:** $< Left, Right >$, a pair of left and right limits for the stability.
    $Left \leftarrow 1;$
    $Right \leftarrow 1;$
    $children \leftarrow$ FindChildren$(\mathbb{K}, C)$ ;        /* $O(|N| \cdot |M|^2$ */
    $minDiffSize \leftarrow \infty;$
    **foreach** $ch \in children$ **do**     /* $O(|M|)$ iterations at most */
        $diffSize \leftarrow |\text{Ext}(C) \setminus \text{Ext}(ch)|;$
        $minDiffSize \leftarrow \min(minDiffSize, diffSize);$
        $Left \leftarrow Left - 2^{-diffSize};$
    $Right \leftarrow 1 - 2^{-minDiffSize};$
    **return** $< Left, Right >;$

---

**Algorithm 1:** An algorithm computing stability bounds according to (2)

---

**Function** FindStabilityLimitsPlusMC
    **Input:** Context $\mathbb{K} = (G, M, I)$; concept $C$; precision $\varepsilon$ and error rate $\delta$ for
        Monte-Carlo.
    **Output:** $< Left, Right >$, a pair of left and right limits for the stability.
    $< Left, Right > \leftarrow$ FindStabilityLimits$(\mathbb{K}, C)$;
    **if** $Right - Left > 2 \cdot \varepsilon$ **then**
        $stabilityMC \leftarrow$ FindStabilityByMonteCarlo$(\mathbb{K}, C, \varepsilon, \delta)$;
        $Left \leftarrow \max(Left, stabilityMC - \varepsilon);$
        $Right \leftarrow \min(Right, stabilityMC + \varepsilon);$
    **return** $< Left, Right >;$

---

**Algorithm 2:** An algorithm based on combination of (2) and Monte-Carlo approach.

**Definition 1.** *Given a concept $c$, concept stability $Stab(c)$ is defined as*

$$Stab(c) := \frac{|\{s \in \wp(Ext(c)) \mid s' = Int(c)\}|}{2^{|Ext(c)|}} \tag{1}$$

*i.e. the relative number of subsets of the concept extent (denoted by $Ext(c)$), whose description (i.e. the result of $(\cdot)'$) is equal to the concept intent (denoted by $Int(c)$) where $\wp(P)$ is the power set of $P$.*

*Example 1.* Figure 1 shows a lattice for the context in Table 1, for simplicity some intents are not given. The extent of the highlighted concept $c$ is $Ext(c) = \{g_1, g_2, g_3, g_4\}$, thus, its power set contains $2^4$ elements. The descriptions of 5 subsets of $Ext(c)$ ($\{g_1\}, \ldots, \{g_4\}$ and $\emptyset$) are different from $Int(c) = \{m_6\}$, while all other subsets of $Ext(c)$ have a description equal to $\{m_6\}$. So, $Stab(c) = \frac{2^4 - 5}{2^4} = 0.69$.

Stability measures the dependence of a concept intent on objects of the concept extent. More precisely this intuition behind stability can be described by the following proposition originally introduced in [11, 14].

**Proposition 1.** *Let* $\mathbb{K} = (G, M, I)$ *be a formal context and* $c$ *a formal concept of* $\mathbb{K}$. *For a set* $H \subseteq G$, *let* $I_H = I \cap H \times M$ *and* $\mathbb{K}_H = (H, M, I_H)$. *Then,*

$$Stab(c) = \frac{|\{\mathbb{K}_H \mid H \subseteq G \text{ and } Int(c) \text{ is closed in } \mathbb{K}_H\}}{2^{|G|}}$$

The proposition says that stability of a concept $c$ is the relative number of subcontexts where there exists the concept $c$ with intent $Int(c)$. A stable concept can be found in many such subcontexts, and therefore is likely to be found in an unrelated context built from the population under study. This "likely" was never studied and one of the goals of this paper is to check if stability is useful to find significant patterns within the whole population.

It was shown that, given a context and a concept, the computation of concept stability is #P-complete [8, 9]. One of the fastest algorithm for processing concept stability using a concept lattice $L$ is proposed in [11], with a worst-case complexity of $O(L^2)$, where $L$ is the size of the concept lattice. This theoretical complexity bound is significantly higher than that of algorithms computing all formal concepts and in practice computing stability may take much more time than lattice building algorithms [15]. Moreover, this algorithm needs the lattice structure to be computed, requiring additional computations and memory usage. Thus, finding a good estimate of concept stability is an important question. Here we present an efficient way for such an estimate.

## 2.2   Estimation of Stability

Given a concept $c$ and its descendant $d$, we have $(\forall s \subseteq Ext(d))(s'' \subseteq Ext(d) \wedge s' \supseteq Int(d) \supset Int(c))$ i.e. $s' \neq Int(c)$. Thus, we can exclude all subsets of the extent of a descendant while computing the numerator of stability in (1). On the other hand only subsets of the extents of descendants should be excluded from the numerator in (1). Thus, if we exclude the subsets of the extents of all immediate descendants, we exclude everything that is needed but probably some subsets can be excluded several times. Hence we obtain a lower bound for stability:

$$1 - \sum_{d \in DD(c)} \frac{1}{2^{\Delta(c,d)}} \leq Stab(c) \leq 1 - \max_{d \in DD(c)} \frac{1}{2^{\Delta(c,d)}}, \tag{2}$$

where $DD(c)$ is a set of all direct descendants of $c$ in the lattice and $\Delta(c, d)$ is the size of the set-difference between extent of $c$ and extent of $d$, i.e. $\Delta(c, d) = |Ext(c) \setminus Ext(d)|$. The pseudo-code for computing this estimate is shown in Algorithm 1. The time complexity of this approach for a concept is equal to the complexity of finding immediate descendants of the concept, i.e. $O(n \cdot m^2)$.

*Example 2.* If we want to compute stable concepts (with stability more than 0.97), then according to the upper bound in (2) we should compute for each concept $c$ in the lattice $\Delta_{\min}(c) = \min\limits_{d \in DD(c)} \Delta(c, d)$ and select concepts obeying $\Delta_{\min}(c) \geq -\log(1 - 0.97) = 5.06$.

The upper bound of the equation can be found in [11], while the lower bound has not been studied yet. We know that given a context $(G, M, I)$, the number of children for any concept is limited by cardinality of $M$. Every summand in the lower bound of stability in (2) is smaller than $2^{-\Delta_{\min}(c)}$. This gives the following estimate.

$$1 - |M| \cdot 2^{-\Delta_{\min}(c)} \leq 1 - \sum_{d \in DD(c)} 2^{-\Delta(c,d)} \leq Stab(c) \qquad (3)$$

This suggests that stability can have an exponential behavior w.r.t. the size of the context and, thus, most of the concepts have stability close to 1 when the size of the context increases. This behavior of stability is also noticed by authors of [16] for their dataset. So, to use stability for large datasets it is worth computing logarithmic stability for every concept $c$:

$$LStab(c) = -\log_2(1 - Stab(c)) \qquad (4)$$

Taking into account the bounds in (2) and in (3), we have the following:

$$\Delta_{\min}(c) - \log_2(|M|) \leq -\log_2 \Big( \sum_{d \in DD(c)} 2^{-\Delta(c,d)} \Big) \leq LStab(c) \leq \Delta_{\min}(c) \qquad (5)$$

This approach is referred as the *bounding method*. It can efficiently bound stability for any concept of the lattice. However, the tightness of this bound cannot be ensured.

In [13] the authors suggest a method for approximating concept stability based on a Monte Carlo approach. Given a concept $c$, the idea is to randomly count the number of "good" subsets $s \subseteq Ext(c)$ of the extent of $c$ such that $s' = Int(c)$. Then knowing the number of iterations $N$ and the number of "good" subsets $N_{good}$, stability can be calculated as the relation between them: $Stab(c) = \frac{N_{good}}{N}$. In their paper authors provide the following approximation of the number of iterations:

$$N > \frac{1}{2\varepsilon^2} \ln \frac{2}{\delta} \qquad (6)$$

where $\varepsilon$ is the precision of the approximation and $\delta$ is the error rate, i.e. if one have computed stability approximation $s$, then the exact value of stability is within the interval $[s - \varepsilon; s + \varepsilon]$ with the probability $1 - \delta$. This method will be later referred as the *Monte Carlo method*.

*Example 3.* In order to approximate stability with precision $\varepsilon = 0.01$ and error rate $\delta = 0.01$, it is necessary to make at least $N = 2.65 \cdot 10^4$ iterations.

Example 3 shows that the number of iterations for one concept can be huge and, thus, the Monte Carlo method should be less efficient than the bounding method. Nevertheless the Monte Carlo method can ensure a certain level of tightness. Consequently the bounding method and the Monte Carlo method

**Table 2.** Datasets used in the experiments. Column 'Shortcut' refers to the short name of the dataset used in the rest of the paper; 'Size' is the number of objects in the dataset; 'Max. Size' is the maximal number of objects in a random subset of the dataset the lattice structure can be computed for; 'Max. Lat. Size' is the size of the corresponding lattice; 'Lat. Time' is the time in seconds for computing this lattice; 'Stab. Time' is the time in seconds to compute stability for every concept in the maximal lattice.

| Dataset | Shortcut | Size | Max. Size | Max. Lat. Size | Lat. Time | Stab. Time |
|---|---|---|---|---|---|---|
| Mushrooms[1] | Mush | 8124 | 8124 | $2.3 \cdot 10^5$ | 324 | 57 |
| Plants[2] | Plants | 34781 | 1000 | $2 \cdot 10^6$ | 45 | $10^4$ |
| Chess[3] | Chess | 3198 | 100 | $2 \cdot 10^6$ | 30 | $7.4 \cdot 10^3$ |
| Solar Flare (II)[4] | Flare | 1066 | 1066 | 2988 | 0 | 0 |
| Nursery[5] | Nurs | 12960 | 12960 | $1.2 \cdot 10^5$ | 245 | 5 |

can be used in a complementary way as follows. First, the stability bounds are computed. Second, if the tightness of the bounding method is worse than the tightness of the Monte Carlo method, the latter should be applied. The pseudo-code of this approach is shown in Algorithm 2. In this paper it is referred as the *combined method.*

Recall that there are three other estimates of stability [8, 9, 11] whose study is out of the scope of the present paper. Two of these estimates are applicable incrementally, i.e. when stability is known for a concept from some context and several objects are added to this context authors estimate the stability of the corresponding concept in the new lattice. For the third estimate no efficient computation is known for the moment.

In the next section we present two types of experiments. In Subsection 3.1 an experiment on the predictability of stability is presented. The discussion continues in Subsection 3.3 with the behaviour of stability thresholds and in Subsection 3.4 with stability ordering ability.

# 3   Experiment on Predictability of Stability

The experiments are run on an "Intel(R) Core(TM) i7-2600 CPU @ 3.40GHz" computer with 8Gb of memory under Ubuntu 12. The algorithms are not paral-lelized. Public datasets available from the UCI repository [17] are used for the experimentation. These datasets are shown in Table 2. With their different size and complexity, these datasets provide a rich experimental basis. Complexity here stands for the size of the concept lattice given the initial number of objects in the corresponding context. For example, Chess is the most complex dataset

---

[1] http://archive.ics.uci.edu/ml/datasets/Mushroom
[2] http://archive.ics.uci.edu/ml/machine-learning-databases/plants/
[3] http://archive.ics.uci.edu/ml/datasets/Chess+(King-Rook+vs.+King-Pawn)
[4] http://archive.ics.uci.edu/ml/datasets/Solar+Flare
[5] http://archive.ics.uci.edu/ml/datasets/Nursery

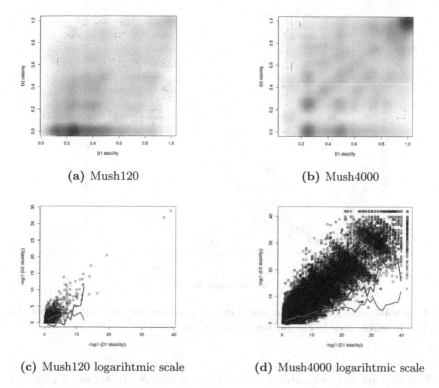

(a) Mush120

(b) Mush4000

(c) Mush120 logarihtmic scale

(d) Mush4000 logarihtmic scale

**Fig. 2.** Stability in the test dataset w.r.t the reference one

as for only 100 objects in the context there are already $2 \cdot 10^6$ of concepts in the concept lattice.

### 3.1   The Experiment Flow

Recall that the stability of a concept $c$ can be considered as the probability for the intent of $c$ to be preserved in the lattice when some objects are removed. However, when computing stability, one wants to know if the intent of a stable concept is a general characteristic rather than an artefact specific for a dataset. For that it is necessary to evaluate stability w.r.t. a test dataset different from the reference one. Reference and test datasets are two names of disjoint datasets on which the stability behaviour is evaluated. In order to do that the following scheme of experiment is developed:

1. Given a dataset $\mathbb{K}$ of size $K$ objects, experiments are performed on dataset subsets whose size in terms of number of objects is $N$. This size is required to be at least half the size of $K$. For example, for a dataset of size $K = 10$ the size of it subset can be $N = 4$.

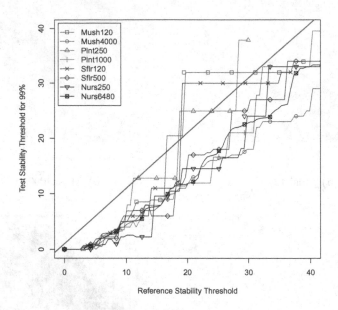

**Fig. 3.** Stability threshold in the test dataset ensuring that 99% of concepts corresponding to stable ones in the reference dataset are stable

2. Two disjoint dataset subsets $\mathbb{K}_1$ and $\mathbb{K}_2$ of size $N$ (in terms of objects) of dataset $\mathbb{K}$ are generated by sampling, e.g. $\mathbb{K}_1 = \{g_2, g_5, g_6, g_9\}$ and $K_2 = \{g_3, g_7, g_8, g_{10}\}$. Later, $\mathbb{K}_1$ is used as a reference dataset for computing stability, while $\mathbb{K}_2$ is a test dataset for evaluating stability computed in $\mathbb{K}_1$.

3. The corresponding sets of concepts $\mathcal{L}_1$ and $\mathcal{L}_2$ with their stability are built for both datasets $\mathbb{K}_1$ and $\mathbb{K}_2$.

4. The concepts with the same intents in $\mathcal{L}_1$ and $\mathcal{L}_2$ are declared as corresponding concepts.

5. Based on this list of corresponding concepts, a list of pairs $S = \{\langle X, Y \rangle, \dots\}$ is built, where $X$ is the stability of the concept in $\mathcal{L}_1$ and $Y$ is the stability of the corresponding concept in $\mathcal{L}_2$. If an intent exists only in one dataset, its stability is set to zero in the other dataset (following the definition of stability). Finally, the list $LS = \{\langle X_{\log}, Y_{\log} \rangle, \dots\}$ includes the stability pairs in $S$ in logarithmic scale as stated in formula (4).

6. Then sets of pairs $S$ and $LS$ are further used to study the behaviour of stability on disjoint (independent) datasets coming from the same general population.

The idea of evaluating stability computed on a reference dataset w.r.t. a test dataset comes from the supervised classification methods. Moreover, this idea is often used to evaluate statistical measures for pattern selection and can be found as a part of pattern selection algorithms with a good performance [18].

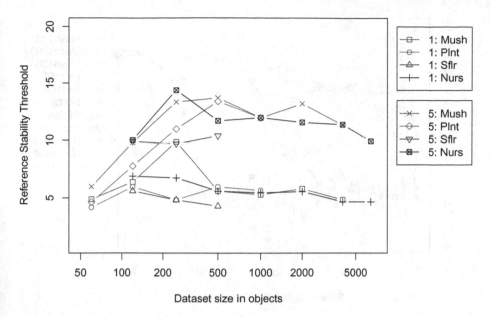

**Fig. 4.** Stability threshold in the reference dataset ensuring that 99% of concepts in the test dataset corresponding to stable concepts in the reference dataset are stable with stability thresholds 1 or 5

## 3.2 The General Behaviour of Stability

Sets of pairs $S$ and $LS$ can be drawn by matching every point $\langle X, Y \rangle$ to a point in a 2D-plot. The best case is $y = x$, i.e. stability for a concept in $\mathcal{L}_1$ is equal to stability of the corresponding concept in $\mathcal{L}_2$, meaning that stability is not dependant on the dataset. However, this is hardly the case in real-world experiments. All relevancy measures depend on the dataset, while any measure should be able to predict its value independently of the dataset. Figures 2a and 2b show the corresponding diagrams for the datasets `Mush120` and `Mush4000`.[6,7] These figures also highlight the fact that many concepts have stability close to 1, and that the larger is the dataset, the larger is the number of concepts with stability close to 1. It is in accordance with the work [16] where most of the concepts have the stability close to 1. However, when the logarithmic set $LS$ is used, a blurred line $y = x$ can be perceived in Figures 2c and 2d. Moreover, selecting the concepts which are stable w.r.t. a high threshold, say $\theta_r$, in the reference dataset $\mathbb{K}_1$, the corresponding concepts in $\mathbb{K}_2$ are stable w.r.t. a lower threshold, say $\theta_t$. Thus, we can conclude that stability is more tractable in the logarithmic scale, and then we only consider this logarithmic scale in the rest of the paper.

---

[6] From here, the name of a dataset followed by a number such as '$NameN$' refers to an experiment based on the dataset $Name$ where $\mathbb{K}_1$ and $\mathbb{K}_2$ are of the size $N$.

[7] See http://www.loria.fr/~abuzmako/stability-meaning/ for other diagrams.

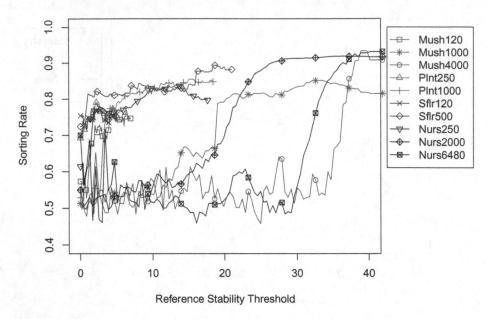

**Fig. 5.** Local sorting rate for different datasets. The rate is computed for the test dataset concepts corresponding to the first 1000 stable concepts in the reference dataset with stability above a given threshold.

## 3.3    Setting a Stability Threshold

The dependency between two thresholds $\theta_r$ and $\theta_t$ of stability are shown in Figure 3. The x-axis corresponds to the stability threshold in the reference dataset $\mathbb{K}_1$, while the y-axis corresponds to the stability threshold in the test dataset $\mathbb{K}_2$. The lines correspond to the 99% level, i.e. given the stability in $K_1$, what should be the stability threshold in the test dataset $\mathbb{K}_2$ such that 99% of stable concepts in $K_1$ are also stable in $K_2$. In this figure one can see that lines begin to grow from 5 meaning that given stability threshold less than 5 in $\mathbb{K}_1$ no stability threshold in the test dataset $\mathbb{K}_2$ can ensure 99% of stable concepts. We can also see two types of lines. The lines with stairs correspond to the datasets with small number of stable concepts, while the others behave nearly the same. This behavior suggests that in order to ensure that a concept remains stable in another dataset with threshold $\theta_{\log}$, its stability in the reference dataset should be within $[\theta_{\log} + 5, \theta_{\log} + 10]$.

Let us consider the behavior of the stability thresholds w.r.t the size of the dataset. The dependency between the size of the dataset and the difference between stability thresholds in the reference ($\mathbb{K}_1$) and in the test ($\mathbb{K}_2$) datasets is shown in Figure 4. The x-axis corresponds to the size of the dataset, the y-axis corresponds to the stability threshold in $K_1$ such that 99% of concepts selected by this threshold are stable in the test dataset $\mathbb{K}_2$ with a certain threshold (1 or 5). For example, the line '5: Mush' corresponds to the stability threshold

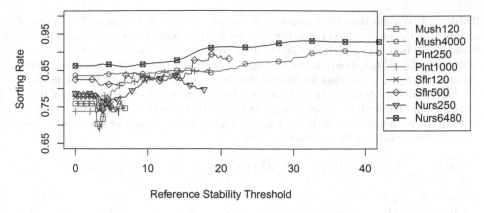

**Fig. 6.** Global sorting rate for different datasets

$\theta$ ensuring that all concepts having stability more than $\theta$ in $\mathbb{K}_1$ correspond to concepts having stability at least 5 in the test dataset $\mathbb{K}_2$. We can see that for large datasets the stability threshold is independent of the dataset, while for small datasets the diversity is higher. Here for large datasets the stability threshold should be set to 5–6 in a reference dataset in order to ensure that 99% of stable concepts have corresponding concepts in another dataset. This threshold should be set to 12 in order to ensure that 99% of stable concepts correspond to concepts having stability at least 5 in another dataset.

### 3.4 Stability and Ranking

Stability can be used for ranking concepts by decreasing its value. Thus, it is useful to study the linear order corresponding to the ranking relation. A way to study an order of an array $ar$ is to compute its sorting rate $r$, i.e. the relative number of pairs in the array sorted in the ascending order: $r = 2 \cdot \frac{\{(i,j)|i<j \text{ and } ar_i \leq ar_j\}}{|ar| \cdot (|ar|-1)}$.
A sorting rate equal to 1 means that the array is in the ascending order, while 0 means that it is in the descending order; the value 0.5 means that there is no order at all. Figure 5 shows local sorting rate (LSR), i.e. given a threshold the first 1000 stable concepts in $\mathbb{K}_1$ are taken and the sorting rate for the array of stabilities of the corresponding concepts in $\mathbb{K}_2$ is computed. This plot shows that for large datasets, the LSR is high (around 0.8–0.9) only for high stability thresholds in $\mathbb{K}_1$. For the smaller datasets the local sorting rate is around 0.7–0.8 for all thresholds. It means that stability preserves LSR only for the most stable concepts where the difference in stability between concepts is high enough, i.e. an error in order is less likely.

Finally, Figure 6 shows the global sorting rate (GSR) for different datasets, i.e. the sorting rate of stabilities in $\mathbb{K}_2$ for all concepts corresponding to the concepts selected by a threshold in $K_1$. We can see that the GSR for all datasets is slowly increasing and for small thresholds it is higher than the LSR. It shows

**Table 3.** Execution time for different steps on different datasets. `Size` is the number of concepts in the lattice; `Lattice` is the time for lattice computation with its structure; `Stab.` is the time for computing exact stability; `FCbO` is the time for computing the set of concepts by FCbO; `Freq.` is the frequency threshold applied for big datasets; `Est. Method` is the execution time for computing the estimate of stability by the estimate method; `Comb. Method` is the execution time for computing the estmate of stability be the combined method; the percentage here means that the program has been stopped after a certain amount of work; `MC calls` is the number of calls to the Monte-Carlo routine. All times are given in seconds.

| Dataset | Size | Lattice | Stab. | FCbO | Freq. | Est. Method | Comb. Method | MC calls |
|---|---|---|---|---|---|---|---|---|
| Mush8124 | $2.3 \cdot 10^5$ | 324 | 57 | 0.7 | 0 | $2 \cdot 10^3$ | $6 \cdot 10^3$ | $6 \cdot 10^4$ |
| Plnt1000 | $2 \cdot 10^6$ | 45 | $10^4$ | 78 | 0 | 181 | 446 | $3 \cdot 10^3$ |
| Chss100 | $2 \cdot 10^6$ | 46 | $10^4$ | 3.5 | 0 | 90 | 192 | $2.3 \cdot 10^3$ |
| SFlr1066 | 2988 | 0 | 0 | 0 | 0 | 0.7 | 11 | 284 |
| Nurs12960 | $1.2 \cdot 10^5$ | 245 | 5 | 0.2 | 0 | 425 | $1.2 \cdot 10^3$ | $4 \cdot 10^4$ |
| Chss3196 | $4.4 \cdot 10^6$ | – | – | 42 | 1000 | $2 \cdot 10^4$ | $3.5 \cdot 10^4$ (2%) | ? |
| Plnt34781 | $5.8 \cdot 10^6$ | – | – | 795 | 1750 | $4.1 \cdot 10^5$ | $4.6 \cdot 10^5$ (4.7%) | ? |

that stability gives a global ordering of concepts, while the local ordering is not reliable for small thresholds.

# 4   Computing an Estimate of Stability

In this section we study the efficiency of computing various estimates of stability. Table 3 shows computation times for different methods and datasets. The lattice structure is built by our implementation of AddIntent [19] and the set of concepts is computed by FCbO [20][8]. The datasets selected for experiments are the datasets of maximal tractable size (see Table 2) plus **Chess** and **Plants** with all the objects. For the last two datasets the numbers of concepts is huge. Such datasets can be analyzed by finding only frequent concepts, i.e. concepts with significantly large extents. Although an incomplete set of concepts without lattice structure cannot be processed by the algorithm from [11], stability can be estimated using formula (5), by Monte Carlo approach or their combination. For the cases where the estimation of stability takes too much time, the percentage of the processed concepts before termination is shown in the brackets. For the sake of efficiency, an estimation or an approximation of stability for a concept is stopped whenever it is clear that the concept is unstable i.e. stability is less than 3 in the logarithmic scale.

We can see that even the combined method is significantly slower than the bounding method and, hence, there is no reason to only work with the Monte Carlo method as it is slower and does not provide a better precision. Moreover,

---

[8] The implementation is taken from http://icfca2012.markuskirchberg.net.

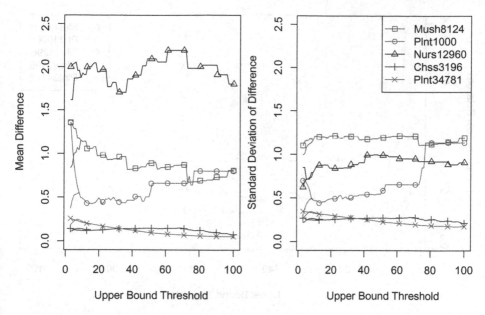

**Fig. 7.** The mean and the standard deviation of the stability estimate interval

although the number of calls to Monte Carlo routine is small in the combined method, the computational efficiency of the stability estimate can dramatically decrease, making the usage of combined method unfeasible. The estimates are more efficient in terms of computational time for large lattices, i.e. lattices with a high number of concepts for one object from the context. We can see that in some cases the estimates for small lattices take much more time than the estimates for large lattices. This can be explained by the fact that the corresponding contexts contain many objects and attributes and that the computational efficiency of the estimates is highly dependent on the size of the context.

The tightness of the estimates is shown in Figure 7. On the x-axis the values of the upper bound stability threshold are plotted while on the y-axis the mean difference in the estimate are plotted. The plots are split in area of [0, 10]; the bottom line corresponds to the improvement achieved by additional use of Monte Carlo in the combined method. According to formula (5) Monte Carlo can give any improvements only in the case where stability upper bound is less than 13 (taking into account that for these datasets there are less than 100 attributes, and Monte Carlo parameters are in accordance with Example 3). In practice, however, this bound is even smaller (less then 10). These plots show that generally mean and standard deviation of the estimate difference do not change w.r.t. the upper bound, however they can significantly depend on the dataset. In our experiments it appears that the well-structured dataset (Mush, Nurs) has higher mean value then the unstructured ones, while the big datasets with only frequent concepts have low mean-values and standard deviations.

If we want to rank concepts w.r.t. stability, how many pairs of concepts become incomparable when we use the estimates? Figure 8 shows the loss rate of the

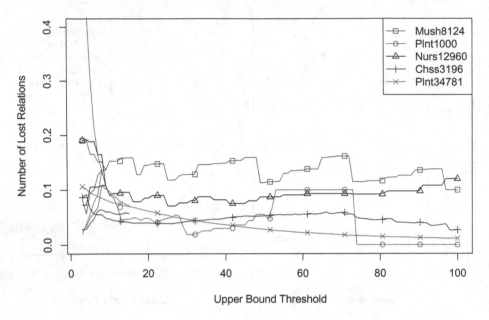

**Fig. 8.** Losing rate of relations for stability estimate

estimates, i.e. the relative number of concept pairs which cannot be compared by the estimate. Although the loss rate for the interval $[0, 10]$ can be high, it can be efficiently reduced by using the combined method.

## 5    Conclusion

In this paper we study concept stability and its estimates on different datasets. It is shown that stability computed in the logarithmic scale is more easy to interpret. Our experiments show that stability of a concept is correlated with the probability that the concept intent occurs in another dataset with high stability, i.e. it is an efficient measure for ranking patterns. However, independently of a dataset, as found experimentally, a concept should have a value of logarithmic stability greater than 5 in order to reflect any property of the population. Moreover, if the stability threshold in a reference dataset is $\theta$, then the stability of the corresponding concept in another dataset is likely to be higher than $\theta - 10$ or even $\theta - 5$. We also remarked that stability is able to sort concepts in two independent datasets with nearly the same order by selecting concepts with stability greater than a certain threshold. However, the sorting rate of the first 1000 concepts from two independent datasets with stability above a certain threshold is high if the threshold is very high.

In the second part of this paper we showed that the introduced estimate is an efficient way for ranking concepts w.r.t. stability. It can be applied for an incomplete set of concepts and, hence, has more potential applications than the exact methods. The introduced approach can be meaningfully combined with a

Monte Carlo method, providing better precision for weakly stable concepts by means of additional computational time. The precision and the sorting rate of the studied approximations are reasonably high and can be efficiently used for the stability computation.

There are many future research directions. One of them is to study other approaches for ranking formal concepts with a similar technique. An interesting question is to adapt the above approach to the comparison of different ranking methods. Next, the properties of stability suggest that interesting concepts can be found by resampling, i.e. analyzing many small parts of a large dataset, thus providing a key to an efficient processing of datasets with Formal Concept Analysis. Finally, the estimate we have proposed in this paper can be combined with an efficient realization, e.g., by means of parallel computation.

**Acknowledgments.** this research was supported by the Basic Research Program at the National Research University Higher School of Economics (Moscow, Russia) and by the BioIntelligence project (France).

# References

1. Ganter, B., Wille, R.: Formal Concept Analysis: Mathematical Foundations, 1st edn. Springer (1999)
2. Ganter, B., Kuznetsov, S.O.: Pattern Structures and Their Projections. In: Delugach, H.S., Stumme, G. (eds.) ICCS 2001. LNCS (LNAI), vol. 2120, pp. 129–142. Springer, Heidelberg (2001)
3. Bělohlávek, R., Vychodil, V.: Formal Concept Analysis with Constraints by Closure Operators. In: Schärfe, H., Hitzler, P., Øhrstrøm, P. (eds.) ICCS 2006. LNCS (LNAI), vol. 4068, pp. 131–143. Springer, Heidelberg (2006)
4. Belohlavek, R., Vychodil, V.: Formal Concept Analysis With Background Knowledge: Attribute Priorities. IEEE Transactions on Systems, Man, and Cybernetics, Part C (Applications and Reviews) 39(4), 399–409 (2009)
5. Dias, S.M., Vieira, N.J.: Applying the JBOS reduction method for relevant knowledge extraction. Expert Systems with Applications 40(5), 1880–1887 (2013)
6. Buzmakov, A., Egho, E., Jay, N., Kuznetsov, S.O., Napoli, A., Raïssi, C.: On Projections of Sequential Pattern Structures (with an application on care trajectories). In: Proc. 10th International Conference on Concept Lattices and Their Applications, pp. 199–208 (2013)
7. Belohlavek, R., Trnecka, M.: Basic Level in Formal Concept Analysis: Interesting Concepts and Psychological Ramifications. In: Proceedings of the Twenty-Third International Joint Conference on Artificial Intelligence, IJCAI 2013, pp. 1233–1239. AAAI Press (August 2013)
8. Kuznetsov, S.O.: Stability as an Estimate of the Degree of Substantiation of Hypotheses on the Basis of Operational Similarity. Automatic Documentation and Mathematical Linguistics (Nauch. Tekh. Inf. Ser. 2) 24(6), 62–75 (1990)
9. Kuznetsov, S.O.: On stability of a formal concept. Annals of Mathematics and Artificial Intelligence 49(1-4), 101–115 (2007)
10. Kuznetsov, S.O., Obiedkov, S., Roth, C.: Reducing the Representation Complexity of Lattice-Based Taxonomies. In: Priss, U., Polovina, S., Hill, R. (eds.) ICCS 2007. LNCS (LNAI), vol. 4604, pp. 241–254. Springer, Heidelberg (2007)

11. Roth, C., Obiedkov, S., Kourie, D.G.: On succinct representation of knowledge community taxonomies with formal concept analysis A Formal Concept Analysis Approach in Applied Epistemology. International Journal of Foundations of Computer Science 19(02), 383–404 (2008)
12. Klimushkin, M., Obiedkov, S., Roth, C.: Approaches to the Selection of Relevant Concepts in the Case of Noisy Data. In: Kwuida, L., Sertkaya, B. (eds.) ICFCA 2010. LNCS, vol. 5986, pp. 255–266. Springer, Heidelberg (2010)
13. Babin, M.A., Kuznetsov, S.O.: Approximating Concept Stability. In: Domenach, F., Ignatov, D.I., Poelmans, J. (eds.) ICFCA 2012. LNCS, vol. 7278, pp. 7–15. Springer, Heidelberg (2012)
14. Roth, C., Obiedkov, S., Kourie, D.: Towards concise representation for taxonomies of epistemic communities. In: Yahia, S.B., Nguifo, E.M., Belohlavek, R. (eds.) CLA 2006. LNCS (LNAI), vol. 4923, pp. 240–255. Springer, Heidelberg (2008)
15. Buzmakov, A., Egho, E., Jay, N., Kuznetsov, S.O., Napoli, A., Raïssi, C.: The representation of sequential patterns and their projections within Formal Concept Analysis. In: Workshop Notes for LML (PKDD), pp. 65–79 (2013)
16. Jay, N., Kohler, F., Napoli, A.: Analysis of Social Communities with Iceberg and Stability-Based Concept Lattices. In: Medina, R., Obiedkov, S. (eds.) ICFCA 2008. LNCS (LNAI), vol. 4933, pp. 258–272. Springer, Heidelberg (2008)
17. Frank, A., Asuncion, A.: UCI Machine Learning Repository, University of California, Irvine, School of Information and Computer Science (2010), http://archive.ics.uci.edu/ml
18. Webb, G.I.: Discovering Significant Patterns. Machine Learning 68(1), 1–33 (2007)
19. van der Merwe, D., Obiedkov, S., Kourie, D.: AddIntent: A new incremental algorithm for constructing concept lattices. In: Eklund, P. (ed.) ICFCA 2004. LNCS (LNAI), vol. 2961, pp. 372–385. Springer, Heidelberg (2004)
20. Krajca, P., Outrata, J., Vychodil, V.: Advances in Algorithms Based on CbO. In: Proc. of the 8th International Conference on Concept Lattices and Their Applications (CLA 2010), pp. 325–337 (2010)

# Factors and Skills

Bernhard Ganter and Cynthia Vera Glodeanu

Technische Universität Dresden,
01062 Dresden, Germany
{Bernhard.Ganter,Cynthia-Vera.Glodeanu}@tu-dresden.de

**Abstract.** Inspired by Knowledge and Learning Spaces, we present a novel framework for explaining the answering patterns of learners through competences and skills. More precisely, we investigate how a given learner-question data may be ascribed by a set of competences such that a learner masters a question if and only if they have a competence that is sufficient for mastering the question. Each competence is some combination of skills, but there may be restrictions on which skills can be combined. In general a question does not require a unique competence.

**Keywords:** Skills, Competences, Knowledge Spaces, Formal Concept Analysis, Learner-question data, Boolean factorisation.

## 1 Introduction

The theory of *Knowledge Spaces*, as it was introduced by Doignon and Falmagne [1], is closely related to Formal Concept Analysis. Several extensions have been studied, among them the "Competence based Knowledge Space Theory" (CbKST) [2], and, more recently, the theory of *Learning Spaces* [3]. Here we present and extend some ideas from CbKST, using the language of Formal Concept Analysis. We illustrate the basic definitions and results by a small example. Random effects, though important, will not be considered in this basic version.

## 2 Competences and Factors

We consider a formal context $(L, Q, \square)$ with the following intended interpretation: The elements of $L$ are called **learners**, those of $Q$ are the **questions**, and $l \square q$ expresses that learner $l$ **masters** question $q$. In the jargon of Formal Concept Analysis the set of questions mastered by learner $l$ then is denoted by $l^{\square}$, and $q^{\square}$ is a shorthand notation for the set of learners who master question $q$.

This interpretation should be understood in a very general manner: $Q$ might, for example, be a set of diseases, $L$ a set of therapies and $l \square q$ indicates that therapy $l$ heals disease $q$. Or $L$ is a set of customers, $Q$ a set of products and $l \square q$ indicates that product $q$ is a possible choice for customer $l$, et cetera.

We investigate how $(L, Q, \square)$ may be explained by a set $\mathcal{C}$ of **competences** in such a way that a learner masters a question $q$ if and only if they have a competence that is sufficient for mastering $q$.

C.V. Glodeanu, M. Kaytoue, and C. Sacarea (Eds.): ICFCA 2014, LNAI 8478, pp. 173–187, 2014.

This leads to the well known problem of finding **Boolean factorisations** [4,5] of $(L, Q, \Box)$. Required for such a factorisation are formal contexts $(L, \mathcal{C}, \circ)$ and $(\mathcal{C}, Q, \models)$ such that

$$l \Box q \iff \exists_{C \in \mathcal{C}} (l \circ C \text{ and } C \models q),$$

which is symbolised by

$$(L, Q, \Box) = (L, \mathcal{C}, \circ) \cdot (\mathcal{C}, Q, \models).$$

Of course, $l \circ C$ is interpreted as "learner $l$ has competence $C$" and $C \models q$ reads as "competence $C$ suffices for mastering question $q$".

It is well understood how this problem must be attacked. The factorisations are in 1-1-correspondence with the coverings of the relation $\Box$ by rectangular subrelations. Their smallest number equals the so-called 2-dimension (see [6] for the definition of $k$-dimension for arbitrary integer $k$) of the complementary context $(L, Q, L \times Q \setminus \Box)$. Determining this dimension is known to be $\mathcal{NP}$-complete. Alternatively, one can show that the factorisation problem is hard by reducing the set basis problem to it, see [5].

There is another approach to Boolean factors which is perhaps more intuitive. For a given formal context one may ask if its attributes can be interpreted as disjunctions of attributes of an other, hopefully simpler context. More formally, let us say that a **disjunctive attribute representation** of $(G, M, I)$ over $(G, N, J)$ is a mapping $\delta : M \to \mathfrak{P}(N)$ such that

$$g \, I \, m \iff \exists_{n \in \delta(m)} \, g \, J \, n.$$

The existence of such an attribute representation leads to a factorisation

$$(G, M, I) = (G, N, J) \cdot (N, M, K) \qquad \text{with } n \, K \, m : \iff n \in \delta(m).$$

Conversely any such factorisation leads to a disjunctive attribute representation via $\delta(m) := m^K$ for all $m \in M$.

*Example 1.* The data that we use is from Korossy [2]. It describes how eleven learners performed for a set $Q := \{a, b, c, d, e, f\}$ of six questions. Only seven distinct answering patterns occurred. These are given in Figure 1.

The concept lattice of the complementary relation (the diagram on the right of Figure 1) contains the information about the possible Boolean factorisations. Its length is five, which gives a lower bound for the 2-dimension (and thereby for the number of competences). But the dimension cannot be larger since there are only five join-irreducible elements. Therefore the incidence relation $\Box$ can be covered by five "rectangles", but not by fewer than five. An example of a covering is

| □ | a | b | c | d | e | f |
|---|---|---|---|---|---|---|
| 02L | × | × | × |   | × | × |
| 03L | × | × | × | × | × | × |
| 05L | × | × | × | × | × |   |
| 08L | × |   | × |   |   |   |
| 11L | × | × | × |   | × |   |
| 13L | × |   | × |   | × | × |
| 20L | × | × | × |   |   |   |

**Fig. 1.** A formal context of learners and questions, and its concept lattice, and the concept lattice of its complementary context (unlabeled)

$$C_1 := \{02L, 03L, 05L, 08L, 11L, 13L, 20L\} \times \{a, c\},$$
$$C_2 := \{02L, 03L, 05L, 11L, 20L\} \times \{a, b, c\},$$
$$C_3 := \{03L, 05L\} \times \{a, b, c, d, e\},$$
$$C_4 := \{02L, 03L, 05L, 11L\} \times \{a, b, c, e\},$$
$$C_5 := \{02L, 03L, 13L\} \times \{a, c, e, f\}.$$

Taking these factors as competences, we get a factorisation of the context in Figure 1 as shown in Figure 2.

| ○ | $C_1$ | $C_2$ | $C_3$ | $C_4$ | $C_5$ |
|---|---|---|---|---|---|
| 02L | × | × |   | × | × |
| 03L | × | × | × | × | × |
| 05L | × | × | × | × |   |
| 08L | × |   |   |   |   |
| 11L | × | × |   | × |   |
| 13L | × |   |   |   | × |
| 20L | × | × |   |   |   |

| ⊨ | a | b | c | d | e | f |
|---|---|---|---|---|---|---|
| $C_1$ | × |   | × |   |   |   |
| $C_2$ | × | × | × |   |   |   |
| $C_3$ | × | × | × | × | × |   |
| $C_4$ | × | × | × |   | × |   |
| $C_5$ | × |   | × |   | × | × |

**Fig. 2.** A factorisation of the context in Figure 1

The concept lattice of the first factorising context is shown in Figure 3.

In view of a desired interpretation, a result like the one presented in Figure 2 may be somewhat disappointing, because it only produces an (ordered) set of abstract "competences" without further explanation. Moreover, the covering with rectangular subrelations is by no means unique. In the above example, we might take as rectangles the columns of the original context, combining columns $a$ and $c$ to one rectangle, and obtain a different factorisation.

In a second step therefore one can investigate competences which comply with a given theoretical **competence model**.

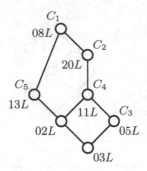

**Fig. 3.** The concept lattice of the first factorising context in Figure 2

Such a model can be given as a formal context $(S, T, *)$, where $S$ is a set of competence "states" which a learner may or may not have, $T$ is a set of competences and $s * t$ indicates that in state $s$ competence $t$ is present.

The basic question then is if the observed learner-question data can be explained by competences from this abstract model. For this, we must associate to every question $q$ the set of those competences from $T$ that are sufficient for mastering the question. Simultaneously, for each learner a suitable competence state from $S$ has to be found that enables the learner to master the questions as observed. A more formal version is given in the following theorem.

**Theorem 1.** *Let formal contexts $(L, Q, \Box)$ and $(S, T, *)$ be given. Then for every mapping $\alpha : L \to S$ the following are equivalent:*

1. *There is a mapping $\sigma : Q \to \mathfrak{P}(T)$ such that*

$$l \,\Box\, q \iff \exists_{C \in \sigma(q)} \, \alpha(l) * C.$$

2. *There is a Boolean factorisation $(L, Q, \Box) = (L, \mathcal{C}, \circ) \cdot (\mathcal{C}, Q, \models)$ together with a mapping $\beta : \mathcal{C} \to T$ such that*

$$l \circ C \iff \alpha(l) * \beta(C).$$

*Proof.* Assuming (1) we let $\mathcal{C} := \bigcup_{q \in Q} \sigma(q)$ and define for $l \in L$, $C \in \mathcal{C}$, and $q \in Q$

$$l \circ C :\iff \alpha(l) * C, \quad \beta := \mathrm{id}, \quad \text{and} \quad C \models q :\iff C \in \sigma(q).$$

The conditions of (2) are now easily verified. Conversely when starting from (2) we get (1) by letting

$$\sigma(q) := \{\beta(C) \mid C \models q\}.$$

## 3    Skills and Competences

Several authors (e.g. Korossy [2], Doignon [7]) have investigated if such competences may be explained by a finite set $S$ of **skills**, which learners may have. For
this they ask for a **skill function**[1], a mapping[2]

$$\sigma : Q \to \mathfrak{P}(\mathfrak{P}(S))$$

with the property that $\sigma(q)$ is an antichain for each $q \in Q$. It is assumed that
the learners have certain skills, as expressed by the **skill context** $(L, S, \bullet)$. The
elements of

$$\mathcal{C} := \bigcup_{q \in Q} \sigma(q)$$

then play the role of the competences. They are ordered by set inclusion $\subseteq$. The
interpretation is that a learner masters a question if they have the necessary
skills, more precisely that

$$l \,\square\, q \iff \exists_{C \in \sigma(q)} C \subseteq l^{\bullet}.$$

The context $(\mathcal{C}, Q, \models)$ is then given by

$$C \models q \iff \exists_{D \in \sigma(q)} D \subseteq C.$$

The above mentioned competence model in this case is $(\mathfrak{P}(S), \mathcal{C}, \subseteq)$.

Each skill function $\sigma$ defines a mapping $p_{\sigma} : \mathfrak{P}(S) \to \mathfrak{P}(Q)$, called the **problem function**, by

$$p_{\sigma}(T) := \{q \in Q \mid \exists_{C \in \sigma(q)} C \subseteq T\}, \qquad T \subseteq S,$$

assigning to each set $T \subseteq S$ the set of problems which can be answered with the
skills in $T$. Equivalent to the above condition is that for each learner $l \in L$ it
holds that

$$l^{\square} = p_{\sigma}(l^{\bullet}),$$

meaning that each learner masters exactly those questions for which they have
the necessary skills.

Problem functions are order preserving maps from $(\mathfrak{P}(S), \subseteq)$ to $(\mathfrak{P}(Q), \subseteq)$, and
indeed, as Düntsch and Gediga [8] have shown, every order preserving function
can be obtained in this way from a unique skill function.

*Example 2.* Continuing the above example we ask how the context in Figure 1
may be explained by skills. It is easier to tackle this problem with respect to a
given factorisation. Consider the first factorising context $(L, \mathcal{C}, \circ)$ in Figure 2.
It displays which competences the individual learners have. In order to express
these competences by subsets of a (yet unknown) set $S$ of "skills" we have to find
mappings

$$\alpha : L \to \mathfrak{P}(S) \quad \text{and} \quad \beta : \mathcal{C} \to \mathfrak{P}(S)$$

---

[1] Skill multiassignment in [7], skill multimap in [3].
[2] We omit some technical conditions which are not necessary for our considerations.

such that a learner $l$ has a competence $C$ if and only if they have all the skills contained in $\beta(C)$, formally

$$l \circ C \iff \alpha(l) \supseteq \beta(C).$$

It is immediate from Proposition 33 in [6] that such mappings can be found if and only if there is an order embedding of $\mathfrak{B}(L, C, \circ)$ into $(\mathfrak{P}(S), \supseteq)$. This is in turn equivalent to the condition that the 2-dimension of $\mathfrak{B}(L, C, \circ)$ is at most the size of $S$, i.e., to

$$\mathrm{fdim}_2(L, C, \circ) \le |S|.$$

The 2-dimension of the lattice in Figure 3 obviously is four, and Figure 4 shows an order embedding into the dual of the power set of $S := \{x, y, z, t\}$.

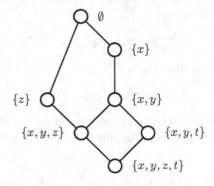

**Fig. 4.** The concept lattice of the first factorising context in Figure 2, embedded into $(\mathfrak{P}(\{x, y, z, t\}), \supseteq)$

A comparison of the labellings in Figures 3 and 4 discloses the skills associated to the learners and to the competences.

| $\bullet$ | $x$ | $y$ | $z$ | $t$ |
|---|---|---|---|---|
| 02L | × | × | × | |
| 03L | × | × | × | × |
| 05L | × | × | | × |
| 08L | | | | |
| 11L | × | × | | |
| 13L | | | × | |
| 20L | × | | | |

$C_1 = \varnothing$
$C_2 = \{x\}$
$C_3 = \{x, y, t\}$
$C_4 = \{x, y\}$
$C_5 = \{z\}.$

**Fig. 5.** The learner-skill context and the competences as sets of skills

The second factorising context in Figure 2 can now be understood as a skill function (Figure 6), however with a slight modification: The attribute intent of each question $q \in \{a, \ldots, f\}$ consists of *all* competences which suffice for mastering the question, not only the minimal ones. We call this an *enriched* skill function. Meagering it for each question to the minimal sufficient competences results in a skill function (Figure 7).

| $\models$ | $a$ | $b$ | $c$ | $d$ | $e$ | $f$ |
|---|---|---|---|---|---|---|
| $C_1 = \varnothing$ | × | | × | | | |
| $C_2 = \{x\}$ | × | × | × | | | |
| $C_3 = \{x, y, t\}$ | × | × | × | × | × | |
| $C_4 = \{x, y\}$ | × | × | × | | × | |
| $C_5 = \{z\}$ | × | | × | | × | × |

**Fig. 6.** The attribute intents define an enriched skill function

| $q$ | $a$ | $b$ | $c$ | $d$ | $e$ | $f$ |
|---|---|---|---|---|---|---|
| $\sigma(q)$ | $\{\varnothing\}$ | $\{\{x\}\}$ | $\{\varnothing\}$ | $\{\{x, y, t\}\}$ | $\{\{x, y\}, \{z\}\}$ | $\{\{z\}\}$ |

**Fig. 7.** The derived skill function

It can now easily be verified that the skill function in Figure 7 together with the learner-skill context in Figure 5 result in the original learner-question data shown in Figure 1.

We summarise our findings in a theorem. This theorem, as well as the next one, may look a little technical, but their content is easy. The first one says, loosely spoken: Given learner-question data, pick a Boolean factorisation and a representation of the first factorising context by sets. Then a skill function is obtained representing the given data.

**Theorem 2.** *Let*

$$(L, Q, \square) = (L, \mathcal{C}, \circ) \cdot (\mathcal{C}, Q, \models)$$

*be a Boolean factorisation and let* $\alpha : L \to \mathfrak{P}(S)$ *and* $\beta : \mathcal{C} \to \mathfrak{P}(S)$, *where* $S$ *is a finite set, be mappings such that*

$$l \circ C \iff \alpha(l) \supseteq \beta(C) \qquad \text{(for all } l \in L, C \in \mathcal{C}\text{)}.$$

*Then the mapping* $\sigma : Q \to \mathfrak{P}(\mathfrak{P}(S))$, *defined by*

$$\sigma(q) := \{\beta(C) \mid \beta(C) \text{ is minimal wrt. } C \models q\},$$

*is a skill function such that*

$$l \square q \iff \exists_{D \in \sigma(q)} D \subseteq \alpha(l).$$

*Proof.* Because of the minimality condition it is clear that $\sigma$ is a skill function. Since we have a Boolean factorisation we get for $l \in L$ and $q \in Q$

$$l \,\square\, q \iff \exists_{C \in \mathcal{C}}\, l \circ C \text{ and } C \models q$$
$$\iff \exists_{C \in \mathcal{C}}\, \alpha(l) \supseteq \beta(C) \text{ and } C \models q$$
$$\iff \exists_{D \in \sigma(q)}\, \alpha(l) \supseteq D.$$

The existence of such a set $D$ follows from the finiteness of $S$.

## 4    From Skills to Factors

In the previous section we have demonstrated how a skill function can be constructed from learner-question data using a two-stage set representation process. It is however not yet obvious that this method always works and, if so, that it leads to a small number of skills.

The latter is indeed not always true. The number of required skills depends on the choice of the Boolean factorisation. In fact, the data of the example can be represented by fewer skills, as we shall show.

Nevertheless is the method general enough to cover all possibilities. Each skill function can be reconstructed, as we shall demonstrate in the next theorem. Informally, it says that when the construction described in Theorem 2 is applied to learner-question data which is based on a skill function, the factorisation and the embedding can be chosen so that this skill function is reconstructed.

**Theorem 3.** *Let finite sets $L$, $Q$, and $S$ (of "learners", "questions", and "skills", respectively) be given together with a mapping $\sigma : Q \to \mathfrak{P}(\mathfrak{P}(S))$ that maps questions to antichains of skill sets (i.e., a skill function) and a mapping $\alpha : L \to \mathfrak{P}(S)$ that assigns to each learner a set of skills. Then for the relation $\square \subseteq L \times Q$, defined by*

$$l \,\square\, q :\iff \exists_{C \in \sigma(q)}\, C \subseteq \alpha(l)$$

*there is a Boolean factorisation $(L, Q, \square) = (L, \mathcal{C}, \circ) \cdot (\mathcal{C}, Q, \models)$ and a bijection $\beta : \mathcal{C} \to \bigcup_{q \in Q} \sigma(q)$, such that*

$$l \circ C \iff \alpha(l) \supseteq \beta(C) \qquad and \qquad C \models q \iff \exists_{D \in \sigma(q)}\, D \subseteq \beta(C).$$

*In particular,*

$$\sigma(q) = \{\beta(C) \mid \beta(C) \text{ is minimal wrt. } C \models q\} \qquad \text{for each } q \in Q.$$

*Proof.* Let $\mathcal{C} := \bigcup_{q \in Q} \sigma(q)$ and $\beta := \mathrm{id}$. Then

$$l \,\square\, q \iff \exists_{C \in \sigma(q)}\, C \subseteq \alpha(l)$$
$$\iff \exists_{C \in \sigma(q)} \exists_{D \in \mathcal{C}}\, \beta(D) = C \subseteq \alpha(l)$$
$$\iff \exists_{C \in \sigma(q)} \exists_{D \in \mathcal{C}}\, C \subseteq \beta(D) \subseteq \alpha(l)$$
$$\iff \exists_{D \in \mathcal{C}}\, l \circ D \text{ and } D \models q.$$

It remains to show that

$$\sigma(q) = \{\beta(C) \mid \beta(C) \text{ is minimal wrt. } C \models q\} \quad \text{for each } q \in Q.$$

To this end let $q \in Q$ and $C \in \mathcal{C}$. We show the two inclusions:

"$\supseteq$" Let $\beta(C)$ be minimal in $\{\beta(C) \mid C \models q\}$. Then, there is $D \in \sigma(q)$ s.t. $D \subseteq \beta(C)$. Since $\beta(C)$ was chosen minimal wrt. $C \models q$, we have $D = \beta(C)$ and thus $\beta(C) \in \sigma(q)$.

"$\subseteq$" Let $D \in \sigma(q)$. There exists $C \in \mathcal{C}$ s.t. $\beta(C) = D$. Hence, $C \models q$. It remains to show that $\beta(C)$ is minimal in $\{\beta(E) \mid E \models q\}$ for $\beta(E) \subsetneq \beta(C)$. Suppose not. Then, there exists $F \in \sigma(q)$ s.t. $F \subseteq \beta(E) \subsetneq \beta(C) \subseteq D$. Thus, $F \subsetneq D$ yielding a contradiction since $\sigma$ is a skill function.

*Example 3.* The learner-question data in Figure 1 can be based on only three skills, as the following tables show. For the three-element skill set $S := \{u, v, w\}$ they define a skill function $\sigma : Q \to \mathfrak{P}(\mathfrak{P}(S))$ and a learner-skill assignment $\alpha : L \to \mathfrak{P}(S)$.

| $q$ | $a$ | $b$ | $c$ | $d$ | $e$ | $f$ |
|---|---|---|---|---|---|---|
| $\sigma(q)$ | $\{\varnothing\}$ | $\{\{v\}, \{w\}\}$ | $\{\varnothing\}$ | $\{\{v, w\}\}$ | $\{\{w\}, \{u\}\}$ | $\{\{u\}\}$ |

| $l$ | $02L$ | $03L$ | $05L$ | $08L$ | $11L$ | $13L$ | $20L$ |
|---|---|---|---|---|---|---|---|
| $\alpha(l)$ | $\{u, v\}$ | $\{u, v, w\}$ | $\{v, w\}$ | $\varnothing$ | $\{w\}$ | $\{u\}$ | $\{v\}$ |

**Fig. 8.** This skill function leads to the learner-question data in Figure 1 if the learner-skill assignment is as given in the second table

The competences are $\mathcal{C} = \{\varnothing, \{u\}, \{v\}, \{w\}, \{v, w\}\}$, the corresponding factors

$$F_i := \{\text{learners that have } C_i\} \times \{\text{questions that are mastered by } C_i\}$$

are as follows:

$$F_1 = \{02L, 03L, 05L, 08L, 11L, 13L, 20L\} \times \{a, c\}$$
$$F_2 = \{02L, 03L, 05L, 20L\} \times \{a, b, c\}$$
$$F_3 = \{03L, 05L\} \times \{a, b, c, d, e\}$$
$$F_4 = \{03L, 05L, 11L\} \times \{a, b, c, e\}$$
$$F_5 = \{02L, 03L, 13L\} \times \{a, c, e, f\}.$$

These rectangular relations indeed cover the "masters"-relation. The corresponding Boolean factorisation of the learner-question context is shown in Figure 9. It differs only slightly from the one given in Figure 2.

However, the 7-element concept lattice of the first factorising context can easily embedded into $(\mathfrak{P}(\{u, v, w\}), \supseteq)$, as Figure 10 shows.

| ∘ | $C_1$ | $C_2$ | $C_3$ | $C_4$ | $C_5$ |
|---|---|---|---|---|---|
| 02L | × | × | | | × |
| 03L | × | × | × | × | × |
| 05L | × | × | × | × | |
| 08L | × | | | | |
| 11L | × | | | × | |
| 13L | × | | | | × |
| 20L | × | × | | | |

| ⊨ | $a$ | $b$ | $c$ | $d$ | $e$ | $f$ |
|---|---|---|---|---|---|---|
| $C_1$ | × | | × | | | |
| $C_2$ | × | × | × | | | |
| $C_3$ | × | × | × | × | × | |
| $C_4$ | × | × | × | | × | |
| $C_5$ | × | | × | | × | × |

**Fig. 9.** Another factorisation of the context in Figure 1. Here the first factorisation context has 2-dimension three.

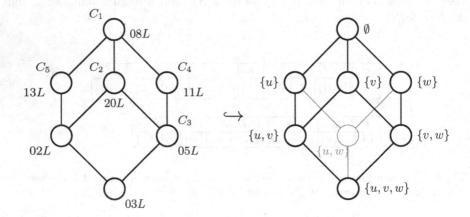

**Fig. 10.** The concept lattice of the first factorising context, embedded into $(\mathfrak{P}(\{u, v, w\}), \supseteq)$

| • | $u$ | $v$ | $w$ |
|---|---|---|---|
| 02L | × | × | |
| 03L | × | × | × |
| 05L | | × | × |
| 08L | | | |
| 11L | | | × |
| 13L | × | | |
| 20L | | × | |

$C_1 = \varnothing$

$C_2 = \{v\}$

$C_3 = \{v, w\}$

$C_4 = \{w\}$

$C_5 = \{u\}.$

**Fig. 11.** The learner-skill context and the competences as sets of skills, for the modified set representation

# 5    Structured Skill Sets

Our approach admits several variations which may be of practical interest. We briefly discuss four of them here.

## 5.1    Graded Skills

The lattice in Figure 4 obviously has an order embedding into a product of two chains, one of size four, the other of size two. This allows to give a more structured interpretation of the four necessary skills: they may be chosen as $\{x, x^+, x^{++}, z\}$, where $x$ is a prerequisite for $x^+$, and $x^+$ a prerequisite for $x^{++}$. The five competences may then be written as

$$C_1 = \varnothing, \quad C_2 = \{x\}, \quad C_3 = \{x^{++}\}, \quad C_4 = \{x^+\}, \quad C_5 = \{z\},$$

with the tacit convention that $x^+$ includes $x$ etc.

This can widely be generalised. A family of competences can be interpreted with skills $\{x_1, x_1^+, x_1^{++}, \ldots, x_k, x_k^+, x_k^{++}\}$ if and only if the 4-dimension of the first factor is at most $k$. But even arbitrarily ordered skill sets can be considered and respective conditions on the factorisations can be formulated.

## 5.2    Propositional Formulae

We may even consider "negative skills". Recall that a skill function encodes that a question $q$ is mastered if and only if at least one competence, i.e., skill combination, from a specified list $\sigma(q)$ is present. So what is required for mastering $q$ is a disjunction of conjunctions of skills, a monotone Boolean term in the language of Propositional Logic.

So why not allow for arbitrary propositional formulae? This can easily be done. Figure 12 shows a representation of our original learner-question data (Figure 1) by propositional formulae in three variables.

But how can this be interpreted? It seems unrealistic that there may be skills which *hinder* a learner mastering a question. However, for other interpretations this may be meaningful. One such case is that of customers selecting goods according to their features. E.g., when buying bread, some customers may prefer one with caraway seeds, while for others this could be a impediment.

## 5.3    The Dichotomic Scale $\mathbb{D}_k$

In the next example we shall make use of the $k$-dimensional **dichotomic scale** $\mathbb{D}_k$, which is one of the standard scales in Formal Concept Analysis (see [6], Lex [9]). It is usually introduced as the $k$-fold semiproduct $\mathbb{D} \times \mathbb{D} \times \cdots \times \mathbb{D}$ of the (one dimensional) dichotomic scale

$$\mathbb{D} := \begin{array}{|c|c|} \hline \cdot & \times \\ \hline \times & \cdot \\ \hline \end{array}.$$

| | $\top$ | $\neg(x \wedge (y \vee z))$ | $\top$ | $\neg y \wedge ((\neg x \wedge z) \vee (x \wedge \neg z))$ | $\neg(y \vee (x \wedge \neg z))$ | $\neg(y \vee (\neg x \wedge z))$ |
|---|---|---|---|---|---|---|
| $\bot\,\bot\,\bot$ | × | × | × | | × | × |
| $\top\,\bot\,\bot$ | × | × | × | × | × | × |
| $\bot\,\bot\,\top$ | × | × | × | × | × | |
| $\top\,\top\,\top$ | × | | × | | | |
| $\bot\,\top\,\top$ | × | × | × | | × | |
| $\top\,\bot\,\top$ | × | | × | | × | × |
| $\bot\,\top\,\bot$ | × | × | × | | | |

**Fig. 12.** Truth value assignments and propositional formulae for the context of Figure 1

For our purposes it is convenient to give another (yet equivalent) description based on a set $V := \{v_1, \ldots, v_k\}$ of symbols[3]. The scale $\mathbb{D}_k$ has $2^k$ objects, $2k$ attributes and $3^k + 1$ formal concepts. As objects we may take the set of all maps from $V$ to $\{+, -\}$. The set of attributes is $S := \{+v_1, \ldots, +v_k, -v_1, \ldots, -v_k\}$. An object $\nu : V \to \{+, -\}$ is incident with an attribute $+s$ (where $s \in V$) iff $\nu(s) = +$, and with the "negative" attribute $-s$ iff $\nu(s) = -$.

A subset of $S$ is called *feasible* if it does not contain a symbol $v$ both in its positive form $+v$ and in its negative form $-v$. The only concept intent that is not feasible is the set $S$, and the corresponding extent is $\emptyset$. Apart from this exception, the concept intents of the dichotomic scale are exactly the feasible subsets of $S$. The concept extent corresponding to a feasible set $T \subseteq S$ consist of those mappings $\nu : V \to \{+, -\}$ that satisfy the condition

$$\text{if } +v \in T \text{ then } \nu(v) = +, \qquad \text{and if } -v \in T \text{ then } \nu(v) = -.$$

The concept extents, apart from the smallest one, therefore can be identified with the partial mappings $\nu : V \to \{\bot, \top, ?\}$.

## 5.4   Incompatible Skills

The propositional approach in Subsection 5.2 is based on the *negation* of skills. In practice however it seems unlikely that a skill is the negation of another one. A more realistic assumption is that skills may be mutually exclusive, but not

---

[3] We avoid naming the elements of $V$ *variables*, because $-v$ is not the negation of $-v$. As a consequence, we later shall work with a modified notion of disjunction.

necessarily exhaustive. In other words: such two skills cannot occur together, but may both be missing. For example, good jockeys usually are not very good high jumpers, because jockeys need to be small, high jumpers to be tall. But most people neither are jumpers nor jockeys.

As in Subsection 5.3 we start with a set $V := \{v_1, \ldots, v_k\}$ of symbols and define $S := \{+v_1, \ldots, +v_k, -v_1, \ldots, -v_k\}$. The elements of $S$ will be the skills, with the intention that for each $i$ the skills $+v_i$ and $-v_i$ are mutually exclusive.

The competence model in this case is introduced as follows: All feasible sets of skills are competences, and all mappings from $V$ to $\{+, -, ?\}$ are possible learner states. The concept intents of the dichotomic scale then are in 1-1-correspondence to the competences, with one exception, which we artificially add: We allow for the set $S$ of *all* skills, though not admissible, as a competence, the *"Chuck Norris competence"*. Similarly, the possible learner states correspond to the concept extents of the dichotomic scale, when we artificially add the possibility of an "almighty" learner that has all skills, negative and positive. The incidence relation in the competence model is the natural one, the one that was discussed in the previous subsection.

Applying Theorem 1 in the case of this slightly artificial competence model yields the following:

**Corollary 1.** *A learner-question context can be interpreted using skills*

$$+v_1, \ldots, +v_k, -v_1, \ldots, -v_k$$

*(where $+v$ and $-v$ are incompatible), iff there is a Boolean factorisation and an order embedding of the concept lattice of the first factorising context into the concept lattice of the $k$-dimensional dichotomic scale.*

*Example 4.* Again we demonstrate this by an example. The concept lattice in Figure 10 (left) can also be embedded into the concept lattice of the 2-dimensional dichotomic scale $\mathbb{D}_2$, see Figure 13, in which we use symbols $x, y$ instead of $v_1, v_2$. Actually, there are several embeddings.

According to the corollary, the learner-question data can be interpreted using two pairs $+x, -x, +y, -y$ of incompatible skills. The competences are mapped to feasible skill sets as follows:

| $C_1$ | $C_2$ | $C_3$ | $C_4$ | $C_5$ |
|-------|-------|-------------|-------|--------|
| $\emptyset$ | $\{+x\}$ | $\{+x, +y\}$ | $\{+y\}$ | $\{-y\}$ |

The observed learner states are the following:

| $02L$ | $03L$ | $05L$ | $08L$ | $11L$ | $13L$ | $20L$ |
|-------------|----------|--------------|-------------|---------|----------|---------|
| $\{+x, -y\}$ | almighty | $\{+x, +y\}$ | $\emptyset$ | $\{+y\}$ | $\{-y\}$ | $\{+x\}$ |

We can also give a skill function based on these skills. It is tempting to do this in propositional form, similar as in Figure 12. However the meaning of disjunction has to be modified, the expression $+v \vee -v$ should not evaluate to $\top$. Instead, we introduce a new symbol depending on $v$ by

$$\delta(v) := +v \vee -v.$$

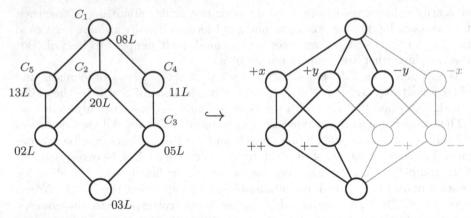

**Fig. 13.** The concept lattice of the learner-competence context in Figure 9 embedded into $\mathfrak{B}(\mathbb{D}_2)$

| □ | $a :: \top$ | $b :: +x \vee +y$ | $c :: \top$ | $d :: +x \wedge +y$ | $e :: \delta(y)$ | $f :: -y$ |
|---|---|---|---|---|---|---|
| 02L: $\{+x, -y\}$ | × | × | × | | × | × |
| 03L: *almighty* | × | × | × | × | × | × |
| 05L: $\{+x, +y\}$ | × | × | × | × | × | |
| 08L: $\emptyset$ | × | | × | | | |
| 11L: $\{+y\}$ | × | × | × | | × | |
| 13L: $\{-y\}$ | × | | × | | × | × |
| 20L: $\{+x\}$ | × | × | × | | | |

**Fig. 14.** A representation of the learner-question data in Figure 1 using two pairs of mutually exclusive skills. $\delta(y)$ is an abbreviation for $+y \vee -y$

The reason is this: If mastering a problem requires $+v$ or $-v$, then one of the two skills $+v$ and $-v$ must be present. This is not necessarily the case, and replacing $+v \vee -v$ by $\top$ therefore leads to errors.

With this notation we obtain from the second factorising context in Figure 10

$$\sigma(a) = C_1 \vee C_2 \vee C_3 \vee C_4 \vee C_5 = \top$$
$$\sigma(b) = C_2 \vee C_3 \vee C_4 = +x \vee +y$$
$$\sigma(c) = C_1 \vee C_2 \vee C_3 \vee C_4 \vee C_5 = \top$$
$$\sigma(d) = C_3 = +x \wedge +y$$
$$\sigma(e) = C_3 \vee C_4 \vee C_5 = \delta(y)$$
$$\sigma(f) = C_5 = -y.$$

Figure 14 finally shows that the combination of these findings indeed represents the learner-question context in Figure 1.

# 6 Conclusion

The combination of Boolean factorisations and of embeddings into standard concept lattices gives promising results for the analysis of learner-question data, in particular for the construction of skill functions according to given competence models. In our presentation we have worked out a few examples. A more general and versatile theory seems possible.

# References

1. Doignon, J.P., Falmagne, J.C.: Spaces for the assessment of knowledge. International Journal of Man-Machine Studies 23(2), 175–196 (1985)
2. Korossy, K.: Modeling knowledge as competence and performance. In: Albert, D., Lukas, J. (eds.) Knowledge Spaces: Theories, Empirical Research, and Applications. Lawrence Erlbaum Associates (1999)
3. Doignon, J.P., Falmagne, J.C.: Learning Spaces. Springer, Heidelberg (2011)
4. Keprt, A., Snásel, V.: Binary factor analysis with help of formal concepts. In: Snásel, V., Belohlávek, R. (eds.) CLA. CEUR Workshop Proceedings, vol. 110. CEUR-WS.org (2004)
5. Belohlávek, R., Vychodil, V.: Formal concepts as optimal factors in boolean factor analysis: Implications and experiments. In: Eklund, P.W., Diatta, J., Liquiere, M. (eds.) CLA. CEUR Workshop Proceedings, vol. 331. CEUR-WS.org (2007)
6. Ganter, B., Wille, R.: Formal Concept Analysis: Mathematical Foundations. Springer, Heidelberg (1999)
7. Doignon, J.P.: Knowledge spaces and skill assignments. In: Fischer, G.H., Laming, D. (eds.) Contributions to Mathematical Psychology, Psychometrics, and Methodology, Recent Research in Psychology, pp. 111–121. Springer, New York (1994)
8. Düntsch, I., Gediga, G.: Skills and knowledge structures. British Journal of Mathematical and Statistical Psychology 48, 9–27 (1995)
9. Lex, W.: Eine Darstellung zur maschinellen Behandlung von Begriffen. In: Ganter, B., Wille, R., Wolff, K.E. (eds.) Beiträge zur Begriffsanalyse, Wissenschaftsverlag, pp. 141–160 (1986)

# Automatized Construction of Implicative Theory of Algebraic Identities of Size Up to 5

Artem Revenko[1,2]

[1] National Research University Higher School of Economics
Pokrovskiy bd. 11, 109028 Moscow, Russia
[2] Technische Universität Dresden
Zellescher Weg 12-14, 01069 Dresden, Germany
artem_viktorovich.revenko@mailbox.tu-dresden.de

**Abstract.** Automation of constructing dependencies between algebraic identities of size up to 5 is investigated. For this purpose a robust active learning technique called Attribute Exploration is used. The technique collects algebra–identity pairs from an expert and builds a concise representation of implicative dependencies (implicative theory) between the identities. It is not possible to accomplish the construction of the implicative theory using only finite algebras and due to this fact heuristics and an algorithm for finding appropriate algebras over an infinite universe are introduced. This allowed for accomplishing the constructing and proving all the obtained implications.

**Keywords:** equational classes, implicative theory, attribute exploration, equational logic, infinite models.

## 1 Introduction

Algebraic identities describe different classes of algebraic structures (equational classes) and therefore play one of the central roles in algebra. Examples include monoids, groups, rings, lattices, and many others. The field of research that studies common patterns of algebraic structures is called universal algebra. One of the main results from which this field of research started – the famous HSP theorem – states that equational description of a class of algebras is equivalent to constructive description (that is operations on algebras which do not lead out of the class) [2]. As noted in [21]: "The role of algebraic equations was pronounced from the start". Therefore, the studying of equational classes is essentially important for mathematics.

A central question one could ask about equational classes is the following: if the class satisfies a given set of identities which other identities are necessarily satisfied by all the members of the class? In other words, which identities are deducible from given ones? The strength and importance of equational deduction can be well appreciated from the words from [5]: "it has even been shown that every problem concerning the derivability of a mathematical statement from a given set of axioms can be reduced to the problem of whether an equation is

C.V. Glodeanu, M. Kaytoue, and C. Sacarea (Eds.): ICFCA 2014, LNAI 8478, pp. 188–202, 2014.

identically satisfied in every relation algebra. One could thus say that, in principle, the whole of mathematical research can be carried out by studying identities
in the arithmetic of relation algebras." It is well known that in general it is not
possible to decide if an identity is deducible from a given set of identities, see
e.g. [20]. Even for a finite set of equations this question can be undecidable [17,
p. 179], [21, p. 28]. However, there are special classes of identities for which the
questions is decidable, for example, groups [8]. The modern field of science called
automated theorem proving has made a big progress in equational deduction (as
a part of deduction in first order logic). To be more precise equational deduction
is semidecidable, meaning that it is not always possible to say if the answer is
negative, i.e. when an identity does not hold. As a counterpart of automatic
theorem provers, automatic model finders are also actively developed. However,
modern tools concentrate on finite models. In this work a pattern and an algorithm for finding infinite models for a particular part of equational logic is
introduced and shown to be enough to find all the counter-examples, therefore,
proving that this part of logic is decidable.

Deductibility is not at all the only question of interest about equational
classes. As pointed out in [4, Recent Developments and Open Problems] finding (finite) bases for equational theories and classification of equational classes
are in scope of current research activities. For the purpose of solving these two
questions in a given set of identities one could find all possible interrelations
between identities inside this set (implicative theory). Up to now no automated
knowledge processing algorithm was offered to automatize the research of the
implicative theory in a given set of identities. In this work this task is addressed
with a robust active learning technique called Attribute Exploration. Attribute
Exploration offers a transparent and useful knowledge acquisition and structur-
ing algorithm. It interactively collects data from an expert and builds a concise
representation of interrelations on the features of the collected data (in the form
of implicative dependencies). A similar investigation was carried out in [13], however, the procedure was not automated, i.e. finding infinite counter-examples,
checking satisfaction of identities in algebras, finding proofs for identities were
performed by hand, therefore, there was no guarantee that a mistake has not
leaked in during the investigation and there is no way to generalize the used
methods for a more general case. The investigation in [13] took several years
and lead to a PhD thesis.

Automation of usage of Attribute Exploration for the exploration of identities and making it efficient issues a number of unique challenges. For example,
though only 70 identities of size up to 5 are under investigation, it turns out
that it is not possible to finish the investigation considering only finite counter-
examples. Proposed pattern for building infinite algebras and an algorithm for
constructing these algebras enabled us to overcome the difficulties and finish the
research. The implicative theory was constructed and proved in automatic mode.
The elaborated methods allow one for designing a flexible knowledge processing
system for a more general case.

The paper is organized as follows. In Section 2 an introduction to Attribute Exploration – the main method used in the current investigation – is given. In the next Section 3 algebraic identities considered in the current work are presented. Algebraic identities are particularly suitable for Attribute Exploration, as they allow a formalization and, therefore, automation of checking their satisfaction in algebras. It is also possible and almost straightforward to generate all finite algebras of given signature over a given universe. Considering only non-isomorphic algebras enables a speed-up of up to $n!$ times, where $n$ is the size of the universe. The set of all finite algebras over a universe of size 2 is the starting point of the investigation. For the next steps finite and infinite counter-examples are needed. The finite counter-examples are found with the help of automatic model finder Mace4 [15]. The existing tools are not suited for finding infinite models, therefore a pattern and an algorithm for finding algebras over an infinite universe satisfying given identities and not satisfying one selected identity are introduced. The question of generating algebras, as well as the proof of the necessity of infinite algebras, are outlined in Section 4. The constructing of implicative theory was finished by means of proving all implications from implication basis with the help of automatic theorem prover Prover9 [15]. Results and conclusion are presented in Section 5.

**Contributions.** All possible pairwise not equivalent identities involving one binary, one unary, and one nullary operations of size up to 5 are considered. There are 70 such identities [13].

- A pattern and an algorithm for generating infinite algebras satisfying a set of identities and not satisfying an identity is developed and implemented;
- An algorithm for checking the satisfaction of identites in algebras is developed and implemented;
- Using only software tools the set of all valid implications between the identities is constructed and proved.

## 2    Attribute Exploration

In what follows we keep to standard definitions of FCA [10]. Let $G$ and $M$ be sets and let $I \subseteq G \times M$ be a binary relation between $G$ and $M$. The triple $\mathbb{K} := (G, M, I)$ is called a *(formal) context*. The set $G$ is called the set of *objects*. The set $M$ is called the set of *attributes*.

Consider mappings $\varphi \colon 2^G \to 2^M$ and $\psi \colon 2^M \to 2^G$:

$$\varphi(X) := \{m \in M \mid gIm \text{ for all } g \in X\},$$

$$\psi(A) := \{g \in G \mid gIm \text{ for all } m \in A\}.$$

Mappings $\varphi$ and $\psi$ define a *Galois connection* between $(2^G, \subseteq)$ and $(2^M, \subseteq)$, i.e. $\varphi(X) \subseteq A \Leftrightarrow \psi(A) \subseteq X$. Usually, instead of $\varphi$ and $\psi$ a single notation $(\cdot)'$ is used. For $X \subseteq G$ the set $X'$ is called the *intent* of $X$. Similarly, for $A \subseteq M$ the set $A'$ is called the *extent* of $A$. If $A$ or $X$ consists of only one element we usually omit the curly brackets.

An object $g$ such that $g' \neq \emptyset$ is called *reducible* in a context $\mathbb{K} := (G, M, I)$ iff $\exists X \subseteq G \setminus g : g' = \bigcap_{j \in X} j'$.

An *implication* of $\mathbb{K} = (G, M, I)$ is defined as a pair $(A, B)$, where $A, B \subseteq M$, written $A \to B$. $A$ is called the *premise*, $B$ is called the *conclusion* of the implication $A \to B$. The implication $A \to B$ is *respected by a set of attributes* $N$ if $A \nsubseteq N$ or $B \subseteq N$. The implication $A \to B$ *holds* (is *valid*) in $\mathbb{K}$ if it is respected by all $g'$, $g \in G$, i.e. every object, that has all the attributes from $A$, also has all the attributes from $B$. New valid implications can be obtained using *Armstrong rules*:

$$\frac{}{A \to A} \quad , \quad \frac{A \to B}{A \cup C \to B} \quad , \quad \frac{A \to B, B \cup C \to D}{A \cup C \to D}$$

A *unit implication* is defined as an implication with only one attribute in its conclusion, i.e. $A \to b$, where $A \subseteq M$, $b \in M$. Every implication $A \to B$ can be regarded as a set of unit implications $\{A \to b \mid b \in B\}$. One can always consider only unit implications without loss of generality.

An *implication basis* of a context $\mathbb{K}$ is defined as a set $\mathfrak{L}_{\mathbb{K}}$ of implications of $\mathbb{K}$, from which any valid implication for $\mathbb{K}$ can be deduced by the Armstrong rules and none of the proper subsets of $\mathfrak{L}_{\mathbb{K}}$ has this property. Reducible objects do not contribute to any implication basis [10], therefore, if one is only interested in an implication basis of the context reducible objects can be eliminated. A context without reducible objects is called *reduced*, the procedure of eliminating reducible objects is called *reducing*.

A minimal in the number of implications basis was defined in [12] and is known as the *canonical implication basis*. In paper [9] the premises of implications from the canonical bases were characterized in terms of pseudo-intents. A subset of attributes $P \subseteq M$ is called a *pseudo-intent*, if $P \neq P''$ and for every pseudo-intent $Q$ such that $Q \subset P$, one has $Q'' \subset P$, where $\subset$ is used in the sense of $\subseteq$ and $\neq$. The canonical implication basis looks as follows: $\{P \to (P'' \setminus P) \mid P$ - pseudo-intent$\}$. The canonical implication basis is used in what follows, however, the investigation could be performed using another implication basis.

*Attribute Exploration* consists in iterations of the following steps until stabilization: computing the implication basis of a context, finding counterexamples to implications, updating the context with counterexamples as new objects, recomputing the basis. Attribute Exploration has been successfully used for investigations in many mostly analytical areas of research. For example, in [14] Attribute Exploration is used for studying Boolean algebras, in [7] lattice properties are studied, in [18] function properties are studied, and there is even a research on Attribute Exploration in fuzzy settings [11].

## 3  Algebraic Identities

In what follows we keep to standard definitions of universal algebra, see e.g. [4].

**Definition 1.** *An algebra $\mathcal{A} = (A, \Phi)$ consists of a set $A$, called* universe *(or* domain, *or* carrier*), and a family of operations $\Phi$ over the set $A$, called* signature. *To every operation $F \in \Phi$ corresponds an arity $s(F) \in \mathbb{N}_0$. The family $\tau$ of arities of all operations is called the* type *of the algebra $\mathcal{A}$.*

*An algebra $\mathcal{A}$ is* finite *if $A$ is finite, otherwise $\mathcal{A}$ is* infinite.

Let $F_k = \{F \in \Phi | s(F) = k\}$, in particular $F_k$ is the set of constants.

**Definition 2.** *Let $X$ be a set of objects called* variables *and $\Phi$ be a set of operations (a signature of an algebra). The set $T_\Phi(X)$ of terms of signature $\Phi$ over $X$ is the smallest set such that*

1. $X \cup F_0 \in T_\Phi(X)$;
2. *If $p_1, \ldots, p_n \in T_\Phi(X)$ and $s(f) = n$ then the term $f(p_1, \ldots, p_n) \in T_\Phi(X))$.*

*The* size *$l(p)$ of a term $p \in T_\Phi(X)$ is the sum of all occurrences of operations and variables in $p$.*

**Definition 3.** *Given a term $p(x_1, \ldots, x_n) \in T_\Phi(X)$ and given an algebra $\mathcal{A}$ of signature $\Phi$ we define a mapping $p^\mathcal{A} : A^n \to A$ as follows:*

1. *if $p$ is a variable $x_i$ then $p^\mathcal{A}(a_1, \ldots, a_n) = a_i$;*
2. *if $p$ is of the form $f(p_1(x_1, \ldots, x_n), \ldots, p_k(x_1, \ldots, x_n))$, where $f \in F_k$, then*

$$p^\mathcal{A}(a_1, \ldots, a_n) = f^\mathcal{A}(p_1^\mathcal{A}(a_1, \ldots, a_n), \ldots, p_k^\mathcal{A}(a_1, \ldots, a_n)),$$

*where $f^\mathcal{A}$ is the operation in algebra $\mathcal{A}$.*

**Definition 4.** *An* identity *of signature $\Phi$ over $X$ is an expression of the form*

$$p \equiv q$$

*where $p, q \in T_\Phi(X)$. An algebra $\mathcal{A}$ satisfies an identity $p(x_1, \ldots, x_n) \equiv q(x_1, \ldots, x_n)$ if for every choice $a_1, \ldots, a_n \in A$ we have $p^\mathcal{A}(a_1, \ldots, a_n) \equiv q^\mathcal{A}(a_1, \ldots, a_n)$.*

*The* size *$l(p \equiv q)$ of an identity $p \equiv q$ is the sum of the sizes of both terms $l(p \equiv q) := l(p) + l(q)$.*

*Example 1.* Consider the identity $(-x) * y * x \equiv y * x$. The size of the left term is $l((-x) * y * x) = 6$, the size of the right term $l(y * x) = 3$, the size of the identity $l((-x) * y * x \equiv y * x) = 9$.

An algorithm of checking the satisfaction of identites in algebras arises from the definition of identities. However, for the aims of constructing infinite algebras we introduce here a more general algorithm capable of processing partial algebras, i.e. algebras with not totally defined operations. $A^X$ denotes the set of all possible mappings from $X$ to $A$ ($A^X := \{f | f : X \to A\}$).

**Input:** $\mathcal{A} = (A, \Phi)$, $p(x_1, \ldots, x_n), q(x_1, \ldots, x_n) \in T_\Phi(X)$.
**Output:** Is an identity $p \equiv q$ satisfied in an algebra $\mathcal{A}$?

```
1  for map in A^X do
2  │   a_1, ..., a_n = map(x_1), ..., map(x_n)
3  │   if p^A(a_1, ..., a_n) is not defined then
4  │   └   return None, p(a_1, ..., a_n)
5  │   if q^A(a_1, ..., a_n) is not defined then
6  │   └   return None, q(a_1, ..., a_n)
7  │   if not p^A(a_1, ..., a_n) = q^A(a_1, ..., a_n) then
8  │   └   return False

9  return True
```

**Algorithm 1:** check_identity_partial

For the signature in the current work we use the notation $(*, -, a)$, where $*$ is a binary operation, $-$ is a unary operation, and $a$ is a nullary operation. We call two identities pairwise equivalent if they are satisfied in the same algebras. Examples of pairwise equivalent identities are: $x * y \equiv y * x$ and $y * z \equiv z * y$; $x \equiv y$ and $x \equiv a$; $a \equiv a$ and $x \equiv x$. In the current investigation all possible pairwise nonequivalent identities of signature $(*, -, a)$ over $\{x, y, z\}$ of size up to 5 are considered. There are exactly 70 such identities [13], the set $M_{\mathrm{id}}$ of all identities is listed below.

Size 2:

$$x \equiv x; \tag{1.1}$$
$$x \equiv y; \tag{1.2}$$

Size 3:

$$a \equiv -a; \tag{1.3}$$
$$a \equiv -x; \tag{1.4}$$
$$x \equiv -x; \tag{1.5}$$

Size 4:

$$a \equiv -(-a); \tag{1.6}$$
$$a \equiv -(-x); \tag{1.7}$$
$$a \equiv a * a; \tag{1.8}$$
$$a \equiv a * x; \tag{1.9}$$
$$a \equiv x * a; \tag{1.10}$$
$$a \equiv x * x; \tag{1.11}$$
$$a \equiv x * y; \tag{1.12}$$
$$-a \equiv -x; \tag{1.13}$$
$$x \equiv -(-x); \tag{1.14}$$

$$x \equiv a * x; \tag{1.15}$$
$$x \equiv x * a; \tag{1.16}$$
$$x \equiv x * x; \tag{1.17}$$
$$x \equiv x * y; \tag{1.18}$$
$$x \equiv y * x; \tag{1.19}$$

Size 5:

$$a \equiv -(-(-a)); \tag{1.20}$$
$$a \equiv -(-(-x)); \tag{1.21}$$
$$a \equiv a * (-a); \tag{1.22}$$
$$a \equiv a * (-x); \tag{1.23}$$
$$a \equiv (-a) * a; \tag{1.24}$$
$$a \equiv (-a) * x; \tag{1.25}$$
$$a \equiv x * (-a); \tag{1.26}$$
$$a \equiv x * (-x); \tag{1.27}$$
$$a \equiv x * (-y); \tag{1.28}$$
$$a \equiv (-x) * a; \tag{1.29}$$
$$a \equiv (-x) * x; \tag{1.30}$$
$$a \equiv (-x) * y; \tag{1.31}$$
$$a \equiv -(a * a); \tag{1.32}$$
$$a \equiv -(a * x); \tag{1.33}$$
$$a \equiv -(x * a); \tag{1.34}$$
$$a \equiv -(x * x); \tag{1.35}$$
$$a \equiv -(x * y); \tag{1.36}$$
$$-a \equiv -(-a); \tag{1.37}$$
$$-a \equiv -(-x); \tag{1.38}$$
$$-a \equiv a * a; \tag{1.39}$$
$$-a \equiv a * x; \tag{1.40}$$
$$-a \equiv x * a; \tag{1.41}$$
$$-a \equiv x * x; \tag{1.42}$$
$$-a \equiv x * y; \tag{1.43}$$
$$x \equiv -(-(-x)); \tag{1.44}$$
$$x \equiv a * (-x); \tag{1.45}$$
$$x \equiv x * (-a); \tag{1.46}$$
$$x \equiv x * (-x); \tag{1.47}$$
$$x \equiv x * (-y); \tag{1.48}$$
$$x \equiv y * (-x); \tag{1.49}$$

$$x \equiv (-a) * x; \tag{1.50}$$
$$x \equiv (-x) * a; \tag{1.51}$$
$$x \equiv (-x) * x; \tag{1.52}$$
$$x \equiv (-x) * y; \tag{1.53}$$
$$x \equiv (-y) * x; \tag{1.54}$$
$$x \equiv -(a * x); \tag{1.55}$$
$$x \equiv -(x * a); \tag{1.56}$$
$$x \equiv -(x * x); \tag{1.57}$$
$$x \equiv -(x * y); \tag{1.58}$$
$$x \equiv -(y * x); \tag{1.59}$$
$$-x \equiv -(-x); \tag{1.60}$$
$$-x \equiv a * a; \tag{1.61}$$
$$-x \equiv a * x; \tag{1.62}$$
$$-x \equiv a * y; \tag{1.63}$$
$$-x \equiv x * a; \tag{1.64}$$
$$-x \equiv x * x; \tag{1.65}$$
$$-x \equiv x * y; \tag{1.66}$$
$$-x \equiv y * a; \tag{1.67}$$
$$-x \equiv y * x; \tag{1.68}$$
$$-x \equiv y * y; \tag{1.69}$$
$$-x \equiv y * z. \tag{1.70}$$

In the general case the equational theory involving identities from above is undecidable, because of, for example, having the identity $x * x \equiv x$ [3, pp. 34–36].

## 4    Algebras

Several important classes of algebras can be defined using all or several operations from chosen signature.

*Example 2.* A groupoid is an algebra $(A, *)$. A groupoid satisfying associativity $(x * y) * z \equiv x * (y * z)$ is called semigroup. If in a semigroup exists an identity element $a$, i.e. $a * x \equiv x * a \equiv x$ the semigroup is called a monoid. A group is a monoid with inverse elements $x * (-x) \equiv (-x) * x \equiv a$, where $-$ is a unary operation on $A$ and $a$ is the identity element of the monoid.

Such classes include finite and infinite algebras. Obviously, it is much easier to work with finite algebras. For example, it is possible to directly use the algorithm check_identity_partial and iterate over all possible mappings in order to check the satisfaction of an identity. However, as was shown in [13], it is not possible to complete the Attribute Exploration considering only finite algebras, therefore, it is necessary to find a way to generate infinite counter-examples.

### 4.1 Generating Finite Algebras

As it is well known that isomorphic algebras satisfy exactly the same identities (see e.g. [4]) there is no need to generate isomorphic algebras. Isomorphic algebras can be obtained by cyclic permutation of elements. As there exist $n!$ such permutations, where $n$ is the size of the universe, it is necessary to generate not more than $(1/n!)$th part of all possible algebras. In the first step of Attribute Exploration all possible 64 non-isomorphic algebras over a universe of size 2 were generated and satisfaction of all chosen identities was directly checked.

An advantage of working with a well developed field of knowledge such as algebraic identities is the existence of well developed software tools such as model finders. This advantage was used in the next steps of Attribute Exploration for the purpose of finding finite counter-examples with the help of Mace4 [15].

### 4.2 Necessity of Infinite Algebras

The necessity of considering infinite algebras follows from the following statements [13].

**Lemma 1.** *If a finite algebra satisfies the identity $x \equiv a * (-x)$ then "$-$" is bijective.*

**Lemma 2.** *If a finite algebra satisfies the identity $x \equiv a * (-x)$ then it satisfies the identity $x \equiv -(a * x)$.*

The statements show that for finite algebras the implication

$$\{x \equiv a * (-x)\} \quad \rightarrow \quad x \equiv -(a * x) \tag{2}$$

is valid.

*Example 3.* A counter-example to Implication (2) is the infinite algebra $\mathcal{A}_\infty = (\mathbb{N}_0, *_\infty, -_\infty, a_\infty)$, defined by ($-_{\mathbb{N}_0}$ stands for binary minus in natural numbers)

$$m *_\infty n = \begin{cases} n, & \text{if } m = 0 \text{ and } n \leq 2; \\ n -_{\mathbb{N}_0} 1, & \text{if } m = 0 \text{ and } n \geq 3; \\ 0, & \text{if } m \geq 1. \end{cases}$$

$$-_\infty n = \begin{cases} n, & \text{if } n \leq 2; \\ n + 1, & \text{if } n \geq 3. \end{cases}$$

$$a_\infty = 0.$$

Although $\mathcal{A}_\infty$ satisfies $x \equiv a * (-x)$ the operation "$-_\infty$" is surjective, but not bijective. This is only possible if universe is infinite.

However, modern automatic model finders like Mace4 [15], E-Darwin [1] and Paradox [6] are only designed for finding finite models. Therefore, we faced the need to develop an algorithm for finding infinite algebras and checking the satisfaction of identities in them.

### 4.3   Generating Infinite Algebras

The task of finding infinite algebras can be stated as follows: given a tuple (a set of identities $P_{ids} \subseteq M_{id}$; an identity $C_{id} \in M_{id}$) find an infinite algebra $\mathcal{A}$ such that all the identities from $P_{ids}$ are satisfied and the identity $C_{id}$ is not satisfied. In other words, it is necessary to find a counter-example to the implication $P_{ids} \to C_{id}$.

For solving this task a pattern of infinite algebra was fixed. The pattern was found heuristically based on personal experience, examples found in [13], and multiple unsuccessful runs of the program. The pattern looks as follows $\mathcal{A} = (\mathbb{N}_0, *, -, 0)$. $\mathbb{N}_{<k}$ stands for the set of all natural numbers less than $k$, $+$ and $\times$ are addition and multiplication defined on natural numbers.

$$
m * n = \begin{cases}
b_{00}, & \text{if } m = 0 \text{ and } n = 0; \\
b_{01}, & \text{if } m = 0 \text{ and } n = 1; \\
b_{02}, & \text{if } m = 0 \text{ and } n = 2; \\
b_{03}, & \text{if } m = 0 \text{ and } n = 3; \\
b_{10}, & \text{if } m = 1 \text{ and } n = 0; \\
b_{11}, & \text{if } m = 1 \text{ and } n = 1; \\
b_{12}, & \text{if } m = 1 \text{ and } n = 2; \\
b_{13}, & \text{if } m = 1 \text{ and } n = 3; \\
b_{20}, & \text{if } m = 2 \text{ and } n = 0; \\
b_{21}, & \text{if } m = 2 \text{ and } n = 1; \\
b_{22}, & \text{if } m = 2 \text{ and } n = 2; \\
b_{30}, & \text{if } m = 3 \text{ and } n = 0; \\
b_{31}, & \text{if } m = 3 \text{ and } n = 1; \\
c_0 \times m + d_0 \times n + e_0, & \text{if } m = 0 \text{ and } n > 3; \\
c_1 \times m + d_1 \times n + e_1, & \text{if } m = 1 \text{ and } n > 3; \\
c_2 \times m + d_2 \times n + e_2, & \text{if } m > 3 \text{ and } n = 0; \\
c_3 \times m + d_3 \times n + e_3, & \text{if } m > 3 \text{ and } n = 1; \\
c_4 \times m + d_4 \times n + e_4, & \text{if } m > 2 \text{ and } m = n; \\
c_5 \times m + d_5 \times n + e_5, & \text{if } m > 1 \text{ and } n = m + 1; \\
c_6 \times m + d_6 \times n + e_6, & \text{if } m > 2 \text{ and } n = m -_{\mathbb{N}_0} 1; \\
c_7 \times m + d_7 \times n + e_7, & \text{if } m > 1 \text{ and } n = m + 2; \\
c_8 \times m + d_8 \times n + e_8, & \text{if } m > 3 \text{ and } n = m -_{\mathbb{N}_0} 2; \\
c_9 \times m + d_9 \times n + e_9, & \text{if } m > 1, n > 1, \text{ and } n \neq m, m \pm_{\mathbb{N}_0} 1, m \pm_{\mathbb{N}_0} 2,
\end{cases}
$$

where $b_{00}, b_{01}, b_{10}, b_{11} \in \mathbb{N}_{<4}$,

$b_{02}, b_{03}, b_{12}, b_{13}, b_{20}, b_{21}, b_{30}, b_{31}, b_{22} \in \mathbb{N}_{<6}$,

$c_{0-9}, d_{0-9} \in \{0, 1\}$, $e_{0-9} \in \{0, 1, -1, 2, -2, 3, -3\}$.

$$-n = \begin{cases} u_0, & \text{if } n = 0; \\ u_1, & \text{if } n = 1; \\ u_2, & \text{if } n = 2; \\ p \times n + q, & \text{if } n \geq 3, \end{cases}$$

where $u_0, u_1, u_2 \in \mathbb{N}_{<5}$, $p \in \mathbb{N}_{<3}$, $q \in \{0, 1, -1, 2, -2, 3, -3\}$.

**Proposition 1.** *To check the satisfaction of an identity of a finite size in an infinite algebra constructed by the pattern it suffices to make only a finite number of substitutions.*

*Proof.* The binary and the unary operations are defined as discrete piecewise linear functions with a finite number of linear regions. Composition of linear functions is a linear function. To check if linear functions of $k$ variables are identically equal it suffices to check in $k + 1$ points. The regions are restricted linearly, therefore their number is at most the product of number of regions of composed functions. Hence, the number of regions is finite. As the size of the identity is finite, the number of variables and the number of composed functions is finite, therefore, it is only necessary to make a finite number of checks in a finite number of regions.                                                                                   □

Therefore, it is only necessary to check an identity over a limited domain. The exact limit depends on the size of the identity and the pattern of operations.

For finding the needed algebras by the pattern an algorithm with backtracking was introduced and implemented, Algorithm 2. The algorithm is presented in Python-like syntax. The keyword **break** exits the innermost loop, the **for**-loop in this example. The clause **else** after the **for**-loop is executed only if the loop ended normally, i.e. without hitting **break**. The keyword **continue** enables the next step of the innermost loop, the **while**-loop in this example.

The function update($\mathcal{A}$, term) determines the cause of a term function being not defined for the given substitution and defines it. Here the possible values of the corresponding operation are ordered and always the first one is taken. For example, let operations be totally undefined, then the result of update($\mathcal{A}$, (5*(-1))) will be that $u_1$ gets the value 0. Notice, that the result of the whole term is still not defined. When the function is called with these arguments the second time the constants $c_2, d_2, e_2$ receive the values 0. Therefore, when searching for counter-examples operations may be left partially defined. However, for checking the satisfaction of identities that are not involved in the input implication the found algebra has to be totally defined. The function **complete** completes the definition.

Function backtrack($\mathcal{A}$) updates the latest changed value to the next possible choice. If possible values for the given operation and arguments are exhausted the function tracks back to the previously assigned operation and arguments and changes that value. If the function has tracked back to the very first operation and arguments then it exists the whole process with failure message. Therefore, this is the only possible way to exit the process without finding the counter-example.

**Input:** $P_{\text{ids}} \rightarrow C_{\text{id}}$, where $P_{\text{ids}} \subseteq M_{\text{id}}$, $C_{\text{id}} \in M_{\text{id}}$.
**Output:** Algebra $\mathcal{A} = (\mathbf{N}_0, *, -, 0)$ satisfying all $P_{\text{ids}}$ and not satisfying $C_{\text{id}}$.

```
1  while True do
2  |  for id in P_ids do
3  |  |    sat, term = check_identity_partial(A, id)
4  |  |    if sat = False then
5  |  |    |    backtrack(A)
6  |  |    |    break
7  |  |    if sat = None then
8  |  |    |    update(A, term)
9  |  |    |    break
10 |  else
11 |  |    sat, term = check_identity_partial(A, C_id)
12 |  |    if sat = True then
13 |  |    |    backtrack(A)
14 |  |    |    continue
15 |  |    if sat = None then
16 |  |    |    update(A, term)
17 |  |    |    continue
18 |  |    if sat = False then
19 |  |    |    return complete(A)
```

**Algorithm 2:** find_algebra

**Proposition 2.** *Algorithm 2 is complete, i.e. if there exists a counter-example that can be constructed by the pattern it will be found by the algorithm, and sound, i.e. the found counter-example satisfies all the identities from the premise and does not satisfy the identity in conclusion.*

*Proof.* Let there be a counter-example, but the algorithm has not found it. There are two possibilities. First possibility is that the algorithm has missed the needed values of the constants in the definition of operations. However, as function **backtrack** first checks through all possible values, this is not possible. Second possibility is that the algorithm has checked the needed values, but did not output the counter-example. This is only possible after a call to function **backtrack** as it is the only exit point without yielding result. However, this function is only called if either not all identities from the premise are satisfied or the conclusion is satisfied, but if so the algebra cannot be a counter-example. Contradiction reached.

Soundness follows from Proposition 1 if the limits for checking are chosen appropriately. □

*Example 4.* Example of running the algorithm on the input $\{x \equiv a * (-x)\} \rightarrow x \equiv -(a * x)$. The process is presented in Table 1 with current values for operations and results of checking identities. The limit of checking is set to 9. Many steps are omitted for the sake of compactness.

**Table 1.** Finding infinite counter-examples for $\{x \equiv a * (-x)\} \rightarrow x \equiv -(a * x)$

| premise satisfied | conclusion satisfied | $*$ | $-$ |
|---|---|---|---|
| None | None | not defined | not defined |
| None | None | not defined | $-0 = 0;$ |
| None | None | $0 * 0 = 0;$ | $-0 = 0;$ |
| None | None | $0 * 0 = 0;$ <br> $0 * 1 = 1;$ <br> $0 * 2 = 2;$ | $-0 = 0;$ <br> $-1 = 1;$ <br> $-2 = 2;$ |
| None | None | $0 * 0 = 0;$ <br> $0 * 1 = 1;$ <br> $0 * 2 = 2;$ | $-0 = 0;$ <br> $-1 = 1;$ <br> $-2 = 2;$ <br> $-n = 3,$ if $n > 2;$ |
| False | None | $0 * 0 = 0;$ <br> $0 * 1 = 1;$ <br> $0 * 2 = 2;$ <br> $0 * 3 = 5;$ | $-0 = 0;$ <br> $-1 = 1;$ <br> $-2 = 2;$ <br> $-n = 3,$ if $n > 2;$ |
| None | None | $0 * 0 = 0;$ <br> $0 * 1 = 1;$ <br> $0 * 2 = 2;$ | $-0 = 0;$ <br> $-1 = 1;$ <br> $-2 = 2;$ <br> $-n = n,$ if $n > 2;$ |
| None | None | $0 * 0 = 0;$ <br> $0 * 1 = 1;$ <br> $0 * 2 = 2;$ <br> $0 * 3 = 3;$ | $-0 = 0;$ <br> $-1 = 1;$ <br> $-2 = 2;$ <br> $-n = n,$ if $n > 2;$ |
| True | True | $0 * 0 = 0;$ <br> $0 * 1 = 1;$ <br> $0 * 2 = 2;$ <br> $0 * 3 = 3;$ <br> $m * n = n,$ if $m = 0,$ $n > 2;$ | $-0 = 0;$ <br> $-1 = 1;$ <br> $-2 = 2;$ <br> $-n = n,$ if $n > 2;$ |
| False | False | $0 * 0 = 0;$ <br> $0 * 1 = 1;$ <br> $0 * 2 = 2;$ <br> $0 * 3 = 3;$ <br> $m * n = n + 3,$ if $m = 0,$ $n > 2;$ | $-0 = 0;$ <br> $-1 = 1;$ <br> $-2 = 2;$ <br> $-n = n,$ if $n > 2;$ |
| False | None | $0 * 0 = 0;$ <br> $0 * 1 = 1;$ <br> $0 * 2 = 2;$ <br> $0 * 3 = 5;$ | $-0 = 0;$ <br> $-1 = 1;$ <br> $-2 = 2;$ <br> $-n = n,$ if $n > 2;$ |
| None | None | $0 * 0 = 0;$ <br> $0 * 1 = 1;$ <br> $0 * 2 = 2;$ | $-0 = 0;$ <br> $-1 = 1;$ <br> $-2 = 2;$ <br> $-n = n + 1,$ if $n > 2;$ |
| True | False | $0 * 0 = 0;$ <br> $0 * 1 = 1;$ <br> $0 * 2 = 2;$ <br> $m * n = n - 1,$ if $m = 0,$ $n > 2;$ | $-0 = 0;$ <br> $-1 = 1;$ <br> $-2 = 2;$ <br> $-n = n + 1,$ if $n > 2;$ |

# 5    Results and Conclusion

Attribute Exploration of the identities was run on computer with Intel Core i5 1.6GHz×4 processor and 6 Gb of RAM running Linux Ubuntu 12.10 x64. FCA package for Python was used for implementation [19], which uses optimized Next Closure algorithm for computing the canonical basis [16]. As mentioned above the initial context contained all nonisomorphic algebras over a universe of size 2, finite counterexamples were found using Mace4, infinite counter-examples were found using implementation of Algorithm 2, proofs were found using Prover9. Below the number of steps and the elapsed time are presented, however, the values may only represent a tentative overall idea as during the investigation the time limits for finding counter-examples were introduced, therefore, not all possible counter-examples were found on every step. When the procedure got stuck the time limits were increased to allow finding new counter-examples. Attribute Exploration took 44 steps (understanding a step as the process of finding the implication basis and trying to find counterexample for each implication from the basis) and approximately 78 hours. After reducing 626 finite algebras and 1529 infinite algebras were left in the context. All 4398 unit implications from the canonical basis were proved. Results of investigation [13] were repeated using only software tools in less than 3 days of processor time.

Preliminary tests show that the used pattern may be used to find counter-examples to implications involving identities of bigger size, however, it is now unclear what the limits for using this pattern are, if the pattern can be extended. The connections between the size and the form of identities and the structure of counter-examples are still to be discovered.

From the point of view of the author it would make sense to attempt to use the elaborated approach for other analytical fields of science where counter-examples could be constructed from the implications.

**Acknowledgements.** The author was supported by German Academic Exchange Service (DAAD).

The author thanks Bernhard Ganter and Sergei O. Kuznetsov for discussions and useful remarks.

# References

1. Baumgartner, P., Fuchs, A., de Nivelle, H., Tinelli, C.: Computing finite models by reduction to function-free clause logic. Journal of Applied Logic (2007)
2. Birkhoff, G.: On the structure of abstract algebras. In: Mathematical Proceedings of the Cambridge Philosophical Society, vol. 31, pp. 433–454. Cambridge Univ. (1935)
3. Bürckert, H.-J., Herold, A., Schmidt-Schauss, M.: On equational theories, unification, and (un) decidability. Journal of Symbolic Computation 8(1), 3–49 (1989)
4. Burris, S., Sankappanavar, H.P.: A course in universal algebra, vol. 78. Springer, New York (1981)

5. Chin, L.H., Tarski, A.: Distributive and modular laws in the arithmetic of relation algebras, vol. 1. University of California Press (1951)
6. Claessen, K., Sörensson, N.: Paradox 1.0, `http://www.cs.miami.edu/~tptp/CASC/19/SystemDescriptions.html#Paradox---1.0`
7. Dau, F.: Implications of properties concerning complementation in finite lattices. In: Dorninger, D., et al. (eds.) Contributions to General Algebra 12, Proceedings of the 58th Workshop on General Algebra "58, Arbeitstagung Allgemeine Algebra", Vienna, Austria, June 3-6, 1999, pp. 145–154. Verlag Johannes Heyn, Klagenfurt (2000)
8. Dehn, M.: Über unendliche diskontinuierliche gruppen. Mathematische Annalen 71(1), 116–144 (1911)
9. Ganter, B.: Two basic algorithms in concept analysis. Preprint-Nr. 831 (1984)
10. Ganter, B., Wille, R.: Formal Concept Analysis: Mathematical Foundations. Springer (1999)
11. Glodeanu, C.V.: Attribute exploration in a fuzzy setting. In: ICFCA 2012 International Conference on Formal Concept Analysis, p. 114 (2012)
12. Guigues, J.-L., Duquenne, V.: Familles minimales d'implications informatives résultant d'un tableau de données binaires. Math. Sci. Hum. 24(95), 5–18 (1986)
13. Kestler, P.: Strukturelle Untersuchungen eines Varietätenverbandes von Gruppoiden mit unärer Operation und ausgezeichnetem Element. PhD thesis, TU Bergakademie, Freiberg (2013)
14. Kwuida, L., Pech, C., Reppe, H.: Generalizations of boolean algebras. An attribute exploration. Mathematica Slovaca 56(2), 145–165 (2006)
15. McCune, W.: Prover9 and mace4 (2005-2010), `http://www.cs.unm.edu/~mccune/prover9/`
16. Obiedkov, S., Duquenne, V.: Attribute-incremental construction of the canonical implication basis. Annals of Mathematics and Artificial Intelligence 49(1-4), 77–99 (2007)
17. Perkins, P.: Unsolvable problems for equational theories. Notre Dame Journal of Formal Logic 8(3), 175–185 (1967)
18. Revenko, A., Kuznetsov, S.O.: Attribute exploration of properties of functions on sets. Fundamenta Informaticae 115(4), 377–394 (2012)
19. Romashkin, N.: Python package for formal concept analysis, `https://github.com/jupp/fca`
20. Tarski, A.: On the calculus of relations. The Journal of Symbolic Logic 6(3), 73–89 (1941)
21. Taylor, W.: Equational logic. University of Houston, Department of Mathematics (1979)

# Closed Patterns and Abstraction Beyond Lattices

Henry Soldano

Université Paris 13, Sorbonne Paris Cité, L.I.P.N UMR-CNRS 7030
F-93430, Villetaneuse, France

**Abstract.** Recently pattern mining has investigated closure operators in families of subsets of an attribute set that are not lattices. In particular, various authors have investigated closure operators starting from a context, in the Formal Concept Analysis (FCA) sense, in which objects are described as usual according to their relation to attributes, and in which a closed element is a maximal element of the equivalence class of elements sharing the same *support*, i.e. occurring in the same objects. The purpose of this paper is twofold. First we thoroughly investigate this framework and relate it to FCA, defining in particular a structure called a *pre-confluence*, weaker than a lattice, in which we can define a closure operator with respect to a set of objects. Second, we show that the requirements allowing us to define abstract concept lattices also allow us to define corresponding abstract Galois pre-confluences.

## 1 Introduction

Until recently searching for *closed* motifs or patterns when exploring data was restricted to lattices as pattern languages. A pattern in some language $\mathcal{L}$ is said closed whenever it can be obtained by applying a *closure operator* to some pattern. This subject has been thoroughly explored both from a mathematical and algorithmical point of view as well in formal concept analysis, Galois analysis and, more recently, in data mining. Most of this work considers *support-closed* patterns. In this case, we also have a set of objects $O$ and a motif may *occur* or not in each object. The *support* of a motif is then the subset of objects in which the motif occurs. The language is a lattice with respect to a *general-to-specific* ordering and each object is described by a motif. A motif then occurs in an object whenever the motif is more general than the object description. Motifs that cannot be specialized without losing some object in their support are said *support-closed*. Clearly, there is some redundancy in enumerating all motifs when we are concerned with properties relative to their support, such as *frequency*, and it is interesting to only consider support-closed motifs. In a lattice, support-closed motifs may be efficiently searched for because there exists a closure operator on the lattice that returns as the closure $f(t)$ of some pattern $t$ the unique support-closed pattern sharing the same support as $t$.

The most investigated pattern language is the power set $2^X$ of some attribute set $X$, ordered following the set-theoretic inclusion order. Formal Concept Analysis [1] as well as Galois analysis [2,3] relies on the relation between objects and attributes. In data mining, these ideas have been investigated under the name of *itemsets mining* and rely on the same relation[4].

C.V. Glodeanu, M. Kaytoue, and C. Sacarea (Eds.): ICFCA 2014, LNAI 8478, pp. 203–218, 2014.

Recently, pattern mining has gone beyond this general framework in two directions. First, various mining problems have been investigated that come down to searching for closed motifs which cannot be considered, strictly speaking, as support-closed motifs, such as convex hulls of subsets of a given set of points, or sequential motifs with wild-cards. Solving the problem then means defining and building the corresponding closure operator [5]. To characterize such closure operators, the authors make use of a well-known theorem stating that in a finite lattice $T$ there is a one-to-one correspondence between the families closed under the *meet* operator and the closure operators on $T$. Second, various mining problems have been addressed in which the pattern language is not a lattice, in particular problems where closed motifs are support-closed motifs with respect to some dataset of objects. A framework has been proposed for that purpose in which the language is a family $F$ included in a host lattice $2^X$. The pair $(F, 2^X)$ is denoted as a *set system*. For instance, consider the set of the subgraphs generated by a subset of the set $X$ of the edges of a given graph $G = (V, X)$. Such a subgraph can be represented as a subset of $X$ and therefore searching for support-closed patterns can be performed as a standard $2^X$ lattice mining problem. However if we want to consider as a language the family $F$ of connected subgraphs of $G$, then $F$ is not a lattice[1]. Still there is a closure operator that relates a connected subgraph to a support-closed connected subgraph. This means that we can use the same kind of algorithm that specializes a closed pattern, computes the support of the new pattern and closes it in the same way we do in the lattice mining case. In their paper [6] M. Boley and coauthors state in particular the necessary and sufficient conditions that the family $F$ of a set system $(F, X)$ has to fulfill in order to guarantee that whatever the dataset $O$ of objects is[2], there exists a closure operator to compute support-closed patterns. These conditions on set systems defines the property of *confluence* that requires a kind of *local union closure* on $F$: given three elements $t, t_1, t_2$ non empty elements of $F$, if $t_1$ and $t_2$ are greater than or equal to $t$ then necessarily $t_1 \cup t_2$ belongs to $F$. This condition is clearly satisfied by the connected subgraphs family in the representation mentioned above: consider two connected subgraphs represented as their edge sets $t_1$ and $t_2$ and each including an edge set $t$, then as the subgraph generated by $t$ is connected and non empty, $t_1 \cup t_2$ also generates a connected subgraph and therefore belongs to $F$.

Our contribution concerns the two directions mentioned above. First, we state sufficient conditions to obtain closed patterns for structures weaker than lattices. This extends the theorem on finite lattices mentioned above to finite partial orders in which there exists a *local meet operator* and that we call *pre-confluences*. We obtain then that the set $f[F]$ of closed elements of a pre-confluence $F$ also is a pre-confluence. The main condition requires that given three elements $t, t_1, t_2$ of $F$, if $t_1$ and $t_2$ belongs to the upset $\uparrow t$ then there exists a greatest lower bound of $t_1$ and $t_2$ in the upset $\uparrow t$ of $F$. This *local* meet element is denoted by $t_1 \wedge_t t_2$. In the case of set systems, the family $F = \{a, b, abc, abd, abcd\}$ is a pre-confluent family. Here we have that $abc \wedge_a abd = a$ and $abc \wedge_b abd = b$ i.e. there are two maximal lower bounds of $abc$ and $abd$ in $F$ because $ab$ does not belong to $F$. The left part of Figure 1, represents a pre-confluent family $F$ where $a, b, c, d$ are the edges of a graph.

---

[1] The intersection of two such connected subgraphs is not necessarily connected.

[2] With some mild restriction we discuss further.

Second, we show that when adding to the pre-confluence property some condition on the elements of a set of objects $O$, there exists a closure operator returning the support-closed elements of $F$ with respect to $O$. When requiring the existence of such a closure operator for any database $O$ whose objects are represented as elements of a lattice $T \supseteq F$, we need a stronger property denoted by *confluence\** which is stronger than confluence on set systems and which scope is extended to any host lattice $T$. The property requires that the local join condition has to be satisfied even when $t$ is the bottom element of $T$ and belongs to $F$. The pre-confluence on the left part of Figure 1 is a confluence\* hosted by the whole set of subgraphs generated by subsets of $\{a, b, c, d\}$.

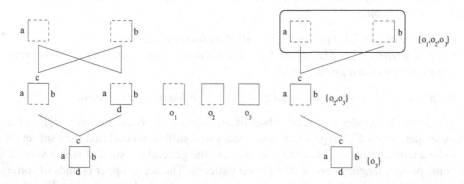

**Fig. 1.** The diagram on the left represents a family $F$ of connected subgraphs each generated by a subset (represented by a word) of the edges $\{a, b, c, d\}$ of the original graph. The subgraphs generated by $a$ and by $b$ are the minimal elements. $F$ is a pre-confluent family in which, for instance, $\{abc, abd\}$ have two maximal lower bounds, $a = abc \wedge_a abd$ greater than or equal to $a$, and the other, $b = abc \wedge_b abd$ greater than or equal to $b$. The diagram on the right represents the support closed pre-confluence $f[F]$ (see Section 3.2) with respect to the set of subgraphs $O = \{o_1, o_2, o_3\}$ represented on the middle part of the figure. The closed patterns $abc$ and $abcd$ represent the greatest connected subgraphs whose extensions are respectively $\{o_2, o_3\}$ and $\{o_3\}$. The thick box around closed patterns $a$ and $b$ indicates that both patterns have the same extension $\{o_1, o_2, o_3\}$. Elements $a$ and $b$ are the closed patterns of the bottom elements of the projected concept lattices built respectively from $(F^a, O^a)$ and $(F^b, O^b)$ and represented respectively as the up sets $f[F]^a$ and $f[F]^b$.

We can also observe in Figure 1 that for any element $x$ of the pre-confluence $F$, the upset $\uparrow x$ is a lattice. This is a general and straightforward result, that allows to link closure operators on pre-confluences to closure operators on lattices, and therefore allows to relate FCA to the analysis of support-closed patterns in pre-confluences.

Finally, a last contribution consists in noticing that when $F$ is a pre-confluence, by applying an interior operator to the extensional space $2^O$, therefore obtaining *abstract supports*, we can build an abstract support closure operator. This means that we extend abstract Galois lattices, as alpha Galois lattices [7], to abstract Galois pre-confluences.

## 2  Closure Subsets of a Partial Order

We are interested here in closed elements of an ordered set. When this ordered set refers to a language for pattern mining, we call patterns the elements of the ordered set.

### 2.1  Preliminaries

We first recall definitions of closure and dual closure operators:

**Definition 1.** *Let $E$ be an ordered set and $f : E \to E$ be an automorphism such that for any $x, y \in E$, $f$ is monotone, i.e. $x \leq y \implies f(x) \leq f(y)$ and idempotent, i.e. $f(f(x)) = f(x)$, then:*

- *if $f$ is extensive, i.e. $f(x) \geq x$, $f$ is called a closure operator*
- *if $f$ is intensive, i.e. $f(x) \leq x$, $f$ is called a dual closure operator or an interior operator, or also a projection.*

*In the first case, an element such that $x = f(x)$ is called a closed element.*

We define hereunder a closure subset of an ordered set $E$ as the range $f[E]$ of a closure operator on $E$. We give then a necessary and sufficient condition for a subset of $E$ to be a closure subset. This condition answers the general question of which subsets of some pattern language are sets of closed patterns. The set of upper bounds of some element $x$ in $E$ is denoted as the up set $\uparrow x = \{y \mid y \geq x\}$ also denoted as $E^x$ when more than one partial order is concerned. In the same way, the set of lower bounds of $x$ is denoted as the down set $\downarrow x = \{y \mid y \leq x\}$ also denoted as $E_x$.

**Definition 2 (T.S. Blyth [8]).** *A subset $C$ of an ordered set $E$ is called a closure subset if there is a closure $f : E \to E$ such that $C = f[E]$.*

**Proposition 1 (T.S. Blyth [8]).** *A subset $C$ of an ordered set $E$ is a closure subset of $E$ if and only if for every $x \in E$ the set $\uparrow x \cap C$ has a bottom element $x_*$. The closure $f : E \to E$ is then unique and defined as $f(x) = x_*$.*

However this property does not give a direct information in which pattern languages closed patterns are to be found and in which conditions closure operators exist. A direct information is provided by a well known result on closure subsets of complete $\wedge$-semilattices [1]. This result states that in such a pattern language, the closure subsets are the subsets closed by the meet operator $\wedge$. When the language is the power set of some set $X$, the meet operator simply is the intersection operator $\cap$.

**Proposition 2.** *Let $T$ be a lattice. A subset $C$ of $T$ is a closure subset if and only if $C$ is closed under meet. The closure $f : T \to T$ is then unique and defined as $f(x) = \wedge_{\{c \in C \cap \uparrow x\}} c$ and $C$ is a lattice.*

All ordered sets considered here are finite, and as all lattices are finite lattices they are also complete lattices: any subset of a lattice $T$ is then closed under arbitrary meet and arbitrary join. Note that when saying that $C$ is closed under meets we intend here that the meet of $\emptyset$ also belongs to $C$. Therefore $\top = \wedge_\emptyset c$ belongs to $C$.

We will also further need the dual proposition which states that a subset $A$ of $T$ is a *dual closure subset*, also denoted as an *abstraction*, whenever $A$ is closed under joins. The interior operator $p : T \to T$ is then defined as $p(x) = \vee_{\{a \in A \cap \downarrow x\}} a$, $A$ is a lattice and $\bot$ belongs to $A$. In particular when $T$ is a powerset $2^K$, $p(x) = \cup_{\{a \in A | a \subseteq x\}} a$.

We are interested now in pre-confluences which are structures weaker than lattices.

## 2.2   Closure Subsets in Pre-confluences

**Definition 3.** *Let $F$ be an ordered set such that for any $t \in F$, $\uparrow t$ is a $\wedge$-semilattice and has a top element. $F$ is called a pre-confluence, $x \wedge_t y$ is a local infimum or local meet, and $\top_t$ a local top.*

**Lemma 1.** *Let $F$ be a pre-confluence, then for any $t$ in $F$ and $x, y \in F \cap \uparrow t$*

1. *$\uparrow t$ is a lattice with as join, denoted as $x \vee_F y$, the least element of $\uparrow x \cap \uparrow y$*
2. *Let $t' \geq t$ then $\uparrow t'$ is a sublattice of $\uparrow t$.*

**Proof**

1. As $F$ is a pre-confluence, $\uparrow t$ is a finite $\wedge$-semilattice (with meet $x \wedge_t y$) and has a top element ($\top_t$). As a consequence of a well known result on lattice theory, $\uparrow t$ is lattice. The join $x \vee_t y$ is the least upper bound of $\{x,y\}$ in $\uparrow t$, i.e. the least element of $\uparrow t \cap \uparrow x \cap \uparrow y$ which is also $\uparrow x \cap \uparrow y$, as both $x$ and $y$ are greater than or equal to $t$. As it does not depend on $t$ we simply denote it as $x \vee_F y$.
2. For any $t' \geq t$ and $x, y$ in $\uparrow t'$, $x, y$ also belong to $\uparrow t$, As a consequence, $x \wedge_{t'} y$ is also a lower bound of $\{x, y\}$ in $\uparrow t$, and therefore $t' \leq x \wedge_{t'} y \leq x \wedge_t y$. But this means that $x \wedge_t y$ belongs to $\uparrow t'$ and therefore is also smaller than or equal to $x \wedge_{t'} y$. As a consequence we have that $x \wedge_{t'} y = x \wedge_t y$. As $\uparrow t'$ has same meet and join as $\uparrow t$, it is a sublattice of $\uparrow t$.

$\square$

Furthermore we only need minimal elements of $F$ to check whether $F$ is a pre confluence: whenever there is a local meet and a local top on the up set of minimal elements, there is also a local meet and a top element in the up set of any element of $F$.

**Lemma 2.** *$F$ is a pre-confluence if and only if for any $m \in \min(F)$, $\uparrow m$ is a $\wedge$-semilattice and has a top element.*

**Proof.** if $F$ is a pre-confluence, as $M \subseteq F$ obviously all $\uparrow m$ are $\wedge$-semilattices and have a Top element. Now suppose that all elements $m$ of $M$ are such that $\uparrow m$ is a $\wedge$-semilattice and has a Top element, then consider some $t \geq m$ and two elements $t_1, t_2 \in \uparrow t$, we have then that $t_1, t_2 \in \uparrow m$. We know that $t_1 \wedge_m t_2$ is the greatest lower bound of $\{t_1, t_2\}$ in $\uparrow m$ and as $t$ is a lower bound of $\{t_1, t_2\}$ and $t \in \uparrow m$, we have that $t_1 \wedge_m t_2 \in \uparrow t$. As a consequence $t_1 \wedge_m t_2$ is also the greatest lower bound of $\{t_1, t_2\} \in \uparrow t$ and so $t_1 \wedge_t t_2$ exists and this means that $\uparrow t$ is a $\wedge$-semilattice. Furthermore, $\top_m$ also belongs to $\uparrow t$ and therefore $\uparrow t$ also has a greatest element. As for any $t \in F$ there exists some $m \in M$ such that $t \geq m$, then $F$ is a pre-confluence. $\square$

**Definition 4.** *A subset $C$ of a pre-confluence $F$ is called closed under local meet whenever for any element $t$ and any $C' \subseteq C \cap \uparrow t$ we have*

$$\bigwedge_{t \, \{c \in C'\}} c \ \ belongs \ to \ C.$$

This means in particular that $\top_t = \bigwedge_{t\emptyset} c$ belongs to any subset which is closed under local meet and then, by definition, $C$ is also a a pre-confluence. The following theorem extends Proposition 2 to pre-confluences:

**Theorem 1.** *Let $F$ be a pre-confluence. A subset $C$ of $F$ is a closure subset if and only if $C$ is closed under local meet. The closure $f : F \to F$ is then defined as $f(t) = \bigwedge_{t\{c \in C \cap \uparrow t\}} c$ and $C = f[F]$ is a pre-confluence.*

**Proof**

We use Proposition 2 and the fact that $\uparrow t$ in a pre-confluence is a lattice.

- $\Rightarrow$ $C$ is a closure subset of $F$ means that there exists a closure operator $f : F \to F$ such that $f[F] = C$.
  As $F$ is a pre-confluence, for any $t \in F$, $C^t = \uparrow t \cap C$ is a lattice with meet operator $\wedge_t$. Furthermore, for any $x \in \uparrow t$, we have that $f(x) \in \uparrow t$ (extensivity of $f$). We can then define $f_t : \uparrow t \to \uparrow t$ such that for any $x \in \uparrow t$, $f_t(x) = f(x)$. It is straightforward that $f_t$ is a closure on $\uparrow t$ as $f$ is a closure on $F$.
  As a result, from Proposition 2 we have that $C^t = f_t[\uparrow t]$ is closed under the meet operator $\wedge_t$ of $\uparrow t$. But, as this is true for any $t$ in $F$, this also means that $C = \cup_{t \in F} C^t$ is by definition closed under local meet.
- $\Leftarrow$ Let $C$ be a subset of $F$ closed under local meet, and let for any $t$ in $F$, $C^t = \uparrow t \cap C$. By hypothesis, for any $x, y \in \uparrow t$, $x \wedge_t y$ belongs to $C$, and as $x \wedge_t y$ is the greatest lower bound of $x$ and $y$ in $\uparrow t$, we have that $x \wedge_t y$ belongs to $C^t$. This means that $C^t$ is a subset of the lattice $\uparrow t$ and is closed under the meet operator. As a result of Proposition 2 we have then that there exists a closure $f_t : \uparrow t \to \uparrow t$ which is such that for any $x \in \uparrow t$, $f_t(x) = \bigwedge_{t c \in \uparrow x \cap C^t} c$. Furthermore, as $x \in \uparrow t$, we have that $\uparrow x \cap C^t = \uparrow x \cap C$ and therefore $f_t(x) = \bigwedge_{t c \in \uparrow x \cap C} c$ and also as $\uparrow x$ is a sublattice of $\uparrow t$, $f_t(x) = f_x(x) = \bigwedge_{x c \in \uparrow x \cap C} c$ . Let then define $f : F \to F$ as $f(x) = f_x(x)$. It is straightforward that $f$ is a closure:
  - $f(x) = f_t(x)$ for any $t \leq x$, therefore as $f_t$ is a closure, $f_t(x) \geq x$. As there always exists such a $t$, then $f(x) \geq x$
  - if $x \geq y$ we have some $t$ such that $x, y \in \uparrow t$, therefore $f(x) = f_t(x)$ and $f(y) = f_t(y)$ and therefore $f(x) \geq f(y)$.
  - We have that $f(x) \geq x$ and there is some $t$ in $F$ such that $f(x), x$ both belong to $\uparrow t$, therefore $f(f(x)) = f_t(f_t(x)) = f_t(x) = f(x)$.

□

As a summary, we have a generalization of the meet operator which is the basis of most work on closed patterns in data mining, as well as all work on formal concept analysis. This generalization, denoted as local meet operator ensures the existence of closure operators whose ranges are subsets closed with respect to the local meet operator. Whenever we consider a pre-confluence as a subset of a finite powerset $2^X$ we call $F$ also a pre-confluent family. A typical example of such a structure is the set of subgraphs generated by the vertices (or edges) of a given graph. We consider here the family $F = \{a, b, abc, abd, abcd\}$ which diagram is represented in the leftmost part of Fgure 1. Here we have that $abc \wedge_a abd = a$ and $abc \wedge_b abd = b$ i.e. there are two maximal lower bounds of $abc$ and $abd$ in $F$ because $ab$ does not belong to $F$. Note that the up sets $F^a$ and $F^b$ are lattices, and share the same join operator, which in this case is the union operator.

## 3   Support Closed Patterns with Respect to a Set of Objects

### 3.1   Support Closures in Lattices

The standard case in which closed patterns are searched for is when the language is a lattice and that closure of a pattern relies on the occurrences of the pattern in a set of objects. In data mining the set of occurrences is known as the support of the pattern whereas in Formal concept analysis the set of occurrences defines the extension of the pattern and the extent of the corresponding concept.

**Definition 5.** *Let $F$ be a partial order and $O$ a set of objects, a relation of occurrence on $F \times O$ is such that if $t_1 \geq t_2$ and $t_1$ occurs in $o$ then $t_2$ occurs in $o$.*

*The* extension *of $t$ in $O$ is defined as $\mathrm{ext}(t) = \{o \in O \mid t \text{ occurs in } o\}$.*

*The* cover *of $o$ is defined as the part of $F$ whose elements occur in the object $o$, i.e. $S(o) = \{t \in F \mid t \text{ occurs in } o\}$.*

*The* cover *of a subset $e$ of objects is defined as the part of $F$ whose elements occur in all objects of $e$, i.e. $S(e) = \bigcap_{\{o \in e\}} S(o)$.*

We will say hereafter indifferently that $t$ belongs to the cover of $o$, or that $t$ occurs in $o$. The intuition here is that the order is a specificity order and whenever a pattern occurs in some object $o$ then a more general pattern will also occur in $o$. This rewrites also as $t_1 \geq t_2 \Rightarrow \mathrm{ext}(t_1) \subseteq \mathrm{ext}(t_2)$.

When $F$ is a lattice, the interesting case is the one in which objects can be described as elements of $F$:

**Proposition 3.** *Let $T$ be a lattice and $O$ a set of objects, then if for any object $o$ the cover of $o$ has a greatest element $d(o)$, denoted as the* description *of $o$ in $T$, then, for any subset $e$ of $O$*

$$int(e) = \bigwedge_{o \in e} d(o)$$

*is the greatest element of the cover $S(e)$ of $e$, denoted as the* intension *of $e$, and $(\mathrm{int}, \mathrm{ext})$ is a Galois connection on $(2^O, T)$.*

**Proposition 4.** $\mathrm{int} \circ \mathrm{ext}$ *and* $\mathrm{ext} \circ \mathrm{int}$ *are closure operators respectively on $T$ and $2^O$ and the corresponding sets of closed elements are anti-isomorphic[3] lattices whose related pair $(t, e)$ form a lattice called a* Galois lattice.

In FCA, the lattice is a powerset $2^X$ of attributes, the description of an object $i$ is the subset of attributes in relation with $i$, the table of this relation is a formal context and the Galois lattice formed by pairs of corresponding closed elements in $2^X$ and $2^O$ ordered following $2^O$ is called a concept lattice. Note that, any Galois connection between two lattices may be rewritten as the connection between two powersets and therefore there is no strict gain in expressive power in the more general setting. However, the direct formulation as sets of closed elements of the lattice $T$ is often useful [9,10,3]. Proposition 3 follows from, for instance, theorem 2 in [3]. Projected or abstract Galois lattices have been recently defined by noticing that applying an interior (or projection) operator on $T$ [10,11] or $2^O$ (or both) [11,7] when there exists a Galois connection between them, we obtain again closure operators and lattices of closure subsets. Because of the one-to-one correspondence between projections (dual closures) and abstractions (subsets closed under joins) the corresponding projected Galois lattices are also called *abstract Galois lattices*[12].

**Proposition 5.** *Let* (int, ext) *be a Galois connection on* $(2^O, T)$.

- *Let $p$ be an interior operator on $T$, then* $(p \circ \text{int}, \text{ext})$ *defines a Galois connection on* $((2^O, p(T))$
- *Let $p$ be an interior operator on $2^O$, then* $(\text{int}, p \circ \text{ext})$ *defines a Galois connection on* $(p(2^O), T)$

*In both cases the closure subsets are anti-isomorphic and form a Galois lattice, denoted respectively as intensional and extensional abstract Galois lattices.*

In the conditions of Proposition 3 when considering two elements as equivalent whenever they share the same extension with respect to $O$, a closed element $f(t) = \text{int} \circ \text{ext}(t)$ is the greatest element of the equivalence class associated to $\text{ext}(t)$. More generally, in data mining extensions are denoted as *supports*, and an element $x$ of a pattern language is said *support closed* with respect to $O$ whenever for any element $y > x$ we have that $\text{ext}(y) \subset \text{ext}(x)$ [6]. In other words, a support-closed element $x$ is a maximal element of the equivalence class associated to its support $\text{ext}(x)$. The previous proposition says that when its conditions are satisfied support-closed elements are obtained using a *support closure* operator and that there is exactly one such support-closed element in each equivalence class.

## 3.2   Support Closures in Pre-confluences

We discuss now under which conditions support closures exist in pre-confluences. We will further denote the up set $\uparrow t$ as $F^t$. First we benefit from the fact that up sets of a pre-confluence are lattices in this straigthforward corollary of Proposition 3:

**Lemma 3.** *Let $F$ be a pre-confluence, $O$ be a set of objects, and consider $O^t = \text{ext}(t)$.*
*If, for any object $o$ and any element $t$ of $F$ that occurs in $o$, $S(o) \cap F^t$ has a greatest element $d_t(o)$, then, for any subset $e$ of $O^t$,*

---

[3] i.e.isomorphic to the dual of $f[T]$.

$$int_t(e) = \bigwedge_{t \ \{o \in e\}} d_t(o)$$

*is the greatest element of the cover of e in* $F^t$, $(int_t, ext)$ *is a Galois connection on* $(2^{O^t}, F^t)$ *and* $int_t \circ ext$ *is the support closure operator on* $F^t$ *with respect to* $O$.

We will further denote $d_t(o)$ as the local description of $o$ with respect to $t$. We have then the following proposition:

**Theorem 2.** *Let F be a pre-confluence, O a set of objects, then*
*   *If for any object o and any element t of F that occurs in o,o has a local description* $d_t(o)$, *then for any subset e of O, and any t belonging to the cover* $S(e)$,

- $int_t(e)$ *is the greatest element of* $S(e) \cap F^t$
- *f defined by* $f(t) = int_t \circ ext(t)$ *is a support closure on F with respect to O.*

*The pairs* $(f(t), ext(t))$ *form a pre-confluence isomorphic to* $f[F]]$ *called a Galois pre-confluence.*

**Proof.** For any $t$ in $F$ that occurs in $e$, we have that the cover of $e$ in $F^t$ is $S(e) \cap F^t$ and as a consequence, $int_t \circ ext(t))$ is the greatest element of $S(e) \cap F^t$. Moreover a consequence of Lemma 3 is that for any $t$ in $F$, $ext \circ int_t(ext(t)) = ext(t)$. The closed element $f(t)$ is the greatest element greater than or equal to $t$ and sharing the same extension as $t$. This means that the *support-closed* elements of $F$ form the closure subset $f[F]$, $f$ is a support closure operator and by Theorem 1 $f[F]$ is a pre-confluence. □

This means that the interesting case regarding pre-confluences is the one in which each object has a local description with respect to any $t$ that occurs in $o$. This also means that the subset $F(e)$ of elements whose extension is some $e$ is partitioned in such a way that each part has a greatest element $t_m$ and contains the elements $F(e)$ smaller than $t_m$, and that $t_m = f(t)$ for all these elements.

The following lemma shows that pre-confluences generalize lattices and Theorem 2 generalizes Proposition 3:

**Lemma 4.** *Whenever a pre-confluence F has a bottom element* $\perp_F$, *then*

1. *F is a lattice*
2. *If the occurrence relation is such that* $\perp_F$ *occurs in all objects of O, then if for any object o and any* $t \in S(o)$, *o has a local description* $d_t(o)$ *with respect to t, then o has a description* $d(o)$.

**Proof.** This is straightforward regarding (1) as any $\uparrow t$ is a lattice and therefore, $\uparrow\perp_F = F$ also is a lattice. Regarding (2), first remark that as $\perp_F$ occurs in any $S(o)$, then $S(o)$ is non empty, then again as $\uparrow\perp_F = F$ we have $\uparrow\perp_F \cap S(o) = S(o)$ and therefore $S(o)$ has a greatest element. □

In the next section we connect these structures to Formal Concept Analysis.

## 4  Galois Pre-confluences as Union of Galois Lattices

We consider now the standard case of a lattice $T$ in which each object $o$ of $O$ has a description $d(i)$ in $T$, and we further consider that any element of $T$ can be such a description. We are then interested in which subsets $F$ of $T$ have support-closures with respect to any $O$. We connect here to the seminal result of M. Boley and co-authors [6] on confluent systems. To avoid confusion, up sets and down sets of $T$ starting from an element $x$ will be denoted respectively as $T^x$ and $T_x$ wherease the notations $F^t$ and $F_t$ will be used for the up sets and down sets of the subset $F$.

We will first need a lemma to characterize how an object, as an element $x$ of $T$, can be represented in $F$, then we add a condition to pre-confluences to obtain confluences* on which support closure operators are defined whatever is the object set $O$.

**Lemma 5.** *Let $F$ be a subset of a lattice $T$.*
*If for any $t \in F$ and any $x \in T^t$, there exists a greatest element $p_t(x)$ in $F^t \cap T_x$, then the mapping $p_t : T^t \to T^t$ is an interior operator on $T^t$ and $p_t(T^t) = F^t$*

**Proof**

1. $p_t(x) \leq x$ ? The hypothesis ensures that $p_t(x)$ belongs to $T_x$ and therefore $p_t$ is intensive.
2. $x \leq y \to p_t(x) \leq p_t(y)$?
   We have $T_x$ is included in $T_y$ and therefore $F^t \cap T_x$ is included in $F^t \cap T_y$ and the greatest element of $F^t \cap T_y$ is greater than or equal to the greatest element of $F^t \cap T_x$. Therefore $p_t(x) \leq p_t(y)$.
3. $d_t(d_t(x)) = d_t(x)$ ?
   First note that $p_t(x) \in F^t$ and therefore in $T^t$. This means that $p_t(p_t(x))$ is defined. Let $q = p_t(x)$, then $p_t(q) = \max F^t \cap T_q$. But as $q$ belongs to $F^t$ and by definition $q = \max T_q$, we conclude that $q$ is the greatest element of $F^t \cap T_q$ i.e. $p_t(q) = q$

By definition of $p_t$, $p_t(T^t) \subseteq F^t$. Furthermore, we have seen that for any $q \in F^t$ we have $q = p_t(q)$ and therefore $p_t(T^t) \supseteq F^t$. As a consequence $p_t(T^t) = F^t$.    □

$p_t(x)$ is *the local description* of $x$ in $F^t$.

**Proposition 6.** *Let $F$ be a subset of a lattice $T$, the following properties are equivalent:*

1. *For any $t \in F$ and any $x \in T^t$, there exists a greatest element $p_t(x)$ in $F^t \cap T_x$*
2. *For any $x, y, t$ in $F$ such that $x$ and $y$ belong to $F^t$, $x \vee y$ belongs to $F$*
3. *$F$ is a pre-confluence with join $\vee_F = \vee$*

*$F$ is then denoted as a confluence* on $T$ and we have that $p_t(x) = \vee_{q \in F^t \cap T_x} q$*

**Proof**
1 implies 2 as $T^t$ is a lattice and $p_t$ is a projection on $T^t$ and therefore $F^t = p_t(T^t)$ is closed under join (as projections are dual of closure operators and closure subsets on lattices are closed under meet). 2 implies 3 as for any $t \in F$ we have that $F^t$ is closed under union and has by definition $t$ as its least element, and therefore $F^t$ is a lattice and have a meet operator $\wedge_t$. Finally 3 implies 1 as when considering two greatest elements

$q$ and $q'$ in $F^t \cap T_x$ we have that $q \vee q'$ is in $F$ and therefore in $F^t$ and is also in $T_x$, and as a result we have that $q = q'$. The definition of the projection $p_t$ on the lattice $F^t$ as a join is a consequence of the dual result of Proposition 2. □

**Theorem 3.** *Let $F$ be a confluence\* of a lattice $T$, $O$ be a set of objects described as elements of $T$ and $p_t$ denote the local description operators on $F$, then we have that:*

$f(t) = p_t \circ \text{int} \circ \text{ext}(t)$ *where* $(\text{int}, \text{ext})$ *is a Galois connection on* $(T, O)$, *is a support closure operator on $F$ with respect to $O$.*

**Proof.** As $F$ is a confluence\*, from Proposition 6 we deduce that the conditions of Proposition 2 are satisfied. Furthermore, recall that $p_t$ is a projection on $T^t$ and by Lemma 5 that $p_t(T^t) = F^t$, we have then that following Proposition 5 $(p_t \circ \text{int} \circ \text{ext})$ is the support closure operator on $F^t$ with respect to $O^t$. □

Conversely, in order to guarantee that such a support closure operator exists for any set of objects $O$ described in $T$, a subset of $T$ has to be a confluence\*:

**Proposition 7.** *Let $F$ be a subset of the lattice $T$, then the support closure operator on $F$ with respect to any set $O$ whose objects are described as elements of $T$ exists if and only if $F$ is a confluence\*.*

**Proof.** We only need here to show that whenever $F$ is not a confluence\*, there is always an object set $O$ such that the support closure operator does not exist. Recall that $F$ is not a confluence\* means that there is some $t$ in $F$ and some $x, y$ in $F^t$ such that $x \vee y$ does not belong to $F$ (and so $x \neq y$). Consider then $O = \{x \vee y\}$, we have that $S(\{x \vee y\} \cap F^t$ has both $x$ and $y$ has maximal elements as any element that occurs in $x \vee y$ has to be smaller than or equal to $x \vee y$ and cannot be greater than both $x$ and $y$ because it would then be greater than $x \vee y$ the lowest upper bound of $x$ and $y$. Now, a support closure operator $f$ should be such that $f(x) = x$ and $f(y) = y$ has they are both maximal elements of the cover of $x \vee y$. Furthermore, consider that $t$ is one of the maximal lower bounds of $x$ and $y$ (if not, we can replace $t$ by one such element, and still have that $F$ is not a confluence\*). Then the support closure of $t$, $f(t)$ should be be either $x$ or $y$ as $t$ is not a maximal element of $S(x \vee y)$. But, whatever is the choice $f$ is then not monotone and therefore $f$ cannot be a closure operator. □

In [6], the lattice $T$ is a powerset $2^X$ and a confluent system $S$ is similar to the latter definition of confluences\* except that $\bot = \emptyset$ belongs to $S$ but $x \cup y$ is only required to belong to $F$ when $x \supseteq t$ and $y \supseteq t$ for any $t! = \emptyset$. Proposition 7 is a straightforward adaptation of the theorem of [6] when $T = 2^X$, confluent systems replaces confluences\*, and which prohibits to have any attribute common to all objects in $O$. A useful result is the following:

**Proposition 8.** *If $F$ is a confluence\*, then if $q \leq t$, and $x \in T^q$, then $p_t(x) = p_q(x)$*

**Proof.** By definition $p_t(x)$ (resp. $p_q(x)$) is the greatest element of $F^t \cap T_x$ (resp. $F^q \cap T_x$). As $F^t \subseteq F^q$, we have also that $F^t \cap T_x \subseteq F^q \cap T_x$. As both sets have greatest elements, the greatest element of $F^t \cap T_x$ is also the greatest element of $F^q \cap T_x$. □

This means that to compute the support closure of some $t$ we only need $p_m$ where $m \in \min(F)$. Implicitly this also means that whether $t$ is greater than two minimal elements $m$ and $m'$ then $p_m(t) = p'_m(t)$. For instance, in our example of connected subgraphs generated by edges of some graph, the minimal elements are the edges. As a consequence, connected subgraphs under some edge $e$ simply are obtained by projecting subgraphs containing $e$ on their connected component containing $e$.

To summarize, first the support closure set $f(F)$ of a confluence* $F$ on some lattice $T$, forms a pre-confluence of $T$, and second, we only need the minimal elements of $F$ and their associated interior operators to characterize the pre-confluence $f[F]$. When considering $T = 2^X$, $T^t$ is $2^{X \setminus t}$ and $p_t$ is an interior operator on $2^{X \setminus t}$.

## 4.1 Implications

Another question regards the definition and construction of an implication basis whose implications have both left part and right part in $F$. An implication $p \to q$ holds on $F$ whenever $\text{ext}(p) \subseteq \text{ext}(q)$ and a basis of such implications is typically made of implications such that both $p$ and $q$ belong to the same equivalence class i.e. $\text{ext}(p) = \text{ext}(q)$. Whenever $F$ is a lattice, the nodes of the concept lattice represents these equivalence classes and $q$ is a closed pattern i.e. the greatest element of the class, and therefore we have $p \leq q$. As an example the *min-max basis* is made of the implications $p \to q$ where $p \neq q$ and $p$ is a minimal element of the class of $q$ [13]. Whenever $F$ is a confluence*, we have seen that each such equivalence class is associated to several closed patterns $q_1 \ldots q_m$ each being the greatest element of a subclass. We have then in the basis both implications of the form $p_i \to q_i$ where $p_i \leq q_i$ and both belong to subclass $i$ together with implications of the form $p_j \to q_i$ where $j \neq i$ and therefore $p_j$ and $q_j$ are unordered. We extend the idea of the min-max basis to confluences* as follows:

**Definition 6.** *Let $F$ be a confluence*, and $F(e) = \{t \in F \mid \text{ext}(t) = e\}$, the min-max basis $B = B_i \cup B_e$ of implications in $F$ is defined as the set*
$$\{p \to q \mid \text{ext}(p) = \text{ext}(q), p \neq q, p \in \min(F(e)), q \in f[F(e)] \}$$
*The internal sub basis $B_i$ is made of the implications of the form $p_i \to q_i$ where $p_i \leq q_i$ and the external sub basis $B_e$ is made of the implications of the form $p_j \to q_i$ where $\{p_j, q_j\}$ are unordered.*

There are other implication basis such as the minimal Guigue-Duquenne basis [14] that can be as well extended to the case of confluences*.

## 4.2 Example

We consider here the example displayed in Figure 1. We have $F = \{a, b, abc, abd, abcd\}$ and $O = \{ab, abc, abcd\}$. To compute the closures in $F$ we take advantage of the fact that $F$ has two minimal elements $a$ and $b$ and that for any $t \geq a$ (resp. $t \geq b$) we can write $f(t) = p_a \circ \text{int} \circ \text{ext}(t)$ (resp. $(f(t) = p_b \circ \text{int} \circ \text{ext}(t))$. We obtain then:

- $f(a) = p_a \circ \text{int}(\{ab, abc, abcd\}) = p_a(ab) = a$
- $f(b) = p_b \circ \text{int}(\{ab, abc, abcd\}) = p_b(ab) = b$

- $f(abc) = p_a \circ \text{int}(\{abc, abcd\}) = p_a(abc) = abc$ (we could have used $p_b$ as $abc \in T^b$ with the same result $abc$)
- $f(abd) = p_a \circ \text{int}(\{abcd\}) = p_a(abcd) = abcd$ (same remark as above)
- $f(abcd) = p_a \circ \text{int}(\{abcd\}) = p_a(abcd) = abcd$ (same remark as above)

Note that the confluence* $F$ is the union of the two lattices $F^a = \{a, abc, abd, abcd\}$ and $F^b = \{b, abc, abd, abcd\}$. Therefore we have $f[F] = \{a, b, abc, abcd\}$ which is a pre-confluence whose minimal elements are $f(a) = a$ and $f(b) = b$. We have that $f[F] = f[F^A] \cup f[F^b]$ where $f[F^a]$ and $f[F^b]$ are the sets of closed patterns from the concept lattices built respectively from $(F^a, O^a)$, and from $(F^b, O^b)$. We have here $f[F^a] = \{a, abc, abcd\}$ and $f[F^b] = \{b, abc, abcd\}$.

Regarding the min-max implication basis we first consider the set of extensions $\text{ext}[F] = \{e_1 = \{ab, abc, abcd\}, e_2 = \{abc, abcd\}, e_3 = \{abcd\}\}$ together with the corresponding equivalence classes $F(e_1), F(e_2), F(e_3)$. Each each equivalence class is divided into subclasses each containing one closed element:

- $F(e_1) = \{a\} + \{b\}$
- $F(e_2) = \{abc\}$
- $F(e_3) = \{abd, abcd\}$

Figure 1 displays on the left the confluence* $F$, on the middle we have the object set $O$, and on the right is represented the pre-confluence $f[F]$ of support closed patterns of $F$. The min-max implication basis is made of the internal basis $B_i = \{abc \to abcd\}$ (this implication holds both in $(F^a, O^a)$ and in $(F^b, O^b)$) plus the external basis $B_e = \{a \to b, b \to a\}$.

## 5   Abstract Closed Patterns in Confluences*

In this section we consider abstract closed patterns as those obtained in *extensionally abstract Galois lattices*, denoted here as abstract Galois lattices for short, by constraining the space $2^O$. The general idea, as proposed in [12] and resulting in Proposition 5 in section 3.1 is that an abstract Galois lattice is obtained by selecting as an extensional space a subset $A$ of $2^O$ closed under union i.e. an abstraction (or dual closure subset) and therefore such that $A = p_A(2^O)$ where $p_A$ where $p_A$ is an interior operator on $2^O$. The intuitive meaning is that the abstract extension $\text{ext}_A(t)$ of some pattern $t$ will then be the union of the elements of $A$ contained in its (standard) extension, i.e. $\text{ext}_A = p_A \circ \text{ext}$ and the corresponding *abstract support closure operator with respect to $A$* is therefore $f_A = \text{int} \circ p_A \circ \text{ext}$. Intuitively, as noticed in [7], this is because the corresponding abstract Galois lattice is isomorphic, and as same support closure subset as the Galois lattice associated to the object set $O(A)$ each object $a$ of which is an element of $A$ and described in $T$ as $\text{int}(a)^4$. It is then straighforward that we obtain that abstract Galois pre-confluences are simply the Galois pre-confluences obtained through this change on object set.

---

4 In fact we just need the $\cup$-irreducible elements of $A$ as objects.

**Theorem 4.** *Let $F$ be a confluence\* of a lattice $T$, $O$ a set whose objects are described as elements of $T$, $A = p_A(O)$ an abstraction of $A$, then:*

*Let $p_t$ denote the local description operators on $F$, we have that*
$$f_A(t) = p_t \circ \mathrm{int} \circ p_A \circ \mathrm{ext}(t), \text{ where } (\mathrm{int}, p_A \circ \mathrm{ext}) \text{ is a Galois connection on } (T, A),$$
*is a support closure operator on $F$ with respect to $A$ and $f_A[F]$ is a pre-confluence.*

We continue here the example of section 4.2 by using the abstraction $A = \{\{o_1, o_2\}, o_1, o_3\}\} = \{\{ab, abc\}, \{ab, abcd\}\}$. Recall that $p_A(e) = \cup_{\{a \in A | a \subseteq e\}} a$. We obtain then:

- $f_A(a) = p_a \circ \mathrm{int} \circ p_A(\{o_1, o_2, o_3\}) = a$ as $p_A(\{o_1, o_2, o_3\}) = \{o_1, o_2, o_3\} = \{ab, abc, abcd\}$
- $f_A(b) = p_b \circ \mathrm{int} \circ p_A(\{o_1, o_2, o_3\}) = b$ (same reason as above)
- $f_A(abc) = p_a \circ \mathrm{int} \circ p_A(\{o_2, o_3\}) = \mathsf{T}_a = abcd$ as $p_A(\{o_2, o_3\}) = \emptyset$ and therefore $p_a \circ \mathrm{int}(\emptyset) = p_a(\mathsf{T}_a) = \mathsf{T}_a$
- $f_A(abd) = p_a \circ \mathrm{int} \circ p_A(\{o_3\}) = \mathsf{T}_a = abcd$ as $p_A(\{o_3\}) = \emptyset$ (as above)
- $f_A(abcd) = p_a \circ \mathrm{int} \circ p_A(\{o_3\}) = abcd$ (same as above)

$F$ is represented on the left of Figure 2. The corresponding abstract support closure pre-confluence $f_A[F]$ is displayed on the right of the figure. What happens here, is that there are only two possible extensions as $\mathrm{ext}_A[F] = \{\emptyset, O\}$. As a result the two minimal elements of $f_A[F]$ share the same abstract extension $O$ whereas the unique maximal element $\mathsf{T}_a = \mathsf{T}_b = abcd$ have an empty abstract extension.

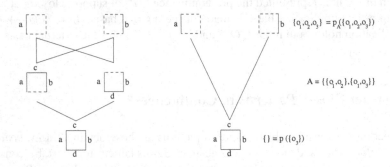

**Fig. 2.** Diagram of the abstract support closed connected subgraphs pre-confluence $f_A[F]$ (on the left part of the figure) with respect to the abstraction $A = \{\{o_1, o_2\}, \{o_1, o_3\}\}$ of $O$. The support closed element $abc$ of $f[F]$ as been projected to the maximal element of $F$, $abcd$, because its extension $\{abc, abcd\}$ is projected on $\emptyset$ as no element of $A$ is included in $\{abc, abcd\}$.

# 6   Algorithmics

An algorithm to build closure support on confluent families on $2^X$ has been proposed in [6] whenever $F$ is *strongly accessible*. This restriction[5] ensures a polynomial delay in outputting support closed elements. This algorithm has further been implemented

---

[5] For $(F, X)$ to be a strongly accessible set system, it is required that between any pair of elements $t_1, t_2$ with $t_1 \leq t_2$ in $F$ there is a path $t_1, t_1 \cup \{x_1\}, ..., t_1 \cup \{x_1, ... x_k\} = t_2$ all elements of which belong to $F$.

as a generic tool and in order to be efficient on multicores architectures particular in PARAMINER [15]. Adapting it to confluences* is straightforward by avoiding computing the support closure of $\emptyset$. Basically, the algorithm performs a depth-first search each step of which consists in adding an attribute $x$ to the current closed pattern $t$, checking whether the resulting pattern $t \cup \{x\}$ is in $F$, and closing the pattern. A SELECT function states whether a pattern belongs to $F$ and closure is only computed if it returns TRUE. The function has an ad hoc implementation according to the problem in hand. In terms of interior operators, SELECT implicitly tests whether $p_t(t \cup \{x\}) = t \cup \{x\}$ is true. A CLOSURE function computes the closure of any $t \in F$ by implicitly applying $p_t$ to $\mathrm{int}(\mathrm{ext}(t))$. Again the implementation is ad hoc, depending of the problem at hand. An open question is the construction and visualisation of the diagram of the pre-confluence of support closed elements and of the corresponding min-max implication basis.

# 7 Conclusion

Motivated by the problem of finding closed patterns in languages as the set of connected subgraphs of a graph, we have investigated an extension of FCA where the pattern language is a pre-confluence, i.e. a partial order defined through the existence of a local meet operator, and that can be expressed as a constrained union of a set of lattices. We have first extended the standard property that relates closure subsets and subsets closed under the meet operator to the case of pre-confluences. Then we have discussed the existence of support-closure operators in pre-confluences, extending a result of [6] and we have called a Galois pre-confluence the pre-confluence of support closed patterns. Related FCA works concern indirect approaches in which a support closure operator is defined on sets of connected graphs, thus resulting in a standard concept lattice whose intents contains several support-closed connected graphs[16]. We have also shown that applying interior operators to the powerset of objects we obtain, as in the lattice case, abstract support closures. The connection to FCA we have attempted to rises some technical questions, as the construction of diagrams of closure subsets, as well as more fundamental questions. For instance, when considering a support closed element as the intensional part of some concept, i.e. an *intent*, we may have two different concepts with the same *extent* which is somewhat disturbing. On the other hand, we could consider that the extension defines the concept, i.e. is an *extent* and in this case, a concept may have several *intents*. Finally, regarding applications, its seems worthwhile to consider such structures, as they are frequent when modeling data using graphs.

**Acknowledgements.** Many thanks to Bernard Monjardet for his invaluable comments and to Sylvie Borne and Sophie Toulouse for their help in the preparation of this manuscript.

# References

1. Ganter, B., Wille, R.: Formal Concept Analysis: Mathematical Foundations. Springer (1999)
2. Caspard, N., Monjardet, B.: The lattices of closure systems, closure operators, and implicational systems on a finite set: a survey. Discrete Appl. Math. 127(2), 241–269 (2003)

3. Diday, E., Emilion, R.: Maximal and stochastic galois lattices. Discrete Appl. Math. 127(2), 271–284 (2003)
4. Pasquier, N., Bastide, Y., Taouil, R., Lakhal, L.: Efficient mining of association rules using closed itemset lattices. Information Systems 24(1), 25–46 (1999)
5. Arimura, H., Uno, T.: Polynomial-delay and polynomial-space algorithms for mining closed sequences, graphs, and pictures in accessible set systems. In: SDM, pp. 1087–1098. SIAM (2009)
6. Boley, M., Horváth, T., Poigné, A., Wrobel, S.: Listing closed sets of strongly accessible set systems with applications to data mining. Theor. Comput. Sci. 411(3), 691–700 (2010)
7. Ventos, V., Soldano, H.: Alpha Galois Lattices: An Overview. In: Ganter, B., Godin, R. (eds.) ICFCA 2005. LNCS (LNAI), vol. 3403, pp. 299–314. Springer, Heidelberg (2005)
8. Blyth, T.S.: Lattices and Ordered Algebraic Structures. Universitext. Springer (2005)
9. Ferré, S., Ridoux, O.: An introduction to logical information systems. Information Processing and Management 40(3), 383–419 (2004)
10. Ganter, B., Kuznetsov, S.O.: Pattern structures and their projections. In: Delugach, H.S., Stumme, G. (eds.) ICCS 2001. LNCS (LNAI), vol. 2120, pp. 129–142. Springer, Heidelberg (2001)
11. Pernelle, N., Rousset, M.C., Soldano, H., Ventos, V.: Zoom: a nested Galois lattices-based system for conceptual clustering. J. of Experimental and Theoretical Artificial Intelligence 2/3(14), 157–187 (2002)
12. Soldano, H., Ventos, V.: Abstract Concept Lattices. In: Valtchev, P., Jäschke, R. (eds.) ICFCA 2011. LNCS (LNAI), vol. 6628, pp. 235–250. Springer, Heidelberg (2011)
13. Pasquier, N., Taouil, R., Bastide, Y., Stumme, G., Lakhal, L.: Generating a condensed representation for association rules. Journal Intelligent Information Systems (JIIS) 24(1), 29–60 (2005)
14. Guigues, J., Duquenne, V.: Famille non redondante d'implications informatives résultant d'un tableau de données binaires. Mathématiques et Sciences Humaines 95, 5–18 (1986)
15. Negrevergne, B., Termier, A., Rousset, M.C., Méhaut, J.F.: Paraminer: a generic pattern mining algorithm for multi-core architectures. Data Mining and Knowledge Discovery, 1–41 (2013)
16. Kuznetsov, S.O., Samokhin, M.V.: Learning closed sets of labeled graphs for chemical applications. In: Kramer, S., Pfahringer, B. (eds.) ILP 2005. LNCS (LNAI), vol. 3625, pp. 190–208. Springer, Heidelberg (2005)

# Mining Videos from the Web for Electronic Textbooks

Rakesh Agrawal[1], Maria Christoforaki[2,*], Sreenivas Gollapudi[1], Anitha Kannan[1],
Krishnaram Kenthapadi[1], and Adith Swaminathan[3,*]

[1] Microsoft Research
[2] Polytechnic Institute of New York University
[3] Cornell University

**Abstract.** We propose a system for mining videos from the web for supplement-
ing the content of electronic textbooks in order to enhance their utility. Textbooks
are generally organized into sections such that each section explains very few
concepts and every concept is primarily explained in one section. Building upon
these principles from the education literature and drawing upon the theory of *Formal Concept Analysis*, we define the *focus* of a section in terms of a few *indicia*,
which themselves are combinations of concept phrases uniquely present in the
section. We identify videos relevant for a section by ensuring that at least one
of the indicia for the section is present in the video and measuring the extent to
which the video contains the concept phrases occurring in different indicia for
the section. Our user study employing two corpora of textbooks on different subjects from two countries demonstrate that our system is able to find useful videos,
relevant to individual sections.

## 1 Introduction

It is inevitable that the traditional paper-based textbook will gradually evolve into electronic textbooks accessible from computing devices connected to the Internet. To enhance the utility of such electronic textbooks, we propose the problem of mining from
the web a few selective videos related to a section in a textbook and present effective
techniques for this purpose. Our techniques can be used to obtain a candidate set of
relevant videos, which can then be used by different stakeholders: by teachers when
preparing lectures on the material, by authors when creating pointers to supplementary
video material for the textbook, and by students for reinforcing their learning from an
alternative exposition.

The problem of finding suitable videos for a textbook section is quite different from
that of finding videos relevant to a stand-alone piece of text. Textbooks are written
following certain organizational principles in order to enable the reader to understand
their content without incurring undue comprehension burden [13, 22]. Two properties
of a well-written textbook of particular relevance to the present work are: (1) *focus* that
says that each section explains a few concepts, and (2) *unity* that implies that for each
concept there is a unique section that best explains the concept. In the presence of the
unity property, the focus of a section can be viewed as the  the unique contribution of

---

* Work done at Microsoft Research.

C.V. Glodeanu, M. Kaytoue, and C. Sacarea (Eds.): ICFCA 2014, LNAI 8478, pp. 219–234, 2014.

the section to the textbook. The conventional information retrieval methods (*e.g.*, TF-IDF [32], LSA [18], and LDA [9]) are not adept at representing the focus of a textbook section (see [3]).

Hence, we take a departure from traditional retrieval methods and present an approach that first infers the focus of each section, taking into account the content of all other sections, and then finds videos relevant to that focus. Our representation of the focus is derived from the theory of *Formal Concept Analysis* [20]. We represent the focus using, what we call, *indicia*. An *Indicium* for a section is a maximal combination of concept phrases that occurs frequently in that section but is not present in any other section. We also associate a score with each Indicium based on the importance of the underlying concept phrases that captures the significance of the Indicium to the section. We identify videos relevant for a section by ensuring that at least one of the indicia is present in the video and measuring the extent to which the video contains the concept phrases occurring in different indicia, after taking into account their significance. We study the efficacy of our system using textbooks on different subjects from two different countries. This extensive user study shows that our system is able to find useful videos relevant to the individual sections of a textbook.

## 1.1 Assumptions and Scope of the Paper

Before delving into details, we offer a few clarifications. We are assuming an evolutionary transformation of the current textbooks to their electronic versions. Undoubtedly, in the future, there will be textbooks written in a way to specifically exploit the functionalities provided by the electronic medium, but that will take time. Meanwhile, we are interested in taking the current books and enhancing the experience of studying from them. In the same vein, one can even question the continued need for textbooks. However, years of educational research have shown that the textbooks are the educational input most consistently associated with improvements in student learning [52]. They serve as the primary conduits for delivering content knowledge to the students and the teachers base their lesson plans mainly on the material given in textbooks [21]. Pragmatically, their importance in educational instruction is unlikely to diminish in the foreseeable future.

We should also clarify why enhancing electronic textbooks with videos has a high payoff. A number of pilot studies have established the importance of using multimedia content in educational instruction. In a recent work [36], Miller showed that the use of multimedia content is "particularly valuable in helping students acquire the initial mental imagery essential for conceptual understanding". Tantrarungroj [50] used a month-long longitudinal study to show that the students have much better content retention when they are presented with multimedia content along with textual material. The visual modality is particularly strong in many people because a child "sees and recognizes before speaking" [8]. The educational pedagogy informs us that any supplementary material is most effective when it is presented in close proximity to the main material [16]. We therefore augment videos at the section level.

We present the technology core for identifying relevant videos, but do not discuss the mechanisms for integrating them into the textbook. Issues such as implications for royalty sharing and intellectual property rights are outside the scope of the paper. It is known that learning outcomes depend not only on the availability of educational materials, but also on how they are used by the teachers and students and how effectively they have been integrated with other interventions [21, 37]. While such deployment issues are critically important, they are beyond the scope of this paper.

When proposing a video, there are multiple considerations that must be taken into account. They can be broadly grouped into aspects related to the video, the viewer, and the presenter respectively. Video considerations include the relevancy of the content of the video to the textbook section, duration of the video, and the video quality [41]. Viewer considerations include the appropriateness of the video to the viewer's prior knowledge of the subject matter and preference for the type of video such as lecture, demonstration, animation, or enactment. Presenter considerations include the presenter speaking style [33], diction, and accent. In this paper, we address the problem of relevancy: how do we augment textbook sections with relevant videos available on the web?

## 1.2 Textbook Corpora

Our study uses publicly available school-level textbooks on different subjects from two different countries. The first corpus consists of books published by the CK-12 Foundation, U.S.A. that are available online from *ck12.org*. The second corpus comprises of books published by the National Council of Educational Research and Training (NCERT), India. These books are also available online from *ncert.nic.in* and they have been used in prior studies related to textbooks [4, 5, 6]. The language of these books is English. We generate augmentations for every section in every chapter of books in our corpora.

Respecting the space constraint, we present and discuss in depth the results for two books. From the CK-12 corpus, we provide results for the middle school Biology textbook. This book introduces various themes in Biology including Molecular Biology and Genetics, Cell Biology, Prokaryotes, Animals, Plants, and Human Biology. The book consists of 26 chapters, spanning 147 sections, and we consider the augmentations for all the sections in our performance evaluation.

From the NCERT corpus, we present results for the Grade XII Physics textbook. The broad theme of this book is electricity and magnetism. It covers electric charges and fields, electrostatic potential and capacitance, current electricity, moving charges and magnetism, magnetism and matter, electromagnetic induction, alternating current, and electromagnetic waves. This book consists of 15 chapters, spanning 200 sections, and again our evaluation considers the results for all the sections.

Hereafter, we refer to these books as Biology and Physics textbooks respectively.

## 1.3 Organization

The rest of the paper proceeds as follows. We start off by discussing the related work in §2. We then describe our model for representing the focus of a textbook section in §3. We describe how we use our representation of the focus of a textbook section for finding videos relevant to it in §4. We present the results of the user study in §5. We conclude with a summary and directions for future work in §6.

## 2   Related Work

*Aboutness:* The problem of formally defining the focus of a textbook section is related to the question of "what a document is about?". The latter has been been extensively investigated in the information retrieval literature from both theoretical (*e.g.*, [10, 24, 27]) as well as pragmatic perspectives (*e.g.*, [28, 32, 39]). However, the conventional information retrieval techniques are not adept at capturing the focus of a textbook section [3]. Our proposed representation of the focus is rooted in the theory of Formal Concept Analysis [20]. It also agrees with the properties of well-written textbooks enunciated in the education literature [13, 22]. We also validate its efficacy through the application of finding relevant videos for different textbook sections.

*Content-based Video Retrieval:* Quite innovative research has been reported in content-based video retrieval where the emphasis is on retrieving videos based on pre-specified physical object categories such as cars and people and their instances [40, 45]. There is also work on recognition and retrieval for certain classes of events for these objects (*e.g.*, human actions such as handshakes and answering phones [51], sporting events [57], or traffic patterns [25]). Retrieval is initiated by providing a textual query, or a representative image, or a region of the image depicting the object of interest. The TREC Video Retrieval Evaluation [38] has played a key role in the development of methods for content-based exploitation of digital videos. These methods have been designed to recognize objects that can be represented using visual pixels, and thus are inapplicable for recognizing abstract concepts such as 'kinetic energy' that are common and central in textbooks.

*Video Search Engines:* Popular search engines such as Google and Bing include support for video search. These search engines work by indexing the associated metadata and matching keyword queries with the stored metadata. The metadata may include textual description and tags, user comments and ratings, and queries that led to the video. One might be tempted to provide the text string of a textbook section as query to a video search engine and obtain the relevant videos. However, it is well known that the current search engines do not perform well with long queries [26, 29]. Indeed, when we ourselves experimented by querying the search engines using the first few lines of a section, we got none or meaningless results. In one major stream of research on information retrieval with long queries, the focus is on selecting a subset of the query, while in another it is on weighting the terms of the query [55]. This body of research however is not designed to work for queries consisting of arbitrarily long textbook sections.

*Textbook Augmentation:* It has been empirically observed that the linking of encyclopedic information to educational material can improve both the quality of the knowledge acquired and the time needed to obtain such knowledge [17]. Motivated by this finding, techniques for mining the web for augmenting textbooks with selective links to web articles and images have been presented in [4, 6]. We extend this line of research by investigating video augmentations. We also introduce new abstractions and techniques.

*Massive Open Online Courses:* Several institutions have made available the videos of the course lectures, and there are websites (*e.g.*, EducationalVideos.com, VideoLectures.net, WatchKnowLearn.org) that aggregate links to them. Massive open online courses (MOOCs) are a relatively new phenomenon to enable teachers to reach a global student population through video-based pedagogy. Coursera, edX, Khan Academy, and Udacity are examples of platforms that have sprung up to support such courses. We view these platforms as video sources for textbook augmentation, as well as potential consumers of our research.

*Crowdsourcing:* It was proposed in [1] to create an education network to harness the collective efforts of educators, parents, and students to collaboratively enhance the quality of educational material. Some websites (*e.g.*, Notemonk.com) allow students to download textbooks and annotate them. Such annotations can include links found interesting by the students, which can then be aggregated. Some allow teachers (*e.g.*, LessonPlanet.com) to find lesson plans, worksheets, and videos to assist them with their classroom presentations. Yet others (*e.g.*, Graphite.org) help educators to use and share apps, games, videos, and websites. One could view the techniques proposed in this paper as providing an initial consideration set of videos that gets refined using crowdsourcing and other manual approaches.

## 3 Focus of a Textbook Section

Our representation of the focus of a section in a textbook is derived from the *Formal Concept Analysis* (FCA) [20]. The theory of FCA has been shown to have connections to the philosophical logic of human thought [54]. We first provide a brief overview of FCA and then formally define the focus of a section in terms of *indicia*. Later, we evaluate the efficacy of our representation through the application of finding relevant videos for different textbook sections.

### 3.1 Formal Concept Analysis: An Overview

FCA postulates that we are given a context $K$ consisting of a set of objects $G$, a set of attributes (properties) $M$, and a relation $I \subseteq G \times M$ specifying which objects have which attributes. A concept is then a pair $(A, B)$ consisting of: i) its extent $A$, comprising all objects which belong to the concept, and ii) its intent $B$, comprising all attributes which apply to all objects of the extension. A formal concept is defined to be a pair of maximal subset of objects and maximal subset of attributes such that every object has every attribute.

The formal concepts are naturally ordered by the subconcept-superconcept relation as defined by: $(A_1, B_1) \leq (A_2, B_2) \Leftrightarrow A_1 \subseteq A_2 \Leftrightarrow B_1 \supseteq B_2$. The set of all concepts together with the above partial order constitutes the *concept lattice* of the given context. For many applications, it is desirable to limit to the top-most part of a concept lattice since this region corresponds to concepts with a minimum support which are relatively stable to small perturbations (noise) in data, and also since the size of a concept lattice can be exponential in the size of the context in the worst case [30]. In [48], iceberg concept lattices, based on frequent itemsets as known from data mining [7], were introduced to address this issue. Let $\mu \in [0, 1]$ be the minimum support. A concept $(A, B)$ is said to be frequent if at least $\mu$ fraction of objects in $G$ individually have every attribute in $B$. The set of all frequent concepts of a context $K$, together with the partial order between them, is called its *iceberg concept lattice*. See [42] for a comprehensive survey of recent advances in FCA and computational techniques. See [11, 14, 43] for overviews of several applications of FCA in information retrieval.

## 3.2 Using FCA to Represent Focus

Assume we have a textbook, consisting of $n$ sections, each of which is subdivided into paragraphs. The sections and paragraphs can be those specified by the author or they can be determined using techniques such as TextTiling [23]. We will use *cphr* to denote a concept phrase present in a text. Let $\mathcal{C}_{book}$ be the set of all *cphrs* in the book.[1]

Since the formal concepts are abstract, we can only observe their manifestations in the form of underlying *cphrs* appearing in various paragraphs. Given a textbook section $s$, treat different paragraphs of $s$ as objects, different *cphrs* occurring in $s$ as attributes, and define the relationship between objects and attributes based on occurrence of a *cphr* in a paragraph. Thus, a pair of maximal set of paragraphs $P_C$ and maximal combination of *cphrs* $C$ such that every *cphr* in $C$ is present in every paragraph in $P_C$ corresponds to a formal concept of the section.

Observe that the pair representation for a formal concept has redundancy built into it. Clearly, given a formal concept $(A, B)$, the attribute set $B$ completely determines the object set $A$, and vice versa. Thus, the iceberg concept lattice of section $s$ can be thought of as corresponding to a partial order over sets of *cphrs* present in $s$. If $B_1 < B_2$ in this partial order then the set of *cphrs* corresponding to $B_1$ will be a superset of $B_2$. For compactness, therefore, we take the leaf nodes of the partial order since they correspond to the most specific sets of *cphrs* (or equivalently maximal combinations of *cphrs*) that are also frequent in the section.

Finally, since we are interested in concepts that are unique to each section, we add a uniqueness constraint to define the focus of the section. More precisely, we only include those leaf nodes that are rare in any other section [49].

**Definition 1 (Indicium of a section).** *A set of cphrs $C$ present in a section $s$ of the textbook constitutes an* Indicium *of $s$ if (1) $C$ is frequent in $s$, (2) $C$ uniquely occurs in $s$ (i.e., there is no other section of the book in which $C$ is frequent), and (3) $C$ is maximal (i.e., there is no superset of $C$ in $s$ which is also frequent in $s$).*

---

[1] The identification of *cphrs* primarily involves detection based on rules or statistical and learning methods [28, 32]. In the former, the structural properties of phrases form the basis for rule generation, while the importance of a phrase is computed based on its statistical properties in the latter. Building upon [19, 34, 47], our implementation defines the initial set of *cphrs* to be the phrases that map to Wikipedia article titles. This set is refined by removing malformed as well as common phrases based on their probability of occurrence on the Web [53]. Our methodology is oblivious to the specific *cphr* identification technique used, though the performance of the system is dependent on it. Our implementation uses author provided sections and paragraphs.

**Table 1.** Indicia for consecutive sections in the Biology textbook

(a) Respiratory system

| |
|---|
| pharynx, cellular respiration, transporting oxygen |
| cardiac muscle, connective tissue, gas transport |
| nasal cavity, connective tissue, gas transport |

| (b) Respiratory diseases | (c) Digestive system |
|---|---|
| pharynx, respiratory system, epiglottis | pharynx, lipid digestion, pepsin |
| emphysema, epiglottis, cigarette | pharynx, large intestine, salivary gland |
| bronchus, cigarette, respiratory system | gall bladder, large intestine, pepsin |

**Definition 2 (Focus of a section).** *The set of indicia of a section s constitutes the* focus *of s, denoted by* $\Psi_s$.

We remark that our derivation of the definition of focus of a section agrees with the properties of well-written textbooks investigated in the education literature [13, 22]. For an author to have introduced a formal concept in a section, the *cphrs* underlying the formal concept must occur frequently across many paragraphs in the section. As a section contributes unique content to the book and introduces very few formal concepts, their underlying *cphr* combinations must appear uniquely in the section, and if not, then infrequently in other sections. We obtain concise representations as a side effect of the maximality constraint. Our implementation sets $\mu$ to require that an Indicium must appear in at least two paragraphs in the section for it to be considered frequent.

We also remark that our notion of an Indicium is related to the idea of a hypothesis for a class present in the FCA literature. Note that Indicium $C$ is a maximal frequent (and hence closed) itemset in class (text section) $s$, which is not frequent in another class (section). As defined in [31], a hypothesis for class $s$ is a closed itemset occurring in $s$ and not occurring in other classes. A minimal hypothesis is an inclusion-minimal hypothesis. An Indicium is thus a "relaxation" of a minimal hypothesis, allowing it to occur in another class, but not frequently. Thus, the focus of a section consists of the set of relaxed minimal hypotheses for the section.

### 3.3 Illustrative Examples

**Biology Textbook:** Table 1 shows top indicia for three consecutive sections in the Biology textbook (wherein the indicia are ordered by their significance score (see §4.1)). Table 1(a) gives indicia for the section on the anatomy of human respiratory system, Table 1(b) for the next section that discusses respiratory diseases, and Table 1(c) for the subsequent section that explains human digestive system.

We see that in each of these sections, there is at least one Indicium that contains the *cphr* 'pharynx'. In human Biology, 'pharynx' refers to a part of the throat that participates in respiration and digestion. Hence, this phrase is discussed in all three sections and is present in the corresponding indicia. However, other *cphrs* occurring in these indicia provide the additional content (respiration or digestion) with which to disambiguate and represent the focus of the corresponding sections.

Table 2. Indicia for adjacent sections in the Physics textbook

| (a) Magnetism & Gauss' Laws | (b) Earth's Magnetism |
| --- | --- |
| field line, magnetic field, monopole | field line, magnetic field, earth |
| field line, magnet, charged particle | equator, meridian, southern hemisphere |
| electrostatics, field line, monopole | earth, solar wind, poles |

The Indicium ⟨pharynx, cellular respiration, transporting oxygen⟩ in the first section captures the working of the respiratory system in which oxygen enters through the mouth and nose and then travels through the pharynx to reach the lungs. In contrast, in the second section, the Indicium ⟨pharynx, respiratory system, epiglottis⟩ captures how the valve, epiglottis, near the pharynx points upwards during respiration in order to enable breathing. In the third section on the human digestive system, the Indicium ⟨pharynx, lipid digestion, pepsin⟩ differentiates the use of 'pharynx' by using digestion related concept phrases.

As another example, consider the Indicium ⟨emphysema, epiglottis, cigarette⟩ in the second section. The *cphr* 'emphysema' refers to a progressive disease of the lungs caused mainly by smoking tobacco. Smoking tobacco also causes inflammation of epiglottis and hence can cause obstruction of oxygen through the 'pharynx'. Similarly, consider the Indicium ⟨gall bladder, large intestine, pepsin⟩ in the third section. The *cphr* 'pepsin' refers to an enzyme that aids digestion of protein in the stomach and the *cphr* 'gall bladder' to the organ that stores bile and then secretes it to aid digestion.

**Physics textbook:** Table 2 shows top indicia for two adjacent sections in the Physics textbook. Table 2(a) shows indicia for the section on magnetism & Gauss' laws, while Table 2(b) shows them for the section on Earth's magnetism. The first section discusses the magnetic field and the physics behind their effects on moving particles. The second section discusses how Earth acts as a magnet. Consider the first rows of the two tables. They both contain *cphrs* 'field line' and 'magnetic field', but the *cphr* 'monopole' is unique to the Indicium for the first section. The *cphr* 'monopole' appearing in the first section distinguishes this section on general magnetism from the section on Earth's magnetism: a magnetic monopole is a hypothetical particle in particle physics that is an isolated magnet with only one magnetic pole, and hence is not discussed in the context of Earth's magnetism as Earth has both poles. The *cphr* 'earth' is rather generic, but the Indicium formed by combining it with 'field line' and 'magnetic field' is very pertinent to the section on Earth's magnetism.

## 4   Augmenting with Videos

A video might be associated with one or more of the following information: (a) images from the visual channel, (b) audio from the auditory channel, (c) video metadata consisting of title, description and any other video related properties such as duration and format, and (d) textual context (*e.g.*, webpage in which the video may have been embedded). One could attempt to match the textual content of a textbook section to the images from the visual channel of the video. However, today's video recognition systems can effectively recognize only the physical objects that are describable using visual pixels [38], whereas we need to be able to find videos relevant to textbook sections containing abstract concepts. Our system, therefore, employs transcript of the spoken words in the video to infer the relevance of the video to the textbook section. Many videos have such transcripts associated with them; otherwise, one can generate transcripts using speech recognition [46].

Our problem now reduces to the following: given a textbook section (a query), search for related documents over the corpus of video transcripts. At a high level, this problem is similar to the query by document work [56] wherein given a news article (a query), techniques were proposed for identifying related documents from a corpus of blogs. However, our approach differs in two respects. We represent the textbook section using indicia which themselves are founded on formal concept analysis and properties of well-written textbooks, whereas their approach represents the given document by extracting key phrases. Our technique for using the representation to query the corpus (see below) also differs from their approach of issuing a conjunctive query of key phrases to a specialized blog search engine.

Given a section $s$ and its set of indicia, $\Psi_s$, the videos relevant to the section are obtained using a two-step process. First, a candidate set of videos is selected by only including videos whose transcripts contain all *cphrs* from at least one Indicium in $\Psi_s$. For each video in the candidate set, we assign a relevance score by measuring the combined significance of the indicia from the section that are present in the corresponding transcript. Let $\Psi_{s,v} \subseteq \Psi_s$ be the set of indicia of section $s$ that are found in the transcript of video $v$. The relevance score for the video $v$ is given by: $relevanceScore(v) := \sum_{C \in \Psi_{s,v}} f(C)$, where $f(C)$ is the significance score of Indicium $C$. The videos are then ranked using this score, and the top $k$ are chosen for augmenting the section.

## 4.1 Significance of an Indicium

An Indicium consists of a combination of *cphrs* that collectively represent the unique content a section, but many such combinations may exist for the same section. However, some Indicium may offer a more significant representation than others. Hence, we associate a score denoting the significance of an Indicium based on the *importance* of the underlying *cphrs*.[2] We first enunciate the desirable properties of significance score.

*Property 1 (MONOTONICITY).* The significance score of an Indicium is a monotonically increasing function of the importance of its constituent *cphrs*.

This property is rooted in the intuitive notion that an Indicium made up of more important *cphrs* is more significant. In particular, inclusion of an additional *cphr* to an Indicium results in a more significant Indicium (the uniqueness requirement is still preserved).

---

[2] Adopting the "keyphraseness" notion from [34, 35], our implementation defines the importance $\phi(c)$ of a *cphr* $c$ in terms of the likelihood that the *cphr* is hyperlinked to the corresponding article in Wikipedia. The intuition is that more important *cphrs* are more likely to be hyperlinked in Wikipedia. Formally, $\phi(c) := n_{link}(c)/n_{all}(c)$, where $n_{link}(c)$ is the number of Wikipedia articles in which $c$ occurs as a hyperlink and $n_{all}(c)$ is the total number of articles in which $c$ appears. See [28, 32] for other possibilities.

*Property 2* (CONCENTRATION). The significance score of an Indicium increases as the importance of its constituent *cphrs* gets concentrated, that is, the importance is shifted from less important *cphrs* to more important *cphrs* retaining the same total importance.

This property stems from the observation that the more important *cphrs* tend to have a broader scope, for example, representing the entire chapter. By themselves, the less important *cphrs* may not represent a section and may even be ambiguous, but their combination with more important *cphrs* helps to narrow down to the focus of the section. The corresponding Indicium can be thought of as anchoring to more important *cphrs*, and then refining their scope using less important *cphrs*.

For example, all three sections shown in Table 1 discuss the *cphr* 'pharynx'. The additional *cphrs* in the respective sections help to refine the scope of this *cphr* to either respiration or digestion as discussed in §3.3.

### 4.2 Characterization of Significance Score for an Indicium

We next show that the significance score of an Indicium can be obtained using a broad category of simple functions that satisfy properties 1 and 2. Let $f(C)$ denote the significance score of Indicium $C$. Let $c_1, c_2, \ldots, c_l$ be the *cphrs* present in $C$, listed in the decreasing order of their importance, that is, $\phi(c_1) \geq \ldots \geq \phi(c_l)$.

*Claim.* Suppose $f$ is defined as the sum of a univariate function of the importance of constituent *cphrs*: $f(C) := \sum_{c \in C} g(\phi(c))$. Then, $f$ satisfies properties 1 and 2 if $g(.)$ is a monotonically increasing non-negative convex function.

*Proof.* See [3].

Our implementation instantiates function $g$ as $g(x) := e^x$. This function satisfies the requirements in Claim 4.2, and favors indicia for which the importance is concentrated in a few *cphrs*.

## 5  Performance

We now present the results of the user studies we conducted to quantify how well our approach is able to find videos relevant to the focus of each section. We first describe the video corpus, and then provide the results.

### 5.1  Video Corpus

The video corpus consists of education-related, short videos obtained from a focused web crawl [12, 44]. The crawler is seeded with educational videos from a few reputed sites. These videos span broad levels of education ranging from school to higher education to lifelong learning and originate from a variety of sources. Many of these videos had accompanying user uploaded transcripts of the video content. In order to remove variability arising out of the quality of speech recognition of the audio from the auditory channel of the videos, our experiments employed only those videos that contained author uploaded transcripts. There were nearly 50,000 such videos.

## 5.2  Experiments

We carried out two sets of experiments to assess how well our techniques are able to find relevant videos. The first experiment evaluates the proposed videos by measuring the precision of retrieval. The second experiment measures the congruence of the retrieval by computing agreement between the section and the retrieved video, in terms of overlap between concept phrases deemed important for the section and for the video by a panel of judges. We measure overlap using a number of similarity measures.

Ideally, we would have liked to have as judges those students who had studied from the textbooks in our test corpus. In the absence of the access to this subject population to us, we carried out our user study on the Amazon Mechanical Turk platform, taking care to follow the best practices [2].

## 5.3  Precision

**Setup:** Taking cue from the relevance judgment literature [15, 38], we asked the turkers to read a section, watch a video, and then judge if the video was relevant to the section. The default choice in the HIT (Human Intelligence Task) was set to 'not-relevant' so that the judges needed to explicitly choose 'relevant' if they indeed found the video to be relevant. Each judge was required to spend a minimum of 30 minutes on a HIT. We rejected any HIT where the time spent was less than the minimum. Each HIT was judged by seven judges. In this manner, we computed the relevance of the top three videos proposed by our system over all sections in four randomly chosen chapters, for both the textbooks.

**Metric:** Our first metric is the commonly used $precision@k$ [32] which measures the fraction of retrieved videos in the top K positions that are judged to be relevant. For a section $s$, let $v_{s,j}$ be the retrieved video at position $j$. Let $rel(v_{s,j})$ be a binary variable that takes a value of 1 if the majority of judges voted $v_{s,j}$ to be relevant for $s$. Then,

$$precision@k = (\sum_{s \in S} \sum_{j=1}^{k} rel(v_{s,j})/k) \, / \, |S|,$$

where $k$ is the number of videos retrieved for each section and $S$ is the set of sections.

We also measure whether the judges found at least $i$ of the videos shown in top $k$ positions for each section to be relevant, and compute the average across all sections:

$$precision@(i, k) = \sum_{s \in S} \delta[\sum_{j=1}^{k} rel(v_{s,j}) \geq i]) \, / \, |S|,$$

where $\delta[x]$ is an indicator variable that evaluates to 1 if $x$ is true, and to 0 otherwise. This metric is useful if the goal of video augmentation is to find a good candidate set of videos from which the final selection is made by an expert.

**Results:** Figure 1a shows the performance of our system under the first metric for $k = 1, 2, 3$. The results are quite encouraging. In 77% of the sections, the top video retrieved

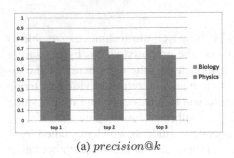
(a) $precision@k$

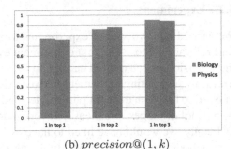
(b) $precision@(1, k)$

**Fig. 1.** Retrieval precision

by our system has been judged relevant. The performance is maintained at 73% even when both first and second videos are required to be judged relevant, and at 63% when all three videos are required to be judged relevant. We can also see that the performance is maintained across both the subjects.

Figure 1b shows the results under the second metric for $i = 1$ and $k = 1, 2, 3$. For 77% of the sections, judges agree with our top augmentation. This number goes up to 86% if we are willing to consider it a success if one of the first two videos is judged relevant. It shoots up to 95% if finding at least one out of three videos to be relevant is treated as success.

### 5.4 Congruence

This experiment measures the agreement between judges' collective understanding of the focus of a section and their collective understanding of the focus of the corresponding video. For this purpose, we designed two HITs, one for the section and the other for the video.

**Setup:** In *SectionHIT* (*VideoHIT*), the judge was asked to read the section (video) and provide top five phrases that best describe the section (video). We converted the phrases from all the judges into unigrams and removed stop words. Let $Y_s$ be the set of unigrams obtained in this manner for section $s$, and $n_s[w]$ be the number of judges that included unigram $w$ in one of the phrases for $s$. Similarly, $Z_v$ and $n_v[w]$ for video $v$.

In this experiment also, judges were required to spend a minimum of 30 minutes on a HIT. The same section (and the corresponding video) was judged by five judges. We selected the judges who took part to be different from those who participated in the experiment reported in §5.3 to remove any biases.

**Metric:** We compute congruence using several similarity measures [32]. For a video $v$ for section $s$, the congruence is computed on the sets $Z_v$ and $Y_s$ of unigrams provided by the judges for video $v$ and section $s$, respectively. We used two symmetric measures: the weighted Jaccard $\left( \frac{\sum_{w \in Z_v \cap Y_s} \min(c_v[w], c_s[w])}{\sum_{w \in Z_v \cup Y_s} \max(c_v[w], c_s[w])} \right)$ and Dice $\left( \frac{2|Z_v \cap Y_s|}{|Z_v| + |Y_s|} \right)$. We also computed asymmetric measures with respect to the section and the video: $|Z_v \cap Y_s| / |Z_v|$ and $|Z_v \cap Y_s| / |Y_s|$ respectively.

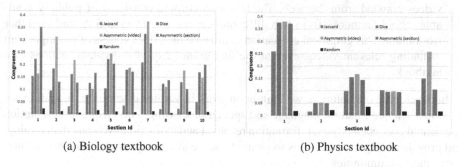

(a) Biology textbook                              (b) Physics textbook

**Fig. 2.** Congruence between section focus and retrieved video

**Results:** Figure 2 shows the results. For each section (shown in X-axis), we selected the top video identified by our approach and computed congruence (shown in Y-axis) between the section and the corresponding top video. For comparison, we also did the following computation. For each section, we randomly sampled as many unigrams as provided by the judges. Similarly, we also randomly sampled unigrams from the matching videos. We used these two sets to compute average congruence over 100 random runs for each ⟨section, video⟩ pair. We can see that the congruence obtained using the unigrams provided by the judges is significantly higher than that of the randomly sampled unigrams under all the measures.

# 6 Conclusions

Motivated by the importance of textbooks in learning, we studied the feasibility of enhancing the predominantly text-oriented textbooks with a few selective videos mined from the web at the level of individual sections. We took an approach that does not view textbook sections as stand-alone pieces of text, but rather part of a logically organized work based on well-founded educational principles in which each textbook section contributes uniquely to the pedagogical objective of the book. Our main contributions are as follows:

- Inspired by the theory of Formal Concept Analysis, we propose that the focus of textbook sections can be defined and identified in terms of a small number of indicia, each of which consists of a combination of concept phrases appearing in the section. Indicia of a textbook section are unique relative to all other sections of the book and can be computed by considering all the sections jointly.
- On the video side, we propose making use of the transcript of the spoken words in the audio from the auditory track of the video. However, videos found on the web are independently produced and without necessarily following the organizational logic of textbooks. We therefore use indicia from a section to identify candidate videos and then score them based on the concept phrases present and their importance.
- We evaluated our video augmentation algorithm through extensive user studies of its performance. The video corpus used in the study consisted of nearly 50,000

videos crawled from the web. The textbook corpora consisted of publicly available school textbooks from two different sources, one from U.S.A. and the other from India. This empirical evaluation confirmed the effectiveness of our algorithm in finding relevant videos even at the fine granularity of individual sections of a textbook.

In developing our solution, we built upon work in various disciplines, including educational sciences, natural language and speech processing, knowledge representation and formal concept analysis, information retrieval and extraction, web and data mining, and crowdsourcing. As such, this work might serve as a bridge for researchers belonging to these communities.

For future, we would like to integrate considerations beyond relevance in our video mining system. We expect incorporating viewer aspects, especially appropriateness to viewer's background and prior knowledge, to be particularly valuable and challenging. It is possible for a video to contain not only content relevant for a particular textbook section, but also additional material. In such cases, we would like to be able to pinpoint the subset of the proposed video. The reader would have noticed that the ideas and techniques we have proposed are quite general and have broader applicability. We would like to explore their effectiveness in augmenting textbooks with other types of content that have been investigated in the past [4, 6].

**Acknowledgments.** We wish to thank Sergei Kuznetsov for introducing us to FCA and and providing insightful feedback.

# References

[1] Improving India's Education System through Information Technology. IBM (2005)

[2] Amazon Mechanical Turk, Requester Best Practices Guide. Amazon Web Services (June 2011)

[3] Agrawal, R., Christoforaki, M., Gollapudi, S., Kannan, A., Kenthapadi, K., Swaminathan, A.: Mining videos from the web for electronic textbooks. Technical Report MSR-TR-2014-5, Microsoft Research (2014)

[4] Agrawal, R., Gollapudi, S., Kannan, A., Kenthapadi, K.: Enriching textbooks with images. In: CIKM (2011)

[5] Agrawal, R., Gollapudi, S., Kannan, A., Kenthapadi, K.: Identifying enrichment candidates in textbooks. In: WWW (2011)

[6] Agrawal, R., Gollapudi, S., Kenthapadi, K., Srivastava, N., Velu, R.: Enriching textbooks through data mining. In: ACM DEV (2010)

[7] Agrawal, R., Mannila, H., Srikant, R., Toivonen, H., Verkamo, A.I.: Fast discovery of association rules. In: Advances in Knowledge Discovery and Data Mining, ch. 12. AAAI/MIT Press (1996)

[8] Berger, J.: Ways of seeing. Penguin (2008)

[9] Blei, D., Ng, A.Y., Jordani, M.: Latent dirichlet allocation. Journal of Machine Learning Research 3 (2003)

[10] Bruza, P.D., Song, D.W., Wong, K.-F.: Aboutness from a commonsense perspective. Journal of the American Society for Information Science 51(12) (2000)

[11] Carpineto, C., Romano, G.: Concept data analysis: Theory and applications. John Wiley & Sons (2004)

[12] Chakrabarti, S., Van den Berg, M., Dom, B.: Focused crawling: A new approach to topic-specific web resource discovery. Computer Networks 31(11) (1999)

[13] Chambliss, M., Calfee, R.: Textbooks for Learning: Nurturing Children's Minds. Wiley-Blackwell (1998)

[14] Cigarrán, J.M., Peñas, A., Gonzalo, J., Verdejo, F.: Automatic selection of noun phrases as document descriptors in an FCA-based information retrieval system. In: Ganter, B., Godin, R. (eds.) ICFCA 2005. LNCS (LNAI), vol. 3403, pp. 49–63. Springer, Heidelberg (2005)

[15] Clarke, C.L.A., Craswell, N., Soboroff, I., Voorhees, E.M.: Overview of the TREC 2011 web track. Technical report, NIST (2011)

[16] Coiro, J., Knobel, M., Lankshear, C., Leu, D. (eds.): Handbook of research on new literacies. Lawrence Erlbaum (2008)

[17] Csomai, A., Mihalcea, R.: Linking educational materials to encyclopedic knowledge. In: AIED (2007)

[18] Deerwester, S.C., Dumais, S.T., Landauer, T.K., Furnas, G.W., Harshman, R.A.: Indexing by latent semantic analysis. JASIS 41(6) (1990)

[19] Gabrilovich, E., Markovitch, S.: Computing semantic relatedness using Wikipedia-based explicit semantic analysis. In: IJCAI (2007)

[20] Ganter, B., Wille, R.: Formal concept analysis: Mathematical foundations. Springer (1999)

[21] Gillies, J., Quijada, J.: Opportunity to learn: A high impact strategy for improving educational outcomes in developing countries. USAID Educational Quality Improvement Program, EQUIP2 (2008)

[22] Gray, W., Leary, B.: What makes a book readable. University of Chicago Press (1935)

[23] Hearst, M.A.: TextTiling: Segmenting text into multi-paragraph subtopic passages. Computational Linguistics 23(1) (1997)

[24] Hjørland, B.: Towards a theory of aboutness, subject, topicality, theme, domain, field, content ... and relevance. Journal of the American Society for Information Science and Technology 52(9) (2001)

[25] Hu, W., Xie, D., Fu, Z., Zeng, W., Maybank, S.: Semantic-based surveillance video retrieval. IEEE Transactions on Image Processing 16(4) (2007)

[26] Huston, S., Croft, W.B.: Evaluating verbose query processing techniques. In: SIGIR (2010)

[27] Hutchins, W.J.: On the problem of aboutness in document analysis. Journal of Informatics 1(1) (1977)

[28] Jurafsky, D., Martin, J.: Speech and language processing. Prentice Hall (2008)

[29] Kumaran, G., Carvalho, V.R.: Reducing long queries using query quality predictors. In: SIGIR (2009)

[30] Kuznetsov, S.O.: On computing the size of a lattice and related decision problems. Order 18(4) (2001)

[31] Kuznetsov, S.O.: Complexity of learning in concept lattices from positive and negative examples. Discrete Applied Mathematics 142(1) (2004)

[32] Manning, C., Raghavan, P., Schütze, H.: Introduction to information retrieval. Cambridge University Press (2008)

[33] Mariooryad, S., Kannan, A., Hakkani-Tur, D., Shriberg, E.: Automatic characterization of speaking styles in educational videos. In: ICASSP (2014)

[34] Medelyan, O., Milne, D., Legg, C., Witten, I.: Mining meaning from Wikipedia. International Journal of Human-Computer Studies 67(9) (2009)

[35] Mihalcea, R., Csomai, A.: Wikify!: Linking documents to encyclopedic knowledge. In: CIKM (2007)

[36] Miller, M.: Integrating online multimedia into college course and classroom: With application to the social sciences. MERLOT Journal of Online Learning and Teaching 5(2) (2009)

[37] Moulton, J.: How do teachers use textbooks and other print materials: A review of the literature. The Improving Educational Quality Project (1994)

[38] Over, P., Awad, G., Fiscus, J., Antonishek, B., Michel, M., Smeaton, A., Kraaij, W., Qunot, G.: TRECVID 2011 – Goals, tasks, data, evaluation mechanisms and metrics. Technical report, NIST (2011)

[39] Paranjpe, D.: Learning document aboutness from implicit user feedback and document structure. In: CIKM (2009)

[40] Patel, B., Meshram, B.: Content based video retrieval. The International Journal of Multimedia & Its Applications (IJMA) 4(5) (2012)

[41] Pinson, M., Wolf, S.: A new standardized method for objectively measuring video quality. IEEE Transactions on Broadcasting 50(3) (2004)

[42] Poelmans, J., Ignatov, D.I., Kuznetsov, S.O., Dedene, G.: Formal concept analysis in knowledge processing: A survey on models and techniques. Expert Systems with Applications 40(16) (2013)

[43] Priss, U.: Formal concept analysis in information science. Annual Review of Information Science and Technology 40 (2006)

[44] Shah, C.: TubeKit: A query-based YouTube crawling toolkit. In: JCDL (2008)

[45] Smoliar, S.W., Zhang, H.: Content based video indexing and retrieval. IEEE MultiMedia 1(2) (1994)

[46] Stolcke, A., Chen, B., Franco, H., Gadde, V., Graciarena, M., Hwang, M., Kirchhoff, K., Mandal, A., Morgan, N., Lei, X., et al.: Recent innovations in speech-to-text transcription at SRI-ICSI-UW. IEEE Transactions on Audio, Speech, and Language Processing 14(5) (2006)

[47] Strube, M., Ponzetto, S.: WikiRelate! Computing semantic relatedness using Wikipedia. In: AAAI (2006)

[48] Stumme, G., Taouil, R., Bastide, Y., Pasquier, N., Lakhal, L.: Computing iceberg concept lattices with TITANIC. Data and Knowledge Engineering 42(2) (2002)

[49] Szathmary, L., Napoli, A., Valtchev, P.: Towards rare itemset mining. In: ICTAI (2007)

[50] Tantrarungroj, P.: Effect of embedded streaming video strategy in an online learning environment on the learning of neuroscience. PhD thesis, Indiana State University (2008)

[51] Tian, Y., Cao, L., Liu, Z., Zhang, Z.: Hierarchical filtered motion for action recognition in crowded videos. IEEE Transactions on Systems, Man, and Cybernetics, Part C: Applications and Reviews 42(3) (2012)

[52] Verspoor, A., Wu, K.B.: Textbooks and educational development. Technical report, World Bank (1990)

[53] Wang, K., Thrasher, C., Viegas, E., Li, X., Hsu, P.: An overview of Microsoft Web N-gram corpus and applications. In: NAACL–HLT (2010)

[54] Wille, R.: Formal concept analysis as mathematical theory of concepts and concept hierarchies. In: Ganter, B., Stumme, G., Wille, R. (eds.) Formal Concept Analysis. LNCS (LNAI), vol. 3626, pp. 1–33. Springer, Heidelberg (2005)

[55] Xue, X., Huston, S., Croft, W.B.: Improving verbose queries using subset distribution. In: CIKM (2010)

[56] Yang, Y., Bansal, N., Dakka, W., Ipeirotis, P., Koudas, N., Papadias, D.: Query by document. In: WSDM (2009)

[57] Zhang, N., Duan, L.-Y., Li, L., Huang, Q., Du, J., Gao, W., Guan, L.: A generic approach for systematic analysis of sports videos. ACM Transactions on Intelligent Systems and Technology 3(3) (2012)

# Automated Enzyme Classification by Formal Concept Analysis

François Coste[1], Gaëlle Garet[1], Agnès Groisillier[2],
Jacques Nicolas[1], and Thierry Tonon[2]

[1] Irisa / Inria Rennes, Campus de Beaulieu, 35042 Rennes cedex, France
jacques.nicolas@inria.fr
http://www.irisa.fr/dyliss
[2] Sorbonne Universités, UPMC Univ Paris 06,
UMR 8227, and CNRS, UMR 8227,
Integrative Biology of Marine Models,
Station Biologique de Roscoff, CS 90074, F-29688, Roscoff cedex, France
http://www.sb-roscoff.fr/umr7139.html

**Abstract.** Enzymes are macro-molecules (linear sequences of linked molecules) with a catalytic activity that make them essential for any biochemical reaction. High throughput genomic techniques give access to the sequence of new enzymes found in living organisms. Guessing the enzyme's functional activity from its sequence is a crucial task that can be approached by comparing the new sequences with those of already known enzymes labeled by a family class. This task is difficult because the activity is based on a combination of small sequence patterns and sequences greatly evolved over time. This paper presents a classifier based on the identification of common subsequence blocks between known and new enzymes and the search of formal concepts built on the cross product of blocks and sequences for each class. Since new enzyme families may emerge, it is important to propose a first classification of enzymes that cannot be assigned to a known family. FCA offers a nice framework to set the task as an optimization problem on the set of concepts. The classifier has been tested with success on a particular set of enzymes present in a large variety of species, the haloacid dehalogenase superfamily.

**Keywords:** bioinformatics, protein classification, FCA application.

# 1 Introduction: Enzyme Classification

This paper presents an application of concept lattices to build a classifier of enzymatic sequences. Enzymes are molecules of living cells with a catalytic activity (they increase reaction rates) that make them essential for biochemical reactions. Enzymes are mainly named and classified according to the reaction they catalyze. A report of a dedicated Nomenclature Committee [1] assigns each enzyme a recommended name and an EC (Enzyme Commission) number made of

---

[1] http://www.iubmb.org/1984

C.V. Glodeanu, M. Kaytoue, and C. Sacarea (Eds.): ICFCA 2014, LNAI 8478, pp. 235–250, 2014.

four hierarchical levels. The first level (indicated by a number from 1 to 6) divides enzymes in six main groups, according to the type of chemical reaction catalyzed (e.g. 3 refers to hydrolases, which involve all reactions decomposing/recomposing molecules by the addition/suppression of water). The second and third levels provide increasing refinements on the mechanism of the reaction. The fourth level is a serial number that is assigned to inform on the specific molecule, *the substrate*, upon which the enzyme acts by forming a transitory complex with it.

An enzyme is a particular type of protein, the main active macro-molecule in cells, which is made of a sequence of linked *amino acids*. It is now easy to obtain the protein sequences contained in various organisms but to find experimentally the function of a protein remains a tedious and expensive task. Biologists are thus interested in automatic approaches that can help them to filter among the numerous possibilities, the most relevant one with respect to the observed sequence. Proteins can also be organized and classified into families and superfamilies based on similarities between their sequences and/or their spatial structures. A number of studies have observed that, whilst relatives within enzyme superfamilies may perform different functions or transform substrates in different ways, they often share some aspects of their chemistry/mechanisms of reactions. Thus, an important step when making hypotheses on the enzyme functional activity is to determine its membership to a structural superfamily and/or family. Two classifications of known protein 3D structures have been developed to capture their evolutionary relationships, CATH [1] and SCOPe [2]. Both of these classifications use elementary structures called *domains*, proteins featuring one or several domains organized in various ways, and often with different functions. There is a relatively small number of superfamilies with respect to the number of domains (e.g. CATH v3.5 contains 2626 superfamilies for 175536 domains) and the issue of predicting the superfamily of a protein from its sequence is relatively easy due to the presence of key domains with some characteristic motifs. In contrast, the family level remains hard to predict from sequences and requires cross-checking of multiple sources of information on the structure or the biochemical characterization of particular sites in the protein.

In this study, given a known superfamily, we consider the issue of classifying a set of new enzyme sequences (the *unlabeled set*) at the family level with respect to a set of sequences that have already been classified (the *labeled set*). We are looking for an explicit classification, with a clear interpretation in terms of the presence of characteristic sites in the sequence. We have addressed this problem in the framework of Formal Concept Analysis and shown it is adapted to the two subproblems that arise in practice: the classification of unlabeled sequences in existing classes (supervised classification) and the creation of new classes (unsupervised classification). The paper is organized as followed: the next section explains how interesting sites have been selected along the sequences, allowing to code them at a domain or subdomain level. Section 3 formalizes the issue in the framework of FCA and gives some account of the literature related to this issue, both in bioinformatics and in the FCA community. Two subsections detail the case of supervised (3.2) and unsupervised (3.3) classification. A last section

(section 4), before conclusion, introduces a real case experiment on a particular enzyme family, which shows the neat interest of the approach in producing meaningful classifications.

## 2   Coding Enzymes Using Multiple Partial Local Alignment

Enzyme functions can be associated to particular positions in their sequences. The corresponding amino acids contribute to shape a specific spatial structure that can interact with the substrate or are directly involved in the catalytic machinery. In practice, short common words extracted from sequences of enzymes sharing a same known activity - i.e. short lists of successive amino acids - can help to point out such active sites. However, two important aspects have to be considered for this task: (1) biochemical knowledge on amino acids, and (2) the divergence of protein sequences through evolution, including point mutations, domain rearrangements and insertion/deletions.

When dealing with protein sequences, it is important, first, to take into account the similarities due to shared physico-chemical properties between letters in the alphabet of the 20 standard amino acids used in proteins: some amino acid substitutions have no impact on the function or the structure of the protein while others have. To consider this knowledge, a standard approach in machine learning consists in directly recoding the proteins on a smaller property-based alphabet, such as the hydropathy index or the Dayhoff encoding ([3], [4]). These coding schemes suffer from being a priori fixed, while the useful properties of a same amino acid may differ from one position to the other. The work described in this manuscript is based on a more specific data-driven approach based on the detection of local conservations shared by labeled and unlabeled sequences.

The second point concerns the identification of putative domains and active sites in the enzyme sequences that relies on the detection of local similarities in the labeled set. It can be achieved by looking for optimal multiple alignment of sequences. In fact, an alignment does not only provide a recoding of sequences, it also keep track of the chaining of elements since the matching edges between characters in the alignment are not allowed to cross. We have extended the standard alignment search by loosening the constraints on admissible alignments in two ways: the alignment is local (involving only substrings) and it is partial (involving only sequences subsets instead of the whole set of sequences as in classical alignment). Altogether, this leads to a partial local multiple alignment (PLMA) of the sequences. Each short strongly conserved region in the PLMA (called *block* in the sequel) will form one of the characters for recoding the sequences: each sequence is represented by the sequence of blocks it is involved in. At this stage, it is important to note that the new sequences to be assigned, the unlabeled sequences, need to be also encoded and are aligned together with the sequences of known class, the labeled sequences. The computation of PLMA has been introduced as the first step performed in Protomata-Learner ([5]), a grammatical inference program aiming at learning finite state automata for the characterization of protein family sequence sets. But even if the choice of the alignment parameters is important in Protomata-Learner to tune the desired level of generalization, we have only used default parameters in this study.

# 3    Class Assignment from Formal Concept Analysis

## 3.1    Formalization of the Classification Problem

The previous section explains how each protein sequence has been converted in a Boolean sequence, i.e. a vector of block presences. The classification task consists, from a set of sequences labeled by a class (a family) and a set of unlabeled sequences, in guessing a class for each unlabeled sequence. This is either a known family class or a novel class never observed in the labeled set but that gains some evidence from the concurrent presence of specific blocks in the unlabeled set.

A natural approach for such an assignment task is to build a classification of all sequences with respect to attributes and to decide the class of the unlabeled sequences from their place among the labeled sequences in the concept tree. This requires to define a similarity measure on the set of attributes, and to set a threshold to discriminate the meaningful clusters. Problems quickly arise when trying to follow this approach: the number of attributes may greatly vary from one superfamily to the other and from one sequence to the other within a same (super)family. Several standard machine learning techniques have been tried for the prediction of enzyme classes or more generally of protein families from sequences. For instance, [6] follows a decision-tree approach (C4.5) to build the classification at the EC level 1 (6 groups), after having extracted 36 features for the description of enzymatic sequences. The same authors have also trained Support Vector Machines [7] for this task. More recently, Kumar et al. [8] addressed the issue of enzyme classification in more depth using Random forest, an approach bagging a number of classification trees (e.g. 200) built on random subsets of features. They used an extended set of 73 sequence-derived features and proposed a classification at 3 levels in the EC hierarchy: level 0 (enzyme/non-enzyme), 1 and 2. Finally, a few authors have tried to distinguish the classification of novel sequences in either known families or in entirely novel families. A nice study is proposed in [9], which uses both a set of Hidden Markov Models trained on each known family to decide the most relevant class of a new sequence and a logistic regression to decide sequences that likely belong to a new class.

In all cases, a decision taken on statistical arguments is useful but not fully satisfactory because it is hard to fix universal values for the necessary parameters and above all, it tends to be a black box. Ultimately a biologist has to check the assignments on the basis of the argumentation logics, his own knowledge, and further biochemical characterization of the sequence(s) of interest. Therefore, it is important to offer an easy access to the way automatic assignments have been decided. Furthermore, we want to be able to distinguish and characterize entirely novel sequence families, since it occurs frequently during the analysis of new organism genomes. Note that we have assumed that all sequences belong to a same superfamily. This way, some aspects of the structure that can be hardly captured at the sequence level are supposed to be present. Since the prediction of the family level is quite good, this is not a hard limitation in practice. Overall, the challenge is thus to check if a more direct and more exploratory approach is possible, where the set of assignment possibilities is made clear to the biologist.

We have thus decided to use a FCA approach to solve this issue: given a relation linking a set of attributes and a set of objects some of which are labeled using a set of class labels, our problem consists in finding a class assignment for unlabeled objects on the basis of the associated concept lattice.

Supervised classification is a relatively common application of concept lattices in the literature. It consists in building a classifier from a concept lattice created with a set of attributes/objects labeled by their classes (learning step) and then predicting the class of new objects by using the generated classifiers (classification step). Published algorithms differ on three points:

1. Object and attribute selection for the creation of the lattice: The vast majority of related papers [10, 11, 12] have used concept lattices built on a learning set of labeled objects to produce a classifier that is used in a second step to assign new objects with unknown class. On the contrary, [13] considers the lattice built on both the labeled and the unlabeled set to focus the search on links between known and unknown objects. Some other papers use a feature selection step to use only "interested" objects or attributes [14].

2. Selection of best concepts: Some methods use a concept selection step before classification, filtering the most relevant ones (for instance in case of missing/noisy data) [15]. To avoid over-fitting, [14, 16] use only the upper lattice to produce most general classifications. Some other measure of significance are used like the coherence or the support of concepts ([11, 12]).

3. Utilization of the lattice as a classifier: After the construction of the lattice and the selection of relevant concepts, there are different ways to use it to classify new objects. Most classifiers use directly the lattice to compute a similarity between objects to be classified and concepts. Various measures exist, including the number of common attributes and/or the support of a class in a concept [11, 14]. For instance, Ikeda [13] estimates the plausibility of each concept to represent a set of objects belonging to a same class. The class label of objects is thus used for scoring. The method selects first the most discriminating concepts for each unlabeled object and classify them with respect to their score. A classifier can also be built by generating rules from the lattice. Indeed, the concept lattice provides a nice ordering for the search of rules [10, 12] directly from selected concepts in the lattice. It is also possible to build a decision tree from the lattice [17], replacing rules by decision nodes. A more complex procedure is possible via the computation of concept intersections in the lattice [18]. Other papers use various classifiers derived from the lattice like nearest neighbors or naive Bayes [16, 19].

In our study, the set of attributes represents the enzyme blocks and there are at least two kinds of objects, the labeled and unlabeled enzyme sequences. The issue is then to introduce the class labels in this framework, in order to handle them directly in the formal concepts. This key point can be solved without changing the formalism, by adding the class value as a particular type of object: each time a block $b$ is observed in a sequence $s$ of class $c$, the pairs $(s, b)$ and $(c, b)$ are added to the formal context relation. Including the classes in the context as objects allows to have the right semantics for the binary relation and the discrimination task: if attribute $b$ appears in a concept with a class $c$, it means that there exists at least one sequence of class $c$ with attribute $b$. If $c$ is the unique class in this concept, then $b$ is characteristic of $c$ and can be used for the classification of unlabeled sequences, otherwise $b$ leads to an ambiguous classification that is also an interesting result for the biologist. Note that using classes as attributes instead of objects would not allow to describe ambiguous classifications. In practice, it is only necessary to produce concepts having at least one unlabeled sequence in the object set, otherwise it is not useful for sequence labeling. The size of the relation remains sufficiently small in this context to produce the whole lattice of formal concepts for this relation. The assignment procedure is based on the exploitation of the lattice.

In a general setting, let $A$ be the attribute set, $C$ the class set, $L$ the labeled set of objects and $U$ the unlabeled set of objects. Let $LUC = (L \uplus U \uplus C)$ and let $\mathcal{I}$ denote the binary relation over $LUC \times A$ and $\mathcal{B}(LUC, A, \mathcal{I})$ the concept lattice. The problem is to find a minimal extension $N$ of $C$ and an argumentation assigning classes of $N \cup C$ to elements of $U$ on the basis of $\mathcal{B}(LUC, A, \mathcal{I})$ .

For this purpose, we propose an iterative scheme where each unlabeled sequence is assigned in turn by looking for its compatible class assignments. A *compatible class assignment* is defined as a class that belongs to some concepts sharing a maximal set of blocks with the unlabeled sequence. Maximality is defined here with respect to set inclusion.

**Definition 1. (compatible class assignment)** *Let* $LUC = (L \uplus U \uplus C)$. *Given a concept lattice* $B = \mathcal{B}(LUC, A, \mathcal{I})$ *and an element* $u$ *of* $U$, *a compatible class assignment is an element* $c \in C$ *such that there exists a concept* $(\{u, c\} \cup X, Y)$ *in* $B$, $X \subset LUC$, *and no* $Y$ *is larger among the possible concepts.*

Another important aspect of the quality of a classification decision is its support with respect to existing labeled sequences. Each class assignment may be associated to a concept that we call *attribute-compatible concept*. This concept gets a support in terms of its number of blocks. Another measure is the support in terms of labeled sequences. However, the compatible concept is not the best one with respect to this measure. It may exist a concept in the lattice, called object-compatible concept, with a larger sequence support:

**Definition 2. (attribute-compatible and object-compatible concept)** *Let* $LUC = (L \uplus U \uplus C)$. *Given a concept lattice* $B = \mathcal{B}(LUC, A, \mathcal{I})$, $u \in U$, *and* $c \in C$, *the attribute-compatible concept and object-compatible concept are concepts* $BC(u, c) = (\{u, c\} \cup X, Y_{max})$ *and* $BC(u, c) = (\{u, c\} \cup X_{max}, Y)$ *of* $B$, *where* $Y_{max} = max\{Y \subset A : (\{u, c\} \cup X, Y) \in B, X \subset LUC\}$ *and* $X_{max} = max\{X \subset LUC : (\{u, c\} \cup X, Y) \in B, Y \subset A\}$.

This way, each class assignment may be scored by the number of blocks of its attribute-compatible concept and the number of sequences of its object-compatible concept.

## 3.2 Supervised Classification

Our method tries to maximize the specificity of the classification decisions and proposes several quality levels for a class assignment towards this end.

At level 1, it checks if some attributes that are specific of a class (i.e. blocks present in sequences belonging to a single class) are also present in the current unlabeled sequence. These attributes, called *characteristic attributes*, are assigned the highest quality value since they do not lead to any ambiguity if present alone. It corresponds to build a characteristic partition that splits $L$ in subsets $L_i, i = 1, m$ with a common class value for each subset and $A$ in $m + 1$ possibly empty subsets $A_i, i = 0, m$ of attributes only present in elements of $L_i$, with $A_0 = A \setminus \cup_{i=1}^{n} A_i$.

If there exists a single compatible class assignment $c$ using only characteristic blocks, the sequence is classified at level 1, with label $c$.

If there are several compatible class assignment $c$ using only characteristic blocks, the sequence has an ambiguous classification and if it cannot be classified at the next level, it is said ambiguous and all its possible classes are displayed.

For sequences that have not been classified at level 1, the method checks at level 2 if some concepts are attribute-compatible with respect to the current unlabeled sequence, irrespective of the specificity of its blocks.

If there exists a single compatible class assignment $c$, the sequence is classified, with label $c$.

If there are several compatible class assignment $c$ , the sequence is said ambiguous and all its possible classes are displayed.

The remaining cases are when no concept is compatible with the unlabeled sequence. It means either that the sequence has no block in common with another sequence and it remains unclassified, or that it is a member of a new family never observed before that use blocks found only in unlabeled sequences.

For instance, figures 1(a), 1(b) and 1(c) represent partial local multiple alignments and in each figure, colored sequences (e.g. $s1$ and $s2$) are labeled sequences while black sequences (e.g. $s3$) are unlabeled and waiting for class assignment. On figure 1(a) the unlabeled sequence $s3$ gets only one compatible class corresponding to the orange concept and can thus be unambiguously classified. However, on figure 1(b) there are two compatible concepts (orange and green), and the unlabeled sequence class assignment is ambiguous. Figure 1(c), provides an example of an unlabeled sequence, $s3$, that remains unclassified because the multiple alignment has found no common block with another sequence. On the same picture a new family is formed with a characteristic concept involving only unlabeled sequences: $\{s4, s5, s6\} \times \{Block1, Block2, Block3\}$. The purpose of the next subsection is to detail the search of such new classes.

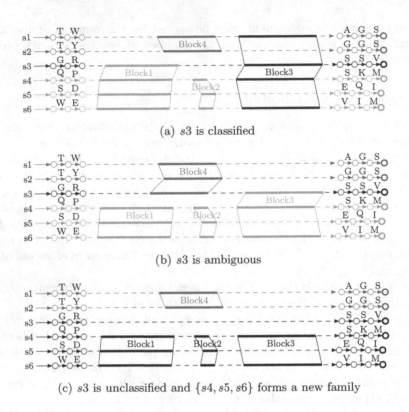

(a) $s3$ is classified

(b) $s3$ is ambiguous

(c) $s3$ is unclassified and $\{s4, s5, s6\}$ forms a new family

**Fig. 1.** Examples of partial local multiple alignments with labeled (colored) and unlabeled (black) sequences

## 3.3 Unsupervised Classification

In terms of FCA, a new family can be characterized like for other families by an associated concept that gathers the sequences of this family and the blocks that form a signature of this family. These blocks are characteristic of unlabeled sequences as is the case for level 1 classification, but this time it is an unsupervised task since the set of classes $N$ is unknown.

This problem is related to biclustering [20]. However, biclustering looks for simultaneous partitioning of the set of objects and attributes. In our case, it is not realistic to expect a partition of both sets. The objects (sequences) share numerous attributes (blocks) and frequently, it is the way they are combined which allow to distinguish different clusters. The issue of object clustering from a formal context is treated in paper [21]. Authors propose a two-step procedures where formal concepts are enlarged to approximate concepts during the first step and then merged in a second step when they overlap sufficiently. This approach draws on the concept lattice as we do in order to find clusters but it shares some common drawbacks with biclustering in relation with our application domain. A partition of objects is useful but not necessary in our case and, furthermore, the method requires careful parameter tuning to get meaningful approximate concepts. In [22], the idea of using the set of formal concepts is further elaborated and no need for thresholds is longer required. Instead of starting from the object×attributes concept lattice, the authors propose to consider the lattice built on the object×concepts context in order to build the object clusters. It seems an interesting idea that could be experimented on the protein classification task. However, the interpretation of clusters becomes more difficult and it is an important preoccupation for the biologist to master the decision process. Another related aspect of all these methods is their heuristic nature. Concept analysis is an exact method and it seems somewhat unfortunate to loose this property in the classification task.

We decided to keep on the idea of associating a concept to each class. We also looked for an exact search of the concepts without parameter tuning, a requirement that implies a neat specification of the target concepts. The issue of deciding the occurrence of new families in $N$ is not trivial due to the conjunction of two difficulties that have to be taken into consideration:

- A given set of sequences participates to a number of concepts. A subset of concepts has to be extracted that covers the set of sequences;
- The set of new families is not necessary a partition: although it should be avoided as much as possible, a given sequence that has evolved to get a bifunctional capacity could belong to two different families.

We have set this issue as the following optimization problem: find an optimal cover of the new family sequences by the set of concepts including characteristic new blocks -only present in unlabeled sequences-. Optimality depends on three criteria of decreasing priority:

1. minimize the number of ambiguous sequences in the concepts (i.e. get closer to a partition);
2. minimize the size of $N$ (i.e. parsimonious hypothesis with a minimum number of necessary new families);
3. maximize globally the support of the new families in terms of number of characteristic blocks.

The two first criteria are the most important but using three criteria ensures to get a single solution in all practical cases we have checked. It would be possible to add other criteria on more complex cases for resolving the ties. The number of sequences of the object-compatible concepts, originally defined as a quality index, could be used for this purpose. All these criteria are coded within a set of logical constraints using Answer Set Programming, a form of declarative programming adapted to combinatorial problems [23]. Once all constraints are expressed as logical formulas, a grounder transform them in a (large) set of boolean formulas. A dedicated solver then looks for possible models of this set (the answers), through a conflict-driven constrained enumeration of admissible solutions [24]. This way, exact optimal concepts can be produced.

## 4    An Experiment with the HaloAcid Dehalogenase Enzyme Superfamily (HAD)

The haloacid dehalogenases superfamily (HADs) represent a large superfamily (120193 sequences reported; http://pfam.sanger.ac.uk/clan/CL0137) of ubiquitous enzymes present in all domains of life. The number of sequences differ between organisms, from around 20 in the *Escherichia coli* bacteria [25] to between 150-200 in the eukaryotic biological models such as *Arabidopsis thaliana* and *Homo sapiens* [26]. HADs serve as the predominant catalysts of organophosphate hydrolysis [27]. Enzymes in this superfamily form covalent enzyme-substrate complexes via a conserved amino acid. They catalyze the cleavage of carbon-halogen bonds (C-halogen), and also feature a variety of hydrolytic activities including phosphatase (CO-P), phosphonatase (C-P) and phosphoglucomutase (CO-P hydrolysis). HAD superfamily enzymes usually function as homodimers (i.e., a complex made of two identical proteins). All structurally characterized superfamily members share a conserved domain, termed the "HAD-like" fold by SCOPe. The typical folds of HAD phosphatases contains three additional structural signatures that contribute to substrate specificity: the "squiggle", "flap", and "cap" domains [28]. HAD have received an increased interest in the last decade since they have the potential to be used in both industrial and pharmaceutical applications, in addition to bioremediation processes [29].

For this experiment, we have worked on the following datasets:

1. 102 sequences from various organisms extracted from the supplementary data of [28]. This set contains 34 families, 3 sequences in each family;
2. 23 sequences from *E. coli* extracted from [25]. This set has 9 families in common with the previous set;
3. 40 sequences from *H. sapiens* extracted from [26];
4. 153 sequences from *A. thaliana* extracted from the TAIR database [2] containing a HAD domain, and additional sequences identified after reviewing the literature [28]. This set includes 23 sequences for which family is unknown.

The first dataset forms the labeled set in our study. The three remaining datasets have been used as unlabeled sets. For some of the sequences contained in these datasets, the real class is known. Indeed, many sequences from *E. coli*, *H. sapiens* and *A. thaliana* have been biochemically characterized and/or have been considered by in silico/in vivo structural analysis, and this provides experimental results on their classification. The sequence family prediction made by FCA can thus be evaluated on this basis.

For all results, we have used the solver Clasp developed in Potsdam University [24].

Figure 2 shows the complete lattice obtained on the smallest context corresponding to the *E. coli* unlabeled dataset. This line diagram has been drawn using the software erca (Eclipse's Relational Concept Analysis [3]) and a reduced

---

[2] http://www.arabidopsis.org/
[3] https://code.google.com/p/erca/

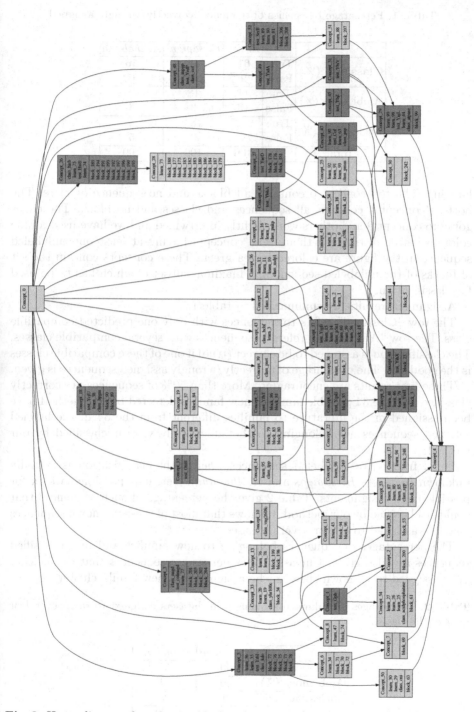

**Fig. 2.** Hasse diagram from lattice blocks x sequences/classes obtained in the experiment with the *E. coli* unlabeled dataset.

**Table 1.** Percentage by species of sequences correctly/wrongly assigned

|  |  | E. coli | H. sapiens | A. thaliana |
|---|---|---|---|---|
| Classified (%) | True | **61** | **65** | **56** |
|  | False | 9 | 3 | 6 |
| Ambiguous (%) | True | 17 | 18 | 18 |
|  | False | 13 | 3 | 8 |
| Unclassified (%) | True | 0 | 8 | 8 |
|  | False | 0 | 3 | 5 |
| Total |  | 100 | 100 | 100 |

labeling. The top concept 0 contains all blocks and no sequence or class. The bottom concept 4 contains all sequences and classes and no block. The edges going to concept 9 and others were slightly intertwined and we have used a blue color to better distinguish them. The concepts having at least one unlabeled sequence in the figure are colored in sea green. These concepts contain the set of blocks of the unlabeled sequences, a maximal subset of which has to be used for classification.

Assignment results are summarized in table 1.

The row "Classified" refers to sequences with only one predicted compatible class. The row "Ambiguous" refers to sequences with several compatible classes. The classification is assumed to be correct (true) if one of these compatible classes is the good one. The percentage of correctly/wrongly assigned sequences is given.

These first results are encouraging. More than 50% of sequences are correctly classified into the 34 possible families, new families detected by the method have been assigned in the literature to families different from the 34 in the labeled set, and sequences not belonging to the superfamily were unclassified by our method.

For a fraction of unlabeled sequences, their right classification is actually unknown (datasets *H. sapiens* and *A. thaliana*). Yet, it is possible to look for possible class assignments. Table 2 give the percentage of such sequences that could be classified by our method. It shows that most of these unknown sequences could be assigned to one or several classes.

The percentages of sequences belonging to new families and of unclassified sequences are also given. Unclassified sequences are sequences that can neither been assigned to a known class nor be assigned to a new family cluster.

**Table 2.** Percentages of unknown sequences in datasets assigned to one, several or none of the classes

|  | H. sapiens | A. thaliana |
|---|---|---|
| Classified (%) | 50 | 54 |
| Ambiguous (%) | 50 | 21 |
| Unclassified (%) | 0 | 25 |
| Total | 100 | 100 |

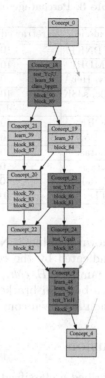

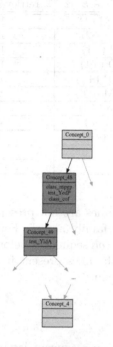

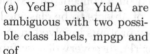

(a) YedP and YidA are ambiguous with two possible class labels, mpgp and cof

(b) YfbT and YcjU can be classified and assigned uniquely with the class bpgm

**Fig. 3.** Different kinds of assignment decisions

For the three datasets, *E. coli*, *H. sapiens* and *A. thaliana*, we find 0, 2 and 11 new subfamilies respectively.

For the *H. sapiens* dataset, sequences predicted to belong to new families are correct: the corresponding families are described in the papers on human HAD [26], and these families are not present in *E. coli* (i.e. the labeled set).

For the *A. thaliana* dataset, it is difficult to know if predicted new families are real because it contains numerous uncertain sequences. Our own review of the literature concludes that 11 unclassified sequences could have been wrongly assigned to the HAD superfamily in the TAIR database.

The specificity of the detection of new families has been tested too. For each known family in the labeled set, a new labeled set has been built that contain all sequences except the sequences belonging to this family. The unlabeled set was made of the *E. coli* dataset and the sequences of the selected family (3 sequences). The selected family should ideally be detected as a new family by our method. We have computed the percentage of retrieved sequences for all families. The

Table 3. Percentage of retrieved sequences within a new family

| new family alone | % retrieved sequences | new family+ *E. coli* | % retrieved sequences |
|---|---|---|---|
| EYA, SPSC, PNKP | 100 | NagD (+1) | 100 |
| SPP, CNII, MDP1 | 100 | HisB (+2) | 100 |
| ATPase, deoxy, HerA | 0 | TPP (+1) | 100 |
| PMM, Yhr100c, s38K | 100 | KDO (+1) | 100 |
| CNI | 67 | MPGP (+1) | 67 |
| Enolase, BCBF | 100 | BPGM (+6) | 67 |
| LPIN, PseT, P5N1 | 100 | Sdt1p (+4) | 43 |
| AcidPhosphatase | 100 | Cof (+6) | 44 |
| Phosphonatase | 100 | PSP (+1) | 75 |
| VNG2608C, dehr | 0 | | |
| Zr25, CTD | 100 | | |

results are shown in table 3. Note that some families are not present in *E. coli* and this is indicated by the column label "new family alone". For the others (column new family + *E. coli*), the number of *E. coli* sequences belonging to the family is given between brackets. On the 34 subfamilies present in the labeled set, the decision has been convincing for 27 of them.

## 5    Conclusion

We have described a classification method based on a concept lattice including both a set of already classified objects and a set of objects to be classified. It has been applied to enzyme sequences, a group of key proteins involved in many biochemical processes and with a high potential for the discovery of new functional molecules. Our results are encouraging and show our classification method to be sensitive and specific. More than half of the unlabeled sequences are correctly classified with respect to the current knowledge for 34 subfamilies and ambiguous sequences represent only one third of the tested sequences, two thirds of them having the correct class assignment. Moreover, each classification decision may be clearly explained and related to known sequences or particular positions in the sequence corresponding to blocks. Ambiguity could be even reduced in practice by looking for sequences that are inherently ambiguous because they are made for instance of two fragments of two proteins of different class. Such potential proteins, which we call chimera, could be automatically extracted during classification.

Another aspect of this work is the unsupervised classification problem for objects with attributes that are characteristic of unlabeled objects. We have suggested a model for solving this problem as an optimization issue taking into account ambiguity, parsimony (number of new classes needed) and intent (number of attributes).

To our knowledge, it is the first time that this issue is properly formalized in bioinformatics. The next step will consist in testing the robustness of the method on species that are very evolutionary distant compared to the other organisms for which test sets were considered. We have selected for this next study the brown alga *Ectocarpus siliculosus*, for which the genome sequence has been recently published [30]. Since attributes describing the sequences have no reason to be limited to blocks, we will try other global features extracted from theses sequences such as amino acid content. We will test if the best in silico assignment within classes correlates with potential substrate specificity. To this aim, a number of algal sequences will also be biochemically characterized.

# References

[1] Sillitoe, I., Cuff, A., Dessailly, B., Dawson, N., Furnham, N., Lee, D., Lees, J., Lewis, T., Studer, R., Rentzsch, R., Yeats, C., Thornton, J.M., Orengo, C.A.: New functional families (funfams) in cath to improve the mapping of conserved functional sites to 3d structures. Nucleic Acids Res. 41(D1), D490–D498 (2013)

[2] Fox, N.K., Brenner, S.E., Chandonia, J.M.: SCOPe: Structural Classification of Proteins-extended, integrating SCOP and ASTRAL data and classification of new structures. Nucleic Acids Res. 42(D1), D304–D309 (2014)

[3] Yokomori, T., Ishida, N., Kobayashi, S.: Learning local languages and its application to protein α-chain identification. In: HICSS (5), pp. 113–122 (1994)

[4] Peris, P., López, D., Campos, M.: Igtm: An algorithm to predict transmembrane domains and topology in proteins. BMC Bioinformatics 9 (2008)

[5] Kerbellec, G.: Apprentissage d'automates modélisant des familles de séquences protéiques. PhD thesis, Université Rennes 1 (2008)

[6] Lee, B.J., Lee, H.G., Lee, J.Y., Ryu, K.H.: Classification of enzyme function from protein sequence based on feature representation. In: Proc. of the 7th IEEE Int. Conf. on Bioinformatics and Bioengineering, BIBE 2007, pp. 741–747 (October 2007)

[7] Lee, B.J., Lee, H.G., Ryu, K.H.: Design of a novel protein feature and enzyme function classification. In: IEEE 8th Int. Conf. on Computer and Information Technology Workshops, CIT Workshops 2008, pp. 450–455 (July 2008)

[8] Kumar, C., Choudhary, A.: A top-down approach to classify enzyme functional classes and sub-classes using random forest. EURASIP Journal on Bioinformatics and Systems Biology 2012(1), 1 (2012)

[9] Brown, D.P., Krishnamurthy, N., Sjölander, K.: Automated protein subfamily identification and classification. PLoS Comput. Biol. 3(8), e160 (2007)

[10] Wang, J., Liang, J., Qian, Y.: Closed-label concept lattice based rule extraction approach. In: Huang, D.-S., Gan, Y., Premaratne, P., Han, K. (eds.) ICIC 2011. LNCS, vol. 6840, pp. 690–698. Springer, Heidelberg (2012)

[11] Carpineto, C., Romano, G.: GALOIS: An order-theoretic approach to conceptual clustering. In: Proceedings of the 10th International Conference on Machine Learning (ICML 1990), pp. 33–40 (July 1993)

[12] Sahami, M.: Learning classification rules using lattices. In: Lavrač, N., Wrobel, S. (eds.) ECML 1995. LNCS, vol. 912, pp. 343–346. Springer, Heidelberg (1995)

[13] Ikeda, M., Yamamoto, A.: Classification by Selecting Plausible Formal Concepts in a Concept Lattice. In: Workshop on Formal Concept Analysis meets Information Retrieval (FCAIR 2013), pp. 22–35 (2013)

[14] Mephu Nguifo, E.: Legal-e: une méthode d'apprentissage de concepts à partir d'exemples, basée sur le treillis de galois. In: Actes du 9ème Congrès Recon. des Formes en Intell. Artificielle (RFIA), Paris, vol. 2, pp. 35–46 (January 1994)

[15] Klimushkin, M., Obiedkov, S., Roth, C.: Approaches to the selection of relevant concepts in the case of noisy data. In: Kwuida, L., Sertkaya, B. (eds.) ICFCA 2010. LNCS, vol. 5986, pp. 255–266. Springer, Heidelberg (2010)

[16] Njiwoua, P.: Améliorer l'apprentissage à partir d'instances grêce à l'induction de concepts: le système cible. In: Science, H., (ed.): Revue d' Intelligence Artificielle, vol. 13, pp. 413–440 (1999)

[17] Kovacs, L.: Generating decision tree from lattice for classification. In: 7th International Conference on Applied Informatics, vol. 2, pp. 377–384 (2007)

[18] Sahami, M.: Learning classification rules using lattices. In: Lavrač, N., Wrobel, S. (eds.) ECML 1995. LNCS, vol. 912, pp. 343–346. Springer, Heidelberg (1995)

[19] Xie, Z., Hsu, W., Liu, Z., Lee, M.L.: Concept lattice based composite classifiers for high predictability. J. Exp. Theor. Artif. Intell. 14(2-3), 143–156 (2002)

[20] Busygin, S., Prokopyev, O., Pardalos, P.M.: Biclustering in data mining. Comput. Oper. Res. 35(9), 2964–2987 (2008)

[21] Gaume, B., Navarro, E., Prade, H.: Clustering bipartite graphs in terms of approximate formal concepts and sub-contexts. International Journal of Computational Intelligence Systems 6(6), 1125–1142 (2013)

[22] Navarro, E., Prade, H., Gaume, B.: Clustering sets of objects using concepts-objects bipartite graphs. In: Hüllermeier, E., Link, S., Fober, T., Seeger, B. (eds.) SUM 2012. LNCS, vol. 7520, pp. 420–432. Springer, Heidelberg (2012)

[23] Brewka, G., Eiter, T., Truszczyński, M.: Answer set programming at a glance. Commun. ACM 54(12), 92–103 (2011)

[24] Gebser, M., Kaufmann, B., Schaub, T.: Conflict-driven answer set solving: From theory to practice. Artif. Intell. 187, 52–89 (2012)

[25] Kuznetsova, E., Proudfoot, M., Gonzalez, C.F., Brown, G., Omelchenko, M.V., Borozan, I., Carmel, L., Wolf, Y.I., Mori, H., Savchenko, A.V., Arrowsmith, C.H., Koonin, E.V., Edwards, A.M., Yakunin, A.F.: Genome-wide Analysis of Substrate Specificities of the *Escherichia coli* Haloacid Dehalogenase-like Phosphatase Family. Journal of Biological Chemistry 281(47), 36149–36161 (2006)

[26] Seifried, A., Schultz, J., Gohla, A.: Human HAD phosphatases: structure, mechanism, and roles in health and disease. FEBS Journal 280(2), 549–571 (2013)

[27] Koonin, E.V., Tatusov, R.L.: Computer analysis of bacterial haloacid dehalogenases defines a large superfamily of hydrolases with diverse specificity: Application of an iterative approach to database search. J. Mol. Bio. 244(1), 125–132 (1994)

[28] Burroughs, A.M., Allen, K.N., Dunaway-Mariano, D., Aravind, L.: Evolutionary Genomics of the HAD Superfamily: Understanding the Structural Adaptations and Catalytic Diversity in a Superfamily of Phosphoesterases and Allied Enzymes. Journal of Molecular Biology 361(5), 1003–1034 (2006)

[29] Janssen, D.B.: Biocatalysis by dehalogenating enzymes. Advances in Applied Microbiology, vol. 61, pp. 233–252. Academic Press (2007)

[30] Mark Cock, J., Sterck, L., Rouz, P., Scornet, D., Allen, A., Amoutzias, G., Anthouard, V., Artiguenave, F., Aury, J., Badger, J.: The Ectocarpus genome and the independent evolution of multicellularity in brown algae. Nature (7298), 617–621 (2010)

# Multilayered, Blocked Formal Concept Analyses for Adaptive Image Compression

Ruairí de Fréin

Telecommunications Software & Systems Group,
Waterford Institute of Technology, Ireland
rdefrein@gmail.com

**Abstract.** Formal Concept Analysis (FCA) decomposes a matrix into a set of sparse matrices capturing its underlying structure. A similar task for real-valued data, transform coding, arises in image compression. Existing cosine transform coding for JPEG image compression uses a fixed, decorrelating transform; however, compression is limited as images rarely consist of pure cosines. The question remains whether an FCA adaptive transform can be applied to image compression. We propose a multi-layer FCA (MFCA) adaptive ordered transform and Sequentially Sifted Linear Programming (SSLP) encoding pair for adaptive image compression. Our hypothesis is that MFCA's sparse linear codes (closures) for natural scenes, are a complete family of ordered, localized, oriented, bandpass receptive fields, predicted by models of the primary visual cortex. Results on real data demonstrate that adaptive compression is feasible. These initial results may play a role in improving compression rates and extending the applicability of FCA to real-valued data.

## 1 Introduction

Sparse Coding (SC) is a class of unsupervised methods for learning *overcomplete bases* to represent data efficiently. An overcomplete basis, $\boldsymbol{\Phi} \in \mathbb{R}^{N \times K}$ –referred to as a *dictionary* from here on– is not necessarily a linearly independent subset of a vector space $V$ (over a field $F$), but it does span $V$. The vectors in the set $\boldsymbol{\Phi}$, denoted $\phi_k \in \mathbb{R}^{N \times 1}$, are called atoms. Important examples are wavelet-related dictionaries (e.g., wavelet packets, stationary wavelets; see Chen, Donoho, and Saunders [1]; Mallat [2]) and learned dictionaries (Lewicki and Sejnowski, [3]; Lewicki and Olshausen [4]; Olshausen and Field [5,6]). The evolution of dictionary design is mapped out in [7], and points to an increased interest in learned dictionaries. The present paper proposes the first application of Formal Concept Analysis (FCA) [8] to SC, specifically, addressing an image compression problem, by: 1) learning an adaptive binary dictionary $\boldsymbol{\Phi} \in \mathbb{Z}^{N \times K}$, via Multilayer Formal Concept Analysis (MFCA); and 2) learning compact coefficients $\boldsymbol{a}$, using Sequentially Sifted Linear Programming on an Ordered Dictionary (SSLPOD), $\boldsymbol{\Phi}$, for an image $\boldsymbol{x}$. SC finds a set of atoms $\phi_k$ such that an input vector $\boldsymbol{x}$ is represented as a linear combination of as few of the atoms as possible –the representation vector, $\boldsymbol{a}$, has lower entropy than the input vector $\boldsymbol{x}$, increasing the compressibility of the image's new representation [9], the system, $\boldsymbol{x} = \sum_{k=1}^{K} a_k \phi_k$.

C.V. Glodeanu, M. Kaytoue, and C. Sacarea (Eds.): ICFCA 2014, LNAI 8478, pp. 251–267, 2014.
© Springer International Publishing Switzerland 2014

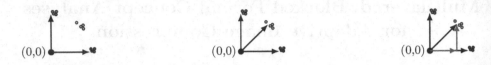

**Fig. 1.** (LHS coordinate system) The canonical basis functions (arrows), $\Phi$, and three sets of data points (red, green, blue dots), $y$, three different image features for example; (Center) an overcomplete basis/dictionary, which is composed of the canonical basis and one additional atom [.7071, .7071]; and finally, (RHS) a comparison of the canonical basis with an overcomplete basis. The additional atom in the center figure captures the green dots with greater sparsity than the canonical basis. This is emphasized in the RHS figure by plotting the weighted (by $a$) sum of the canonical basis functions required to represent the green dots. Using the overcomplete dictionary gives a more compact solution for all data points. However there are many possible representations.

*Related Work:* The advantage of overcompleteness over more traditional analyses for image compression, e.g. Principal Component Analysis (PCA) and the Discrete Cosine Transform (DCT), is that the atoms are better able to capture structures and patterns inherent in input data, $x$, an image. The null space of an overcomplete $\Phi$ introduces extra degrees of freedom in the choice of $a$, which we exploit to improve compressibility. For example, JPEG [10] and JPEG2000 [11] compute *compact* coefficients $a$ by inverting $\Phi$, which is square and nonsingular, and then quantizing the coefficients $\mathcal{Q}_o(\cdot)$: $a = \mathcal{Q}_o(\Phi^{-1}x)$. The DCT and PCA are de-correlating transforms; the DCT [12] is intimately related to the (ideal) Karunen-Loève or Hotelling transform [13]. On the other hand, dictionaries $\Phi$ that are adaptive, or tuned to an image $x$ potentially yield greater compaction of $a$ over the traditional decorrelated transform coefficients. A dictionary constructed using MFCA computes closures [14,15], which are present in the image, whereas the DCT projects the image onto cosine-like atoms (which are generally not present); MFCA is *tuned* to the image. Construction of a transform dictionary using our variant of FCA (see [14,15] for initial results) presents the possibility of greater compaction, along with the traditional advantages of FCA: *lectic* ordering and completeness of the lattice.

Using a MFCA binary dictionary, $\Phi \in \mathbb{Z}^{N \times K}$, is intuitively correct: subtractive atoms (*components*) learned by PCA do not sit well with the additive disjoint building block model, which is popular at present in the computer vision community [16]. For example, an image of a face is composed of atoms (eyes, a nose and a mouth, etc.); subtractive *eye* components have no physical meaning. A binary representation implies features are present, or not present. Once an atom's presence is established, a coefficient $a$ represents *how much* an atom is present –its intensity or color. Binary dictionaries are appealing because they are typically more compactly compressed than a real-valued adaptive dictionary with the same support: a binary dictionary may be encoded by only listing the positions of ones, real-valued dictionaries must also encode the magnitudes [17].

The disadvantage of overcompleteness over an invertible transform, e.g. the DCT [10] or wavelets [11], is that the coefficients, for a given set of atoms $\phi_k$, are no longer unique. Degeneracy is introduced; however, ambiguity due to non-uniqueness allows us to select the type of solution that we want, out of the family of all valid solutions. Sparse solutions generally have appealing properties –they generally indicate compressibility [9]. It is important to also note the role of factors other than *sparsity* (quantization and entropy coding [10,11]).

The sparsity assumption implies that the solution vector, $a = [a_1, a_2, \ldots a_K]$, has as few non-zero components as possible. In practice we equivocate and we desire that $a$ has as few as possible components which are far from zero. Fig. 1 illustrates the difference between the coefficients $a$ generated by the canonical dictionary, and one of the many possible solutions generated by using an over-complete dictionary. A sparse solution gives a compact representation of the data points. The choice of sparsity as a desired representation characteristic is driven by the observation that most sensory data such as natural images may be described as the superposition of a small number of atomic elements such as surfaces or edges (See Fig. 2). Other justifications such as comparisons with the properties of the primary visual cortex have also been advanced [7,6,5]. We draw on a second related concept that has been advanced by the Signal Processing community [18,19]: parts-based representations are Disjoint Orthogonal ([18] discusses the application of disjoint orthogonality in time-frequency analysis), and are generally found by identifying a sparse representation, $a$ [15,14,19,1]. SSLPOD finds the amount by which these sparse linear codes are present.

**Problem 1 *MFCA & SSLPOD:*** *The combined MFCA & SSLPOD problem is solved by minimizing the SC cost function on a set of $M$ input vectors (The $\ell_2$-norm is denoted $|| \cdot ||$):*

$$\min_{a_k^{(j)}, \phi_k} \sum_{m=1}^{M} ||x^{(m)} - \sum_{k=1}^{K} a_k^{(m)} \phi_k||^2 + \lambda \sum_{k=1}^{K} S(a_k^{(m)}) \tag{1}$$

The function $S(a_k^{(m)})$ is a cost function that penalizes $a_k$ for being far away from zero. The first term is the reconstruction term, which tries to minimize the error in the representation. The constant $\lambda$ determines the importance of both terms. For lossless compression, $x^{(m)}$ is an image, and the $\ell_2$-norm (squared) cost is zero. A trade-off between compressibility and error may be introduced by reformulating (Eqn. 1). Here, $\epsilon$, is the approximation target.

$$\min_{a_k^{(j)}, \phi_k} \sum_{m=1}^{M} \sum_{k=1}^{K} S(a_k^{(m)}) \text{ such that } ||x^{(m)} - \sum_{k=1}^{K} a_k^{(m)} \phi_k||^2 \leq \epsilon \tag{2}$$

***FCA's Appeal for SC:*** Using FCA to generate a dictionary providing compaction is novel. We use FCA to provide a binary dictionary for generic compression. Compared to JPEG (where a fixed dictionary is shared by the encoder and decoder) FCA learns content adaptive dictionaries that must be compressed and transmitted also; however this dictionary is binary and has few ones.

The most direct measure of sparsity is the $\ell_0$-norm. It is non-differentiable.

$$S(a_k) = \mathbf{1}_K^T \mathbf{1}(|a_k| > 0), \quad \mathbf{1}(x) = \begin{cases} 1 & \text{if } x \equiv true \\ 0 & \text{otherwise} \end{cases} \quad (3)$$

The matrix $\mathbf{1}_K \in \mathbb{R}^{K \times 1}$ is a matrix of $K$ ones. When the $\ell_0$-norm form of (Eqn. 2) is used, (Eqn. 2) is called the sparse approximation problem [20]. It is in general NP-hard. It is this problem we would ideally like to solve. Common choices for sparsity functions that provide a good approximation for the $\ell_0$-norm are the $\ell_1$-norm $S(a_k) = |a_k|_1 = \sum_{k=1}^K |a_k|$ (See [21,1]). Because MFCA produces binary dictionaries, the $\ell_0$-norm may be computed for the dictionary. Depending on the application, considering a $\ell_0$-cost on the dictionary $S(\boldsymbol{\Phi})$ may be equivalent to considering a $\ell_0$-cost on the coefficients.

Most adaptive compression techniques solve an $\ell_1$-norm approximation of the SC problem. It is possible to make the sparsity penalty arbitrarily small by scaling down $a_k$ and scaling $\phi_k$ up by some large constant when the $\ell_1$-norm is used. One approach is to constrain each $||\phi_k||^2 = 1$. Sparse coding solves:

$$\min_{a_k^{(m)}, \phi_k} \sum_{m=1}^M ||\boldsymbol{x}^{(m)} - \sum_{k=1}^K a_k^{(m)} \phi_k||^2 + \lambda \sum_{k=1}^K S(a_k^{(m)}) \text{ s.t. } ||\phi_k||^2 = 1, \forall k \quad (4)$$

This paper is organized as follows: § 2 defines and justifies the MFCA generative model for 2-D images. § 3 introduces an entropy increasing Binary Layering Tree for image quantization and the MFCA algorithm that learns an adaptive linear transform. § 4 introduces the sequentially sifted linear programming encoding algorithm. The focus of the empirical evaluation is on the new MFCA linear transform (and not entropy coding and quantization). § 5 evaluates the new MFCA-SSLPOD transform coding and encoding scheme on three test images from the USC-SIPI Image Database.

## 2    FCA Generative Model for Images Analysis

***Part-based Assumption and Blocking:*** We apply FCA to the task of learning binary dictionary structures for generic content adapted image compression and justify MFCA's generative model. A grayscale image, for example the cameraman in Fig. 2 (upper Left Hand Corner Figure), is described by a matrix $I \in \mathbb{R}^{R \times G}$, $R \times G$ pixels, which records an intensity value in each element $I(r, g)$. The parts-based assumption implies that each image consists of a number of non-overlapping edges, ridges, or shapes; parts are expressed as 2-D polygons.

**Definition 1.** *The $p^{th}$ part can be described by the chain of $s = 1, \ldots, S$ vertices (pixel coordinates), $v_s^p$ that trace-out its polygon. The first vertex equals the last vertex, $v_0^p = v_S^p$. $\mathcal{P}^p = \{v_s^p\} = \{\{r, g\} | \{r, g\} \text{ is a vertex of the part }\}$*

**Fig. 2.** The original Cameraman is illustrated in the upper Left Hand Side (LHS). Binary masks of the layers of the Cameraman are plotted in increasing intensity from left to right (e.g., $\{(row, column)\} = \{(1,2),(1,3),(1,4),(2,1),(2,2),(2,3),(2,4)\}$). 7 of 16 layers are illustrated. White denotes a pixel is present in that mask; black denotes it is not present. The lower pixel intensity values form the cameraman; cameraman parts are low frequency –the pixels are regionally actived in plot $(1,2)$ & $(1,3)$. The cameraman outline is captured in layer $(1,4)$, which captures edges. Plot $(2,1)$ captures the cameraman's trousers, which are a low frequency region. The remaining layers capture high frequency background.

**Definition 2.** *The face of the $p^{th}$ part, $\mathcal{F}^p$, is the set of image pixels in the part. We express each face set $\mathcal{F}^p$, in matrix form by generating the matrix*

$$\boldsymbol{F}^p(r,g) = \begin{cases} \boldsymbol{I}(r,g) & if \ \{r,g\} \in \mathcal{F}^p \\ 0 & otherwise. \end{cases} \tag{5}$$

**Definition 3.** *Blocking operates on an intensity matrix, $\boldsymbol{I}$, using the arguments $N_b \in \mathbb{Z}$ where $N_b \leq R$ and $N_b \leq G$ and coordinates $(x,y)$. These arguments specify the block-size and the position of the upper-lefthand pixel of the block. It returns a row-vector $\overleftarrow{i}_{xy}$, which is a row-vectorized form of the sub-matrix supported by the block (cf. the definitions above).*

$$\overleftarrow{i}_{xy} = blk\{\boldsymbol{I}, N_b, (x,y)\} = [\boldsymbol{I}(x,y), \boldsymbol{I}(x+1,y), \ldots, \boldsymbol{I}(x+N_b-1,y), \boldsymbol{I}(x,y+1),$$
$$\boldsymbol{I}(x+1,y+1), \ldots, \boldsymbol{I}(x+N_b,y+1), \ldots] \tag{6}$$

We assume that blocks are square in this paper to simplify our notation; rectangular blocks are a valid choice too. In addition, blocks may overlap depending on the application. For notational convenience we may overload blocking by invoking $\overleftarrow{\boldsymbol{I}} = blk\{\boldsymbol{I}, N_b\}$ and blocking-out an entire image $\boldsymbol{I}$, producing the matrix $\overleftarrow{\boldsymbol{I}} \in \mathbb{R}^{\lfloor \frac{R}{N_b} \rfloor \lfloor \frac{G}{N_b} \rfloor \times N_b^2}$. To give a concrete example, the application of $blk\{\boldsymbol{I}, N_b\}$ to the matrix $\boldsymbol{I} \in \mathbb{R}^{512 \times 512}$ yields the blocked form of $\overleftarrow{\boldsymbol{I}} \in \mathbb{R}^{85^2 \times 36}$. The number of blocks is $M = \lfloor \frac{R}{N_b} \rfloor \lfloor \frac{G}{N_b} \rfloor$. We deal with the transpose of the blocked matrix.
*Remark:* Blocking is the appropriate method for partitioning images; each block captures local correlations in the values of intensities [6]. Naturally the block-size

Overloaded Blocking     $\overleftarrow{I} = \overleftarrow{I} = \text{blk}\{I, N_b\} =$

$$\begin{bmatrix} \text{blk}\{I, N_b, (1,1)\} \\ \text{blk}\{I, N_b, (1, N_b + 1)\} \\ \vdots \\ \text{blk}\{I, N_b, (R - N_b + 1, G - N_b + 1)\} \end{bmatrix}$$

Overloaded Deblocking     $I = \text{dblk}\{\overleftarrow{I}, N_b\} =$

$$\begin{bmatrix} \text{dblk}\{i_{1,1}^{\leftarrow}, N_b\} & \text{dblk}\{i_{1,1}^{\leftarrow}, N_b\} & \cdots & \text{dblk}\{i_{1,1}^{\leftarrow}, N_b\} \\ \text{dblk}\{i_{xy}^{\leftarrow}, N_b\} & \text{dblk}\{i_{1,1}^{\leftarrow}, N_b\} & & \text{dblk}\{i_{1,1}^{\leftarrow}, N_b\} \\ \vdots & \vdots & \ddots & \vdots \\ \text{dblk}\{i_{xy}^{\leftarrow}, N_b\} & \text{dblk}\{i_{1,1}^{\leftarrow}, N_b\} & \cdots & \text{dblk}\{i_{1,1}^{\leftarrow}, N_b\} \end{bmatrix}$$

Deblocking     $\overleftarrow{i_{xy}} = \text{dblk}\{i_{xy}^{\leftarrow}, N_b\} =$

$$\begin{bmatrix} I(x, y) & I(x, y+1) & \cdots & I(x, y + N_b) \\ I(x+1, y) & I(x+1, y+1) & \cdots & I(x+1, y+N_b) \\ \vdots & \vdots & \vdots & \cdots \\ I(x + N_b - 1, y) & I(x+N_b, y+1) & \cdots & I(x+N_b, y+N_b) \end{bmatrix}$$

$N_b$ influences the performance of subsequent SC. The selection of $N_b$ causes the size of potential image parts to be less than or equal to, in size, than an $N_b \times N_b$ block; setting $N_b$ to be too large may cause the problem of multiple parts being learned as one part; setting $N_b$ to be too small, may cause parts to be partitioned across multiple blocks. The MPEG video codec blocks frames into $8 \times 8$ blocks, as does JPEG. We define deblocking, the *inverse* of blocking, to reconstruct $I$. The overloaded deblocking operation is also given.

***MultiLayering and the Multilayer Generative Model:*** We formally introduce image layering and parts-based representations.

**Property 1** *The parts-based assumption implies that an image is a linear combination of its composite parts.*

$$I = \sum_{p=1}^{P} F^p \approx \sum_{p=1}^{P} \gamma_p 1(F^p > 0). \tag{7}$$

As the part $1(F^p > 0)$ is a binary matrix, it is scaled by an intensity parameter: $\gamma_p$. A polygon is considered to be a part because the intensity of all elements in the polygon is approximately the same, e.g. $b_t \leq F^p(r,g) < b_u$. This assumption implies that we may partition images based on ranges of pixel intensity, and exploit the element-wise disjointness of parts to segment images. We introduce the idea of layers, which are defined by the bounds on the part weights $b_t \leq I(r,g) < b_u$, because it is unreasonable to assume that each part has exactly uniform intensity. The binary masks associated with the layers plotted in $(1,2), (1,3), (1,4), (2,1), (2,2), (2,3), (2,4)$ in Fig. 2 are good examples of the localized, oriented, bandpass receptive fields, described by [6,5], as being the underpinning properties of the primary visual cortex.

**Property 2** *We define a layer $L^l$ and its associated binary mask $M^l$ to be*

$$L^l(r,g) = \begin{cases} I(r,g) & \text{if } b_t \leq I(r,g) < b_u \\ 0 & \text{otherwise,} \end{cases} \qquad M^l(r,g) = \begin{cases} 1 & \text{if } b_t \leq I(r,g) < b_u \\ 0 & \text{otherwise.} \end{cases} \tag{8}$$

*where $b_u, b_l$ are the upper and lower bounds on pixel intensities that are in the mask, and $l$ is the layer index.*

**Property 3** *Given the parts-based and the layer property we may now express the generative multi-layered mixing model as follows:*

1. **Images are Parts-based:** $M^l = \sum_q 1(F^q > 0), L^l = \sum_q F^q \approx \sum_q \gamma_q 1(F^q > 0) \quad \forall l = 1, \ldots L;$
2. **Images are Element-wise Disjoint:** $\odot_{l=1}^L M^l = 0, \odot_{l=1}^L L^l = 0$ *and* $\odot_{p=1}^P F^p = 0;$
3. **Images are composed of layers:** $I = \sum_{l=1}^L L^l.$

*Element-wise multiplication is denoted by* $\odot$*. In short, layers are linear combinations of parts (1); all layers, layer masks and parts are element-wise disjoint (2); images are linear combinations of layers, and therefore parts (3).*

**Definition 4.** *MFCA Generative Model: By appealing to blocking, the MFCA generative model may be formulated as a new form of SC*

$$x^{(m)} = \sum_{k=1}^K a_k^{(m)} \phi_k = blk\{I, N_b, (x, y)\}^T = \sum_{l=1}^L \sum_{k=1}^K a_k^{(m,l)} \phi_k^l. \tag{9}$$

*The vector* $x^{(m)}$ *is the* $m^{th}$ *block which is positioned at coordinates* $(x, y)$ *of the image* $I$*. The vector* $\phi_k^l \in \mathbb{Z}^{N \times 1}$ *is the* $k^{th}$ *atom of the* $l^{th}$ *layer; it is a segment of* $M^l$*, one of the binary masks, present in the block positioned at* $(x, y)$*. The value* $a_k^{(m,l)} \in \mathbb{R}$ *is the intensity of the presence of the* $k^{th}$ *atom, of the* $l^{th}$ *layer in the* $m^{th}$ *block. We re-write this model by re-ordering the summation and blocking to illustrate how the MFCA SC problem is composed of layer-based sub-problems.*

$$x^{(m)} = \sum_{l=1}^L blk\{L^l, N_b, (x, y)\}^T = \sum_{l=1}^L \sum_{k=1}^K a_k^{(m,l)} \phi_k^l. \tag{10}$$

**Problem 2** *The MFCA-SSLPOD problem is defined as:*

1. $\min_{a_k^{(m)}, \phi_k} \sum_{l=1}^L \sum_{m=1}^M \| blk\{L^l, N_b, (x, y)\}^T - \sum_{k=1}^K a_k^{(m,l)} \bar{\phi}_k^l \|^2 + \lambda \sum_{k=1}^K S(a_k^{(m,l)})$
2. *subject to: the atoms are normalized* $\|\bar{\phi}_k\|^2 = 1, \forall k = 1, \ldots K;$
3. *the binary counterparts of each normalized atom* $1(\bar{\phi}_k^l)$ *are closures of the formal context* $\underset{\smile}{L^l};$
4. *the closures* $1(\bar{\phi}_k^l)$ *of the formal context* $\underset{\smile}{L^l}$ *are lectically ordered;*
5. *the formal contexts (layers) are ordered* $\underset{\smile}{L^1} \leq \underset{\smile}{L^2} \ldots \leq \underset{\smile}{L^l} \leq \ldots;$

*and the constraints and properties specified by the generative model are satisfied.*

## 3   MFCA via Image Quantization: Binary Layering Tree

MFCA-SSLPOD uses entropy-increasing, binary tree layering image quantization (Alg. 1). The ordering of the quantization and linear transform is swapped compared to the traditional DCT approach for JPEG compression [10]. We call these quantizers the outer $\mathcal{Q}_o(\cdot)$ and inner $\mathcal{Q}_i(\cdot)$ quantizers respectively.

$$\text{DCT: } a = \mathcal{Q}_o(\boldsymbol{\Phi}^{-1}\boldsymbol{x}), \quad \text{vs. MFCA-SSLPOD: } a^l = \left(\boldsymbol{\Phi}^l\right)^* \mathcal{Q}_i(\boldsymbol{x}) \forall l. \tag{11}$$

$(\cdot)^*$ denotes the appropriate inverse. The inner quantizer is now described. Alg. 1 first computes a linear fine-grained histogram $\boldsymbol{h} \in \mathbb{Z}^{\beta \times 1}$ with $\beta$ bins of the image pixel intensities, $\boldsymbol{I}$, it then computes a coarse histogram with $L$ bins (with non-uniform decision levels) of the histogram $\boldsymbol{h}$. The number of elements in each bin of the coarse histogram is more equal than for the fine-grained histogram $\boldsymbol{h}$ –the entropy of the histogram is greater; however no quantization error is introduced.

## 3.1   Binary Layering Tree (BLT) Parametrization

The parametrization of the BLT is explained as follows. The constants $V_{\min}$ and $V_{\max}$ are typically the minimum and maximum image pixel intensity, $\min\{\boldsymbol{I}\}$ and $\max\{\boldsymbol{I}\}$ respectively. The constant $\beta$ is the number of bins used to create the fine-grained histogram, $\boldsymbol{h}=$ hist$((\boldsymbol{I}), V_{\min}, V_{\max}, \beta)$, $\beta$ is chosen to be the resolution of the pixel intensities supported by the image file format. Each pixel in the cameraman is an integer value between 0 and 255. Image layers are generated using the sequential binary range partitioning algorithm in Alg. 1. The number of layers $L$ can be expressed as a base-2 number with an integer exponent, $L = 16$. The quantization ranges, $\boldsymbol{b}$, list the indices of the fine-grained histogram $\boldsymbol{h}$ where layer bounds should start and end.

## 3.2   MultiLayer Formal Concept Analysis: MFCA

We present a method for constructing a dictionary using FCA for each layer generated by Alg. 1. MFCA is defined in Alg. 2 and described below in detail. First, we compute the binary mask for layer $l$ using Alg. 1:

$$[\boldsymbol{M}^l, \boldsymbol{L}^l, \boldsymbol{b}(l)] = \mathcal{Q}_i(\boldsymbol{I}, \log_2 L, V_{\min}, V_{\max}, \beta, l), \quad \text{where } \boldsymbol{I} \in \mathbb{R}^{R \times G}. \tag{12}$$

We apply overloaded blocking to $\boldsymbol{M}^l \in \mathbb{Z}^{R \times G}$ which yields, $\overleftarrow{\boldsymbol{M}^l} \in \mathbb{Z}^{N_b^2 \times M}$. Let $O$ and $P$ denote a finite set of $N_b^2$ rows and $M$ column labels respectively for the matrix $\overleftarrow{\boldsymbol{M}^l}$, the blocked binary mask computed for layer $l$. The set of labels $P = \{1 \ldots m \ldots M\}$ denotes the block index, and the set of labels $O = \{1 \ldots n \ldots N_b^2\}$ denotes the elements of each block. The value one is entered in a row-column position to denote that pixel is in that layer; block index $m$ has that pixel; and finally, the pixel is present in the block; a zero entry denotes that the pixel is not present, etc. Therefore, the matrix $\overleftarrow{\boldsymbol{M}^l}$ describes the binary relation between the label sets $O$ and $P$ (we use binary matrix and binary relation subset notation interchangeably for convenience even though this is not strictly correct.); FCA looks to learn structures present in the blocks. We say the row label set $X$ is associated with the column label set $Y$ if $(X, Y) \in \overleftarrow{\boldsymbol{M}^l}$, $X \in O$ and $Y \in P$. The

---

**Algorithm 1.** BLT: $[\boldsymbol{M}^l, \boldsymbol{L}^l, \boldsymbol{b}(l)] = \mathcal{Q}_i(\boldsymbol{I}, \log_2 L, V_{\min}, V_{\max}, \beta, l)$

---

**Input:** $\log_2 L, \boldsymbol{I}, V_{\min}, V_{\max}, \beta, l$, index of the layer required

**Output:** $\boldsymbol{b}(l), \boldsymbol{L}^l(r, g), \boldsymbol{M}^l(r, g)$.

1. Initialization with fine-scaled histogram: $\boldsymbol{h}=$ hist($\text{vec}\{\boldsymbol{I}\}, V_{\min}, V_{\max}, \beta$).
2. Compute $\hat{\boldsymbol{h}}(c) = \sum_{i=1}^{c} h_i, \quad \forall c = 1, \ldots \beta$. Initialize the partition bounds $\boldsymbol{b} = []$.
3. **for** $r = 1 : \log_2 L$ **do**
4.    $\boldsymbol{t} = []$;
5.    **if** ranges==1 **then**
6.       $\boldsymbol{b} = \frac{1}{2}\sum_i h_i$; {Start with the mid-point.}
7.       $i_m = \min(\text{find}(\hat{\boldsymbol{h}} > \boldsymbol{b}))$; {Find transition index.}
8.       $\boldsymbol{b} = [1, i_m, \beta]$; {First set of quantization bounds.}
9.    **else**
10.       **for** $p = 1 : \text{length}(\boldsymbol{b}) - 1$ **do**
11.          Compute valid histogram range and store in $\boldsymbol{v}$: $v_j = h_i$ if $\boldsymbol{b}(p) \leq h_i < \boldsymbol{b}(p+1), \forall i,$; compute $\hat{\boldsymbol{v}}(c) = \sum_{j=1}^{c} v_j, \quad \forall c$.
12.          $m = \frac{1}{2}\sum_j v_j$;
13.          Find Transition Index: $i_m = \boldsymbol{b}(p) + \min(\text{find}(v_c > m))$;
14.          $\boldsymbol{t} = [\boldsymbol{t}, i_m]$;
15.       **end for**
16.    **end if**
17.    $\boldsymbol{b} = [\boldsymbol{b}, \boldsymbol{t}]$; Sort in ascending order $\boldsymbol{b} = sort(\boldsymbol{b})$;
18. **end for**
19. $\boldsymbol{M}^l(r, g) = \begin{cases} 1 & \text{if } \boldsymbol{b}(l) \leq \boldsymbol{I}(r, g) < \boldsymbol{b}(l+1) \\ 0 & \text{otherwise.} \end{cases}, \quad \boldsymbol{L}^l(r, g) = \boldsymbol{M}^l(r, g)\boldsymbol{I}(r, g).$

---

triple $(O, P, \boldsymbol{M}^l)$ is called a formal context of the image layer $l$. Derivation on $X$ and $Y$ where $X \subseteq O, Y \subseteq P$ is defined as

$$X' = \{n \in P \mid \forall m \in O : (n, m) \in \boldsymbol{M}^l\}, \quad Y' = \{m \in O \mid \forall n \in P : (n, m) \in \boldsymbol{M}^l\}. \quad (13)$$

Therefore $X'$ generates the set of columns which are shared by all rows in $X$. Similarly, $Y'$ generates the set of all rows which are common to all columns in $Y$.

A pair $\langle X, Y \rangle$ is called a FC of $(O, P, \boldsymbol{M}^l)$ if and only if $X \subseteq O, Y \subseteq P, X' = Y$, and $Y' = X$. Given a FC, $\langle X, Y \rangle$, $X$ and $Y$ are called its extent and intent. The crucial property of a FC is that the mappings $X \subseteq X''$ and $Y \subseteq Y''$, hereafter known as *closures*, hold. The closure operator can be used to calculate the extent and intent that form a FC; building blocks of the formal context are revealed, by applying the closure mechanism methodically. Establishing a sub/super-concept hierarchy allows for thorough, systematic FCA [8]. Given $X_1, X_2 \subseteq O$ and $Y_1, Y_2 \subseteq P$ the concepts of a context are ordered as follows:

$$\langle X_1, Y_1 \rangle \leqslant \langle X_2, Y_2 \rangle :\Longleftrightarrow X_1 \subseteq X_2 \Longleftrightarrow Y_2 \subseteq Y_1 \quad (14)$$

an ordering which is interesting, because it facilitates the iterative formation of a complete lattice which is called the concept lattice of the context [8]. To define the dictionary generation process, we appeal to the *rank-1* property in [15].

**Algorithm 2.** Multi-layer FCA (MFCA)

**Input:** $I, L, N_b$: Image, number of layers, and block size.
**Output:** $\Phi^l, \forall l$. Ordered dictionaries computed from the closures of each layer
1. **for** l=1,...L **do**
2.    Quanitize $I$: $[M^l, L^l, b(l)] = Q_i(I, \log_2 L, V_{\min}, V_{\max}, \beta, l)$.
3.    Apply overloaded blocking to each layer's mask: $M^l = \text{blk}\{M^l, N_b\}$.
4.    Compute all closures in layer: $\{E\} = \text{AllClosure}(M^l, O, P)$.
5.    Generate dictionary for layer: $\Phi^l = \Phi^l = [\text{full}(E_1), \text{full}(E_2), \ldots, \text{full}(E_{\dot{K}})]$
6. **end for**

**Property 4** *Closures are rank-1 approximations of the formal context and their extents may be concatenated to form ordered dictionaries. If $X \subseteq O$, $Y \subseteq P$, $X' = Y$, and $Y' = X$, we construct vectors by defining the function $\phi_k^l = full(X)$*

$$\phi_k^l(n) = \begin{cases} 1, & if\ n \in X \\ 0, & if\ n \notin X, \end{cases}, \quad y(m) = \begin{cases} 1, & if\ m \in Y \\ 0, & if\ t \notin Y, \end{cases}, \quad then,\ \text{rank}\, y(\phi_k^l)^T = 1. \quad (15)$$

The Nextclosure function defined in [15] generates the set of all extents for $M^l$ using RRCFCA: $E = \text{AllClosure}(M^l, O, P)$.

**Definition 5.** *We may construct an ordered dictionary from the set of extents, E of each ordered closure computed from the binary mask of layer l, $M^l$.*

$$\Phi^l = FullDict(E) = [\phi_1^l, \phi_2^l, \ldots, \phi_k^l, \ldots \phi_K^l] = [full(E_1), full(E_2), \ldots, full(E_K)] \quad (16)$$

We desire that closures are generated iteratively using lectic ordering which is defined *ab initio* by the blocking of images. NextClosure, and its parallel variant RRFCA, generates closures once: a complete exposition of NextClosure is given in [8,15]. Note that dictionaries may not be overcomplete if the layer is very sparse; this issue is addressed in the following section.

## 4    Sequentially Sifted Linear Programming

We introduce a Sequentially Sifted Linear Programming encoder to solve the lossless form of Problem 1, where $S(\cdot)$ is the $\ell_1$-norm and the error term is zero. Recall, the atoms learned by MFCA are ordered lectically; they are generated by NextClosure. We posit that successive atoms are highly correlated. This idea is illustrated in Fig. 3: MFCA atoms $\phi_k^k$ are ordered in this toy exemplar by degree of elevation, e.g. $0°$ increasing to $90°$.

MFCA generates large dictionaries which poses the problem of solving a large Linear Program. We present SSLP of ordered dictionaries as a good but suboptimal solution. SSLP runs successive LP on ranges of atoms (smaller problems), and uses these intermediate solutions to select $\alpha$ atoms of interest via a function called $\hat{\Phi}^l = \text{top}(a_{opt1}^l, \ldots, a_{opts}^l, \ldots, \hat{\Phi}_1^l, \ldots, \hat{\Phi}_s^l, \ldots, \alpha)$. We then concatenate atoms of interest into a super-dictionary $\hat{\boldsymbol{\Phi}}^l$ and encode the images by solving a smaller LP where the constraints are expressed in terms of the super-dictionary. This process is illustrated in Fig. 3 for the case where we have 8 atoms. To define the SSLP algorithm we consider the LP primitive $a^l_{opt,s} = \text{L1opt}(\boldsymbol{x}, \boldsymbol{\Phi}^l_s)$, which is equivalent to solving the following problem where the data $\boldsymbol{x} = \text{vec } \boldsymbol{L}^l$ is vectorized, and the dictionary is expanded appropriately:

$$\min_a \mathbf{1}^T \boldsymbol{a}, \text{ such that } \boldsymbol{a} \geq 0, \text{ and } \boldsymbol{\Phi}^l_s \boldsymbol{a} = \boldsymbol{x}. \tag{17}$$

In what follows, each atom is normalized so that scaling does not affect the selected solution, $\|\boldsymbol{\phi}^l_k\| = 1$. For each range of atoms, we choose intervals of 1000 atoms and we generate the $s^{\text{th}}$ sifting dictionary positioned at $k$:

$$\hat{\boldsymbol{\Phi}}^{l,k}_s = [\boldsymbol{\phi}_k, \dots, \boldsymbol{\phi}_{k+1000} | \boldsymbol{I}_{N^2_b}], \tag{18}$$

where $\boldsymbol{I}_{N^2_b}$ is the canonical basis. This dictionary is called a sifted dictionary as each local dictionary used may not span the positive orthant, and thus the linear program may not have a feasible solution. If the atoms in $\hat{\boldsymbol{\Phi}}^l_s$ do not provide a sparse solution the coefficient energy is captured by the canonical basis. The sifted dictionary behaves in a similar manner to the Dirac delta function in Signal Processing, which serves to sift out components of a signal. The top($\cdot$) function discards atoms based on the indices of the coefficient energy. If the canonical basis coefficients have the $\alpha$ largest coefficients, the atoms may be discarded as they do not represent the image more compactly than the identity.

SSLP may be applied hierarchically in a tree like formation –the size of the super-dictionary depends on $\alpha$. In this paper, SSLP is applied to generate a dictionary for each layer using the 2-step process illustrated in Fig. 3. Alternatively, we can run an LP when 1000 atoms have been generated for a layer, by FCA, and then discard the results so that storage usage during runtime is minimized. The $\ell_1$ penalty encourages the solver to only choose a few good atoms ensure than many atoms are removed and memory is preseved. Note, the SSLP solution is sub-optimal, but lectic ordering ensures that good local minima are found due the local correlation in lectically sorted atoms.

## 5   Empirical Evaluation

We evaluate the MFCA-SSLPOD transform coding and encoding scheme. Subsequent quantization and entropy coding of coefficients is out of the scope. We describe the indicative performance of the MFCA-SSLPOD transform coding and encoding method using $512 \times 512$ pixel images from the USC-SIPI Image Database: Cameraman, Peppers and Barbara. The block-size is $N_b = 8$ and the layer size is $L = 16$. These images are standard test images used for compression.

**BLT Quantization:** Fig. 4 illustrates the fine-grained and rough histograms generated by the BLT layer construction algorithm for the Cameraman. The LHS

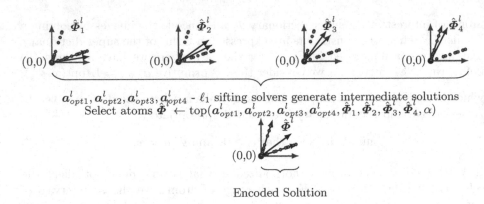

$a^l_{opt1}, a^l_{opt2}, a^l_{opt3}, a^l_{opt4}$ - $\ell_1$ sifting solvers generate intermediate solutions
Select atoms $\hat{\boldsymbol{\Phi}}^l \leftarrow \text{top}(a^l_{opt1}, a^l_{opt2}, a^l_{opt3}, a^l_{opt4}, \hat{\boldsymbol{\Phi}}^l_1, \hat{\boldsymbol{\Phi}}^l_2, \hat{\boldsymbol{\Phi}}^l_3, \hat{\boldsymbol{\Phi}}^l_4, \alpha)$

Encoded Solution

**Fig. 3.** Illustration of SSLP: An FCA-like routine generates 8 atoms in layer $l$, $[\phi^l_1, \ldots, \phi^l_8]$. Sifting dictionaries are generated that contain 2 of these atoms each: $\hat{\boldsymbol{\Phi}}_s$ for $s = 1, 2, 3, 4$, for example $\hat{\boldsymbol{\Phi}}^l_1 = [e_1, e_2, \phi^l_1, \phi^l_2]$. These atoms are plotted in black in the first row of coordinate systems. The Canonical atoms, $e_1, e_2$ are plotted as blue arrows. The test data points $L^l$ are plotted as red dots in each of the coordinate systems. We solve the LP associated with each system in row one and identify the top $\alpha$ coefficients (by magnitude). When $\alpha = 1$ atoms $\phi^l_2, \phi^l_4, \phi^l_8$ are retained in the super-dictionary. The MFCA transform coefficients are computed solving a Linear Program using the super-dictionary, $\hat{\boldsymbol{\Phi}}^l = [e_1, e_2, \phi^l_2, \phi^l_4, \phi^l_8]$ as linear constraints.

is the fine-grained histogram generated from the Cameraman $\boldsymbol{I}$. The numbers of pixels present in each layer, once quantization has been performed, is plotted on the RHS. There is a significant reduction in the dynamic range of the RHS compared to the LHS. The black vertical lines on the LHS denote the computed decision boundaries (histogram bin positions). BLT provides no guarantees on the number of closures (complexity) that are mined from each layer. For example, Fig. 7 lists the number of closures produced by the layers of the Cameraman. In many cases the effective size of $M^l$ is less than $N^2_b \times M$ as layering produces zero rows and columns (the empty extent is of no interest).

**Atoms learned:** The atoms learned by MFCA are closures and are more informative than the DCT atoms; they are adapted to the structures in each of the layers. Fig. 5 illustrates 5 atoms, which are chosen based on their expressiveness and stacked, one column per layer ($l = 1, \ldots, 16$ left to right), for each of the 16 layers $M^l$ of the Cameraman. Comparison with the original cameraman and layer binary masks in Fig. 2 demonstrates that, for example, triangular parts of the cameraman's coat are captured by some of the closures.

**Image Compression:** The number of closures generated by MFCA for the dictionary $\boldsymbol{\Phi}^l$ of each layer is listed for the first three layers in Fig. 7. These numbers are indicative for all layers. We terminate NextClosure when 9999000 closures have been generated. We posit that the large number of layers that generate greater than 9999000 closures is caused by the block-size $N_b$ and the

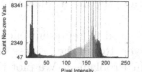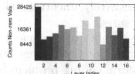

**Fig. 4.** Cameraman layer histogram: The count of the number of active pixels in each layer's range is non-uniform. We form 16 layers by attempting to increase the entropy of this *histogram*, by smoothing it out.

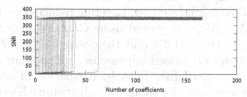

**Fig. 5.** Cameraman Atoms: Each column illustrates five of the parts learned for one of the 16-layers generated for the cameraman. The leftmost column illustrates parts, which constitute the cameraman's coat. Rows $1, 4$ and $5$ of column $1$ contribute triangular parts that represent the outline of the coat.

**Fig. 6.** Rate Distortion of MFCA & SSLPOD: The SNR of each encoding is plotted against the number of sorted coefficients used to generate the approximation. For $\approx$ 4000 blocks (the $512 \times 512$ cameraman image), typically less than 10 coefficients are used per block to achieve a SNR which is greater than 20dB.

| (1,4999000) | (2,37963) | (3,815032) | (4,4999000) |
|---|---|---|---|
| (5,4999000) | (6,4999000) | (7,4999000) | (8,2329900) |
| (9,4999000) | (10,4999000) | (11,4999000) | (12,124886) |
| (13,4999000) | (14,1153000) | (15,4999000) | (16,4999000) |

| Layer | 1 | 2 | 3 |
|---|---|---|---|
| Cameraman | 1999000 | 37963 | 815032 |
| Barbara | 9999000 | 9999000 | 593000 |
| Peppers | 9999000 | 9999000 | 9999000 |

**Fig. 7.** Number of concepts computed per layer for the Cameraman (LHS table) in (layer,number of concepts) form. The RHS compares the number of concepts for the first three layers for the Cameraman, Barbara and Peppers.

image content. If fewer image blocks were used in multiple MFCA routines, fewer closures would be learned. In general the larger the block and image, the larger the number of potential closures. In future work we will partition the image block matrix for each layer and run NextClosure on each partition to reduce the size of the dictionaries. We posit that partitioning may lead to fewer closures being generated. In § 4 we introduced *SSLP* as a method for generating an small ordered dictionary from the complete MFCA ordered dictionary. The super-dictionaries generated from each layer have on the order of 100 atoms. The maximum number of atoms in the super-dictionaries in this paper is 166 atoms.

These atoms are relatively uncorrelated as they are a selected subset from the complete set of closures.

Fig. 6 plots the number of coefficients used to represent each of the image's blocks in the super-dictionary for that layer. Recall the lossless and lossy formulation of the transform coding problem in Problem. 1. We plot the SNR for the encoding for each block as the number of sorted coefficients is increased from 1 to $K$. An SNR of 20dB is good-quality (approximately lossless) compression. Fig. 6 illustrates that most blocks have an SNR which is greater than 20dB when they are represented by less than 10 coefficients. This illustrates that the linear transform compactly represents the image blocks –compression is achieveable. Outer quantization may now be applied to these coefficients, along with entropy encoding, to generate the compressed image in its traditional format. In future work we will consider the role of $\alpha$, which sets the number of maintained atoms in the superdictionary. The compression rates in Fig. 6 are adversely affected by the need for the identity basis to fill-in some troublesome image blocks. We posit that this is because too few atoms were maintained in the super-dictionaries.

**Complexity Reduction and Information Content:** BLT is motivated by the fact that FCA's computational complexity, and thus its run-time, is a function of the density of the formal context. Layering reduces the density of the data passed to FCA. FCA may be computed using Ganter's algorithm [22], Lindig's algorithm [23], CloseByOne [24,25] and their variants [26,27]. The theoretical and empirical complexity of various approaches, which is an important consideration for MFCA, is compared for FCA by Kuznetsov in [28]. Computational complexity is the main measure for comparing algorithms: Kuznetsov and Obiedkov focus on the properties of the data ensemble, namely sparsity, the primary complexity-inducing characteristic of the decomposition. As the number of layers $L$ increases the complexity (sparsity) of each FCA reduces; the number of FCs in each layer reduces, and thus the expressive power of MFCA decreases. The choice $L = 16$ is motivated by the resolution of the pixel intensities in our experiments -8 bits per pixel - increasing $L$ should reduce MFCA complexity.

Aside from sparsity, FCA's main bottlenecks are memory and processing constraints. Ganter's algorithm computes concepts iteratively based on the previous concept, without incurring exponential memory requirements, by exploiting lectic ordering. In some preliminary work [14,15] we introduced RRFCA, which exploits 1) the fact that rank-1 approximations are closures, and 2) the lectic ordering of a set of representative closures to sub-divide FCA into a set of parallel mining tasks (that invoke negligible communication costs). We used RRFCA to speed-up the mining process in this paper. Alternative methods include: CloseByOne produces many concepts in each iteration; Bordat's algorithm, described in [29], introduces a data structure to store previously found concepts, which results in considerable time-savings. This approach is made more efficient in [30] by removing the need for a structure of exponential size.

*Remark:* Incremental approaches for FCA have been made popular by Norris in [31], Dowling in [32], Godin et al. in [33], Carpineto and Romano in [34], Valtchev et al. in [35] and Yu et al. in [36] the authors update the lattice structure when a new object is added to the database. Note that these methods may have a role to play in online adaptation of overcomplete dictionaries (in video compression). To address the memory and computation challenge, we considered using rank reduction method and *disjointness* to select good starting-intents for FCA in [14]. Other approaches that may make MFCA more computationally practical for image compression include: Krajca et al. proposed a parallel version based on CloseByOne in [27]; the first distributed algorithm [37] was developed by Krajca and Vychodil in 2009 using the Map-Reduce framework [38]; and finally, the authors of [39] proposed an efficient, distributed FCA implementation using the Twister Map-Reduce framework [40].

*Conclusions and Future Directions:* We introduced a first FCA-based linear transform that compactly encodes images by converting an 8-bit representation into a number of 1-bit layers and learning the closures for each layer. In future work it would be interesting to assess the effect on dictionary size of taking L-way factorizations of the multilayer representation using a technique similar to [41], which exploits latent components across the layers of MFCA. In addition, we will consider the effect of the parameters $N_b, L, \alpha$ on the compactedness of the derived coefficients and complexity.

**Acknowledgments.** This work was supported by grant 08/SRC/I1403 FAME SRC.

# References

1. Chen, S.S., Donoho, D., Saunders, M.A.: Atomic decomposition by basis pursuit. SIAM J. Scientific Computing 20, 33–61 (1998)
2. Mallat, S.: A Wavelet Tour of Signal Processing, The Sparse Way, 3rd edn. Academic Press (2008)
3. Lewicki, M.S., Sejnowski, T.J., Hughes, H.: Learning overcomplete representations. Neural Computation 12, 337–365 (1998)
4. Lewicki, M.S., Olshausen, B.A.: A probabilistic framework for the adaptation and comparison of image codes. J. Opt. Soc. Am. A 16, 1587–1601 (1999)
5. Olshausen, B.A., Field, D.J.: Emergence of simple-cell receptive field properties by learning a sparse code for natural images. Nature 381(6583), 607–609 (1996)
6. Olshausen, B.O., Fieldt, D.J.: Sparse coding with an overcomplete basis set: a strategy employed by V1. Vision Research 37, 3311–3325 (1997)
7. Rubinstein, R., Bruckstein, A.M., Elad, M.: Dictionaries for Sparse Representation Modeling. Proceedings of the IEEE 98(6), 1045–1057 (2010)
8. Ganter, B., Wille, R.: Formal Concept Analysis: Mathematical Foundations. Springer, Heidelberg (1999)
9. Horev, I., Bryt, O., Rubinstein, R.: Adaptive image compression using sparse dictionaries. In: 19th Int. Conf. Sys., Sig. and Im. Process., pp. 592–595 (2012)
10. Pennebacker, W.B., Mitchell, J.L.: JPEG still image data compression standard. Springer, New York (1993)

11. Taubman, D.S., Marcellin, M.: JPEG2000: image compression fundamentals, standards and practice. Kluwer Academic Publishers, Norwell (2001)
12. Ahmed, N., Natarajan, T., Rao, K.R.: Discrete Cosine Transform. IEEE Trans. Computers C-32, 90–93 (1974)
13. Jolliffe, I.T.: Principal Component Analysis, 2nd edn. Springer Series in Statistics (October 2002)
14. de Fréin, R.: Ghostbusters: A Parts-based NMF Algorithm. In: 24th IET Irish Signals and Systems Conference, pp. 1–8 (June 2013)
15. de Fréin, R.: Formal concept analysis via atomic priming. In: Cellier, P., Distel, F., Ganter, B. (eds.) ICFCA 2013. LNCS, vol. 7880, pp. 92–108. Springer, Heidelberg (2013)
16. Bryt, O., Elad, M.: Compression of facial images using the K-SVD algorithm. Journal of Visual Communication and Image Representation 19(4), 270–283 (2008)
17. Howard, P.G., Vitter, J.S.: Arithmetic coding for data compression. Proceedings of the IEEE 82(6) (June 1994)
18. de Fréin, R., Rickard, S.T.: The synchronized short-time-Fourier-transform: Properties and definitions for multichannel source separation. IEEE Trans. Sig. Proc. 59(1), 91–103 (2011)
19. Lee, D.D., Seung, H.S.: Algorithms for non-negative matrix factorization. In: NIPS, pp. 556–562. MIT Press (2000)
20. Elad, M.: Sparse and redundant representations - from theory to applications in signal and image processing. Springer (2010)
21. Mallat, S., Zhang, Z.: Matching pursuits with time-frequency dictionaries. IEEE Trans. Sig. Proc. 41(12) (1993)
22. Ganter, B.: Two Basic Algorithms in Concept Analysis. In: Kwuida, L., Sertkaya, B. (eds.) ICFCA 2010. LNCS, vol. 5986, pp. 312–340. Springer, Heidelberg (2010)
23. Lindig, C.: Fast Concept Analysis. Working with Conceptual Structures-Contributions to ICCS, pp. 235–248 (2000)
24. Kuznetsov, S.O.: A Fast Algorithm for Computing All Intersections of Objects in a Finite Semi-Lattice. Auto. Doc. and Math. Linguistics 27(5), 11–21 (1993)
25. Andrews, S.: In-Close, a Fast Algorithm for Computing Formal Concepts. In: The Seventeenth International Conference on Conceptual Structures (2009)
26. Vychodil, V.: A New Algorithm for Computing Formal Concepts. In: Cybernetics and Systems, pp. 15–21 (2008)
27. Krajca, P., Outrata, J., Vychodil, V.: Parallel Recursive Algorithm for FCA. In: CLA 2008, vol. 433, pp. 71–82 (2008)
28. Kuznetsov, S.O., Obiedkov, S.A.: Comparing Performance of Algorithms for Generating Concept Lattices. J. Exper. & Th. Artif. Intell. 14, 189–216 (2002)
29. Bordat, J.-P.: Calcul pratique du treillis de Galois d'une correspondance. Mathématiques et Sciences Humaines 96, 31–47 (1986)
30. Berry, A., Bordat, J.-P., Sigayret, A.: A Local Approach to Concept Generation. Annals of Mathematics and Artificial Intelligence 49(1), 117–136 (2006)
31. Norris, E.M.: An Algorithm for Computing the Maximal Rectangles in a Binary Relation. Rev. Roum. Math. Pures et Appl. 23(2), 243–250 (1978)
32. Dowling, C.E.: On the Irredundant Generation of Knowledge Spaces. J. Math. Psychol. 37, 49–62 (1993)
33. Godin, R., Missaoui, R., Alaoui, H.: Incremental Concept Formation Algorithms Based on Galois (Concept) Lattices. Computational Intelligence 11, 246–267 (1995)
34. Carpineto, C., Romano, G.: A Lattice Conceptual Clustering System and Its Application to Browsing Retrieval. Machine Learning, 95–122 (1996)

35. Valtchev, P., Missaoui, R., Lebrun, P.: A Partition-based Approach Towards Constructing Galois (concept) Lattices. Discrete Math., 801–829 (2002)
36. Yu, Y., Qian, X., Zhong, F., Li, X.R.: An Improved Incremental Algorithm for Constructing Concept Lattices. In: Soft. Eng., World Congress, vol. 4, pp. 401–405 (2009)
37. Krajca, P., Vychodil, V.: Distributed Algorithm for Computing Formal Concepts Using Map-Reduce Framework. In: Adams, N.M., Robardet, C., Siebes, A., Boulicaut, J.-F. (eds.) IDA 2009. LNCS, vol. 5772, pp. 333–344. Springer, Heidelberg (2009)
38. Dean, J., Ghemawat, S.: MapReduce: Simplified Data Processing on Large Clusters. In: OSDI, p. 13 (2004)
39. Xu, B., de Fréin, R., Robson, E., Ó Foghlú, M.: Distributed Formal Concept Analysis Algorithms Based on an Iterative MapReduce Framework. In: Domenach, F., Ignatov, D.I., Poelmans, J. (eds.) ICFCA 2012. LNCS, vol. 7278, pp. 292–308. Springer, Heidelberg (2012)
40. Ekanayake, J., Li, H., Zhang, B., Gunarathne, T., Bae, S.H., Qiu, J., Fox, G.: Twister: a Runtime for Iterative MapReduce. In: HPDC 2010, pp. 810–818. ACM (2010)
41. Belohlavek, R., Glodeanu, C., Vychodil, V.: Optimal factorization of three-way binary data using triadic concepts. Order 30(2), 437–454 (2013)

# Attribute Exploration
# with Proper Premises and Incomplete Knowledge
# Applied to the Free Radical Theory of Ageing

Johannes Wollbold[1,*], Rüdiger Köhling[2], and Daniel Borchmann[3,**]

[1] University of Rostock, Germany
johannes.wollbold@uni-rostock.de
[2] University Medicine, Rostock, Germany
[3] Technische Universität Dresden, Germany
borch@tcs.inf.tu-dresden.de

**Abstract.** The classical free radical theory of ageing assumes that oxidative damage by reactive oxygen species (ROS) accumulates with age in a self-enhancing process. The theory has been confirmed by many experiments in various species. However, it is seriously challenged since several years. In this ambiguous situation, we collected and ordered existing knowledge, with a focus on the integration of conflicting findings.

We developed a specific method of knowledge base construction and give a first example of its application. Data reported in literature or generated by our experimental partners is formalized as Ripple Down Rules (RDR), a structure of general rules and exceptions. This rule set is validated and completed by the attribute exploration algorithm: Several, most specific RDR are accepted as background implications for an exploration starting from the examples collected during the RDR knowledge base growth.

The RDR classify biological cases, which are defined by attributes like organism, cell type or stimulation experiment. The classes are different and chosen according to leading questions. We focus on low/high ROS concentration in age and on lifespan. Implications with proper premises are suited for such disjoint basic sets of premises and conclusions. We implemented an easily understandable exploration algorithm in conexp-clj, furthermore an extension of this algorithm to incomplete counterexamples. The correctness and completeness of both algorithms is proven.

**Keywords:** attribute exploration, proper premises, incomplete knowledge, knowledge base, ripple down rule, free radical theory of ageing.

## 1    Introduction

The aim of the present work is to provide a specific methodology to structure and validate knowledge and data. It is applied to a central hypothesis of ageing research

\* Supported by the German Federal Ministry of Education and Research (BMBF), FKZ 0315892A (ROSAGE).
\*\* Supported by DFG Graduiertenkolleg 1763 (QuantLA).

C.V. Glodeanu, M. Kaytoue, and C. Sacarea (Eds.): ICFCA 2014, LNAI 8478, pp. 268–283, 2014.

related to free radicals or – more generally – *reactive oxygen species (ROS)*.[1] ROS are mainly produced in the mitochondria of an eucaryotic, e.g. animal cell. Proteins of the inner mitochondrial membrane build an electrical circuit (*electron transport chain, ETC*) pumping protons out of the inner part of the mitochondrion. Their reflux drives the motion of a protein complex producing adenosine triphosphate (ATP), one of the most important energy storing molecules. ROS have a high potential to oxidize other molecules. Therefore, they can damage proteins or the DNA strain of the mitochondria coding mainly for the proteins of the ETC. [1]

Now, the free radical theory of ageing (FRTA) [11] developed since the 1950s assumes a positive feedback with negative consequences. Damaged proteins of the ETC are supposed to produce more ROS, oxidative damage accumulates with age and progressively disturbs cell and organ functions. Whereas the theory has been confirmed by manifold experiments in various species, also exceptions were found, and every part of the sketched argumentative chain is challenged since several years. Moreover, important physiological roles of ROS are now better understood. As signaling molecules they trigger, for instance, an immune response or apoptosis (a controlled form of cell death). Hence, a more realistic picture emerges, and ROS do not represent the "axis of evil" any more, within the scientific story about social interactions of an organism.

In order to contribute to this dialectic process, we use a related logical structure of rules and exceptions, so-called *Ripple Down Rules* (RDR) [7]. Data reported in the literature or collected by us and our experimental partners is formalized as a set of RDR. This rule set and the corresponding examples are further validated and completed by the attribute exploration algorithm of formal concept analysis (FCA). Implications proposed by the algorithm are accepted or counterexamples are given, if necessary after supplementary literature or data queries. Thus, a rule base is defined systematically. From this rule base, all implications valid according to the available knowledge can be derived logically.

However, RDR are not implications, but more general clauses. In [8], attribute exploration has been generalized to the most general case of cumulated clauses. But then, three comfortable features of the rule base no longer hold without further assumptions: its minimality, uniqueness and the decidability in linear time, if a given implication follows from the rule base – indeed, this inference problem is $\mathcal{NP}$-complete [8, p. 10].

Therefore, in a first step we adopt here the following strategy: If the RDR knowledge base is sufficiently developed and approved, the examples corresponding to all RDR are assembled into a formal context, and the implicational logic of the underlying domain is explored. Secure RDR without exception are entered as background knowledge.

---

[1] Here, free radicals denote molecules containing oxygen with one unpaired electron (i.e. the electron at the same energy level, but with inverse spin is missing), e.g. $O_2^{-}$ (superoxide) or $OH^{\cdot}$ (hydroxyl radical). They belong to the larger class of reactive oxygen species (ROS) comprising also $H_2O_2$ (hydrogen peroxide), where oxygen is only partially reduced. Since different ROS can be rapidly converted to each other, the general term of free radical is often used for all kinds of ROS.

It is the purpose of this paper to develop this first step in detail. To this end, we shall introduce the necessary notions from FCA in the following section. Thereafter, we shall develop a modified version of attribute exploration which includes both *proper premises* and *incomplete counterexamples* in Section 3. Thereafter, we shall apply our algorithm to the FRTA in a small example in Section 4.

## 2    Methods: Mathematical and Logical Background

### 2.1    Ripple Down Rules

The RDR scheme [7, 14] is a formalism used for the progressive extension of a knowledge base. A new case – for instance occurring in medical diagnosis – is classified following one or more paths in a tree of if-then-rules defined as a pair of formula $(\alpha, \beta)$. The case is evaluated according to the most specific rule(s) that is/are applicable to the case (the rule *fires*). If an expert detects a wrong classification, the wrongly firing rule can be understood as a general rule to which an exceptional case has been detected. Then, a new child rule $(\alpha, \beta)$ with supplementary attributes of the premise is defined and stored in the knowledge base, together with the example (*cornerstone case*) that required the exception.

**Table 1.** Cornerstone cases

| | AntiOx1.+ | AntiOx1.− | AntiOx2.+ | AntiOx2.− | ROS.old.+ | ROS.old.− |
|---|---|---|---|---|---|---|
| **1.** | | | | | × | |
| **1.1** | × | | | | | × |
| **1.1.1** | × | | × | × | | |

Since RDR are only the motivation for the present work, we give a rather intuitive description starting from an example derived from the rules in Section 4. Consider the tree with a single branch (brackets for attribute sets are omitted, then ',' signifies $\wedge$):

$$\emptyset \longrightarrow \text{ROS.old.+} \tag{1}$$

$$\text{AntiOx1.+} \longrightarrow \text{ROS.old.−} \tag{1.1}$$

$$\text{AntiOx1.+, AntiOx2.−} \longrightarrow \text{ROS.old.+} \tag{1.1.1}$$

The root rule (1) represents one of the basic hypothesis of the FRTA, the accumulation of ROS during ageing. If the production of an antioxidant enzyme is increased permanently, for instance by a mutation, the cellular ROS concentration in age is reduced (1.1). Other experiments are in conflict with this rule, hence misclassified, and the exceptional rule (1.1.1) is added to the knowledge base: If the concentration of a second antioxidant directed against a different type of ROS is too low, the protective effect of AntiOx1 is minimal. With the rules, the cornerstone cases of Table 1 are stored.

As an RDR only holds if the conditions of its child rules are not fulfilled, they meet the following general definition.

**Definition 1.** *Let $\alpha_i, \alpha_j, \beta$, $i \in I$, $j \in J$ be propositional or first-order formula. Then a ripple down rule is a clause with the following structure:*

$$\bigwedge_{i \in I} \alpha_i \wedge \bigwedge_{j \in J} \neg \alpha_j \longrightarrow \beta.$$

For *single classification RDR (SCRDR)*, the tree of rules and exceptions is binary. We shall use *multiple classification RDR (MCRDR)*, which allow an arbitrary number of child rules. For further details see [14, Definitions 1, 3, 5 and 8].

## 2.2   Formal Concept Analysis

Our considerations rely on notions from the mathematical field of formal concept analysis [10], a subfield of mathematical order theory. We assume a certain familiarity of the reader with fundamental notions such as *formal contexts*, *derivation operators* noted by ′, and *implications*. Here and throughout the article, we shall introduce specific notions which are relevant for our purpose, in particular *proper premises* and formal contexts expressing incomplete knowledge. Basically, we also assume that the reader is familiar with *attribute exploration*, but we give an intuitive introduction with an example, the exploration of the formal context in Table 1.

**Attribute Exploration.** This algorithm supposes that we are interested in the implicational dependencies of a certain *domain*, of which we can think of as a collection of instances which may or may not have particular attributes. In other words, we assume that this domain is representable by a formal context $(G,M,I)$, and we are interested in the implicational theory of that context, and we shall call this formal context the *background context* $\mathbb{K}_{back}$ of the domain. We shall furthermore assume that the set of attributes of the background context is finite. The implicational theory is representable by a rule set from which all valid implications can be logically derived (*completeness*). Classical attribute exploration generates such a rule set, namely implications with *pseudo-intents*[2] as premises. This *stem base* is minimal and unique. It is computed in interaction with a domain expert or a computer program.

Starting, for instance, from the *initial working context* $\mathbb{K} \subseteq \mathbb{K}_{back}$ of Table 1, the algorithm first proposes the implication ROS.old.− ⟶ AntiOx1.+. It is accepted and added to the still empty stem base, since an elevated antioxidant level is a necessary condition for a decrease of ROS. However, the subsequent implication

$$AntiOx2.- \;\longrightarrow\; AntiOx1.+, \; ROS.old.+$$

is rejected. Obviously, not only the case 1.1.1 can occur, but the concentration of both antioxidants can be low. Therefore, a *counterexample* is added to the *current working context* as a new object (row) with the attributes AntiOx2.−, AntiOx1.− and ROS.old.+. Then, the algorithm "defends" the remaining part of the conclusion, and the expert complies with AntiOx2.− ⟶ ROS.old.+. With

---

[2] A pseudo-intent $\tilde{P} \subseteq M$ is defined recursively, starting from $\emptyset$: It is not closed, but contains the closure $Q''$ (regarding the current working context) of all pseudo-intents $Q \subsetneq \tilde{P}$. Attribute exploration computes sets $\tilde{P}$ which in addition are closed under the background knowledge. For these sets, $\tilde{P} \longrightarrow \tilde{P}''$ does not follow from the currently known implications, but is also not invalidated by the current working context.

termination of the algorithm, the stem base contains 7 implications, and the *final working context* one further counterexample.

Note that we have omitted a lot of details here, in particular the method on how to compute sets pseudo-intents as mentioned above. Moreover, in our algorithms and the extended example of Section 4.1, the set of already known implications has not to be empty in the beginning, but *background knowledge* $\mathcal{B}$ can be entered into the algorithm, expressing, e.g., the exclusion of ROS.old.+ and ROS.old.−. A more detailed overview can be found in [16]. For an extensive discussion of attribute exploration we refer to [9, 10].

We finish this section with two technical remarks. First, we assume the expert to *not make errors*, i. e. to give correct answers and counterexamples. If an expert is implemented in terms of an automatic system, then this assumption may be acceptable, however if the expert is a human expert, then the assumption is quite naive. There has been some research into this direction [2], but we shall mostly ignore those issues here due to space restrictions.

Second, we shall make use of a notational idiosyncrasy: If $\mathbb{K} = (G, M, I)$ is a formal context and $A \subseteq M$, then we occasionally shall denote the derivations $A'$ and $A''$ of $A$ in $\mathbb{K}$ by $A'_{\mathbb{K}}$ and $A''_{\mathbb{K}}$. This is due to the fact that we have to deal with situations where more than one formal context is present, and where we need to make clear in which formal context the derivations are conducted.

**Proper Premises.** For our application to the validation of an RDR knowledge base, it will be necessary to classify objects into certain classes $C \subseteq M$ which are described by attributes in $M \setminus C$. Therefore, we shall make use of implications with premise in $M \setminus C$ and conclusion in $C$. It shall turn out in Section 3 that *proper premises* are a helpful notion for this.

**Definition 2.** *For a given formal context* $(G, M, I)$ *and a set of attributes* $P \subseteq M$, *define* $P^{\bullet}$ *to be the set of those attributes in* $M \setminus P$ *that follow from* $P$ *but not from a strict subset of* $P$, *i. e.*

$$P^{\bullet} = P'' \setminus \left( P \cup \bigcup_{S \subsetneq P} S'' \right)$$

*P is called a* proper premise *if* $P^{\bullet}$ *is not empty. It is called a* proper premise for *m if* $m \in P^{\bullet}$.

The set of implications $\mathcal{L}_{\mathcal{P}} := \{P \longrightarrow P^{\bullet} \mid P^{\bullet} \neq \emptyset\}$ is a sound and complete implicational base for $\mathbb{K} := (G, M, I)$ [10]. By definition, a proper premise is $\subseteq$-minimal regarding the property of implying an attribute $m \in M$. Pseudo-intents $\tilde{P} \in M$, however, tend to be large, since they are closed by all implications but $\tilde{P} \longrightarrow \tilde{P}''$. For instance, the premise of

$$\text{AntiOx1.+, AntiOx1.−, AntiOx2.−, ROS.old.+} \rightarrow \text{AntiOx2.+, ROS.old.−} \quad (1)$$

contains the proper premise AntiOx1.+, AntiOx1.−. With $\bot := M$, the respective implication AntiOx1.+, AntiOx1.− $\longrightarrow \bot$ expresses that no cases exist where both AntiOx1.+ and AntiOx1.− occur together, because $M' = \emptyset$. The premise of the implication in (1) contains additional attributes because it is closed under the valid implication AntiOx1.$-$ $\longrightarrow$ AntiOx2.$-$, ROS.old.$+$.

On the other hand, the stem base is always minimal regarding the number of implications, whereas $\mathcal{L_P}$ is often much larger. In the next section, we shall mention a further difference. A more elaborated comparison of attribute exploration with pseudo-intents and proper premises can be found in [16].

# 3 Attribute Exploration for Implications with Proper Premises

In our application in Section 4 we want to know all implicational dependencies between attribute combinations from $M \setminus C$ and attributes from $C$. More precisely, we are interested in investigating the set

$$\mathrm{Th}_C(\mathbb{K}) := \{A \longrightarrow B \mid A \subseteq M \setminus C, B \subseteq C, \mathbb{K} \models (A \longrightarrow B)\},$$

i.e. we want to compute a *base* for this set, which is a subset of $\mathrm{Th}_C(\mathbb{K})$ which is complete for $\mathrm{Th}_C(\mathbb{K})$.

Because all implications imply elements in $C$ from elements in $M \setminus C$, every base of $\mathrm{Th}_C(\mathbb{K})$ will be *iteration-free* (or *direct*), i.e. logical derivations are achieved by unique application of implications of the base. It is known that, in contrast to the stem base, the set

$$\mathcal{L} := \{(P \longrightarrow \{m\}) \in \mathrm{Th}_C(\mathbb{K}) \mid P \text{ proper premise for } m \text{ in } \mathbb{K}\}$$

is such a base, and minimal as well [3]. We shall discuss two exploration algorithms which make use of proper premises, one using completely specified counterexamples, and the other one allowing for incompletely specified counterexamples. Both algorithms were implemented in `conexp-clj` [4].

## 3.1 Complete Counterexamples

To explore all proper premise implications allowing only completely specified counterexamples, we shall use a very simple approach: Given an initial working context $\mathbb{K}$, we will consider all proper premises $P$ of $\mathbb{K}$ for some attribute $m$ and ask whether the implication $P \longrightarrow \{m\}$ is true or not. If the expert confirms, we continue with the next proper premise until no more are left. If the expert rejects, we add a counterexample to $\mathbb{K}$ and start over. A pseudocode-listing of such an algorithm is given in Listing 1.1.

Since our manually curated contexts are rather small, it is not a performance problem to recompute all proper premises as soon as a counterexample is added. Thus, we could keep the algorithms simple, well understandable and could easily implement the distinction between basic sets $C$ and $M \setminus C$ for conclusions and premises. However, the computation can be optimized by using ideas from [16] based on results from the theory of hypergraphs, which allows for computing the proper premises of an updated context from the proper premises of the original one.

**Listing 1.1.** Attribute exploration using proper premises with disjoint basic sets for conclusions and premises

```
0   define algorithm-1(𝕂 = (G,M,I), C ⊊ M, B ⊆ Th_C(𝕂))
1     ℒ := B
2     forall m ∈ C do
3       𝒫 := {P ⊆ M\C | P is proper premise for m in 𝕂}
4       while there exists P ∈ 𝒫 with ℒ ⊭ (P ⟶ {m}) do
5         if expert confirms P ⟶ {m} then
6           ℒ := ℒ∪{P ⟶ {m}}
7         else
8           augment 𝕂 by valid counterexample from the expert
9           𝒫 := {P ⊆ M\C | P is proper premise for m in 𝕂}
10        end
11      end
12    end
13    return ℒ\B
14  end
```

**Theorem 3.** *With termination of* `algorithm-1`, $\mathcal{L} \setminus \mathcal{B}$ *is a sound and complete implicational base with background knowledge* $\mathcal{B}$ *for* $\mathrm{Th}_C(\mathbb{K}_{\mathrm{back}})$, *where* $\mathbb{K}_{\mathrm{back}} = (G_{\mathrm{back}}, M, I_{\mathrm{back}})$ *is the background context of the exploration. Furthermore, it is true that*

$$\mathcal{L} \setminus \mathcal{B} \subseteq \{P \longrightarrow \{m\} \mid m \in C, \ P \text{ is proper premise for } m \text{ in } \mathbb{K}, \ P \subseteq M \setminus C\}.$$

Note that parts of the proof are already contained in [16], but we shall repeat them here for the sake of completeness.

*Proof.* By construction, all implications contained in $\mathcal{L} \setminus \mathcal{B}$ have been confirmed by the expert, and all counterexamples contained in $\mathbb{K}$ are valid counterexamples provided by the expert. Therefore, these counterexamples do not invalidate implications confirmed by the expert, and hence all implications in $\mathcal{L} \setminus \mathcal{B}$ hold in $\mathbb{K}_{\mathrm{back}}$.

Suppose that $\mathcal{L} \setminus \mathcal{B}$ is not complete for $\mathbb{K}_{\mathrm{back}}$ with background knowledge $\mathcal{B}$. Since the set

$$\{P \longrightarrow \{m\} \mid m \in C, \ P \subseteq M \setminus C \text{ proper premise for } m \text{ in } \mathbb{K}_{\mathrm{back}}\}$$

is complete for $\mathrm{Th}_C(\mathbb{K}_{\mathrm{back}})$, there must exist for some attribute $m \in C$ a proper premise $P$ of $m$ in $\mathbb{K}_{\mathrm{back}}$ such that

$$\mathcal{L} \not\models (P \longrightarrow \{m\}). \tag{2}$$

Consider now in `algorithm-1` the iteration for the attribute $m$ just after it has finished the inner **while**-loop, and denote with $\bar{\mathcal{L}}$ the set of currently known implications. Then $\bar{\mathcal{L}} \subseteq \mathcal{L}$, and therefore $\bar{\mathcal{L}} \not\models (P \longrightarrow \{m\})$. On the other hand,

$P \longrightarrow \{m\}$ is valid in $\mathbb{K}_{\text{back}}$, so it is also valid in the current working context $\bar{\mathbb{K}}$. Therefore, there exists a proper premise $\bar{P} \subseteq P$ for $m$ in $\bar{\mathbb{K}}$. Since $\bar{\mathcal{L}} \not\models (P \longrightarrow \{m\})$, it is also true that $\bar{\mathcal{L}} \not\models (\bar{P} \longrightarrow \{m\})$, and thus the implication $\bar{P} \longrightarrow \{m\}$ must be asked to the expert, and we could not have reached the end of the **while**-loop, a contradiction.

It remains to show that all implications in $\mathcal{L} \setminus \mathcal{B}$ have proper premises of $\mathbb{K}_{\text{back}}$ as premises. By contradiction assume that this is not the case and let $(P \longrightarrow \{m\}) \in \mathcal{L} \setminus \mathcal{B}$ such that $P$ is not a proper premise for $m$ in $\mathbb{K}_{\text{back}}$. But this then means that there exists $Q \subsetneq P$ such that $Q \longrightarrow \{m\}$ is valid in $\mathbb{K}_{\text{back}}$. As the final working context of the exploration is a subcontext of $\mathbb{K}_{\text{back}}$, the implication $Q \longrightarrow \{m\}$ is also valid in the final working context. In particular, $P$ is not a proper premise of the final working context.

Therefore, to show the claim it is enough to show that all premises in $\mathcal{L} \setminus \mathcal{B}$ are proper premises of the final working context. To this end it is enough to show that for attributes $m \in C$ and proper premises $P$ of $m$ in some formal context $\mathbb{K}_1$ stay proper premises if we add counterexamples which do not invalidate $P \longrightarrow \{m\}$. However, this is easy to see: Suppose that $P$ would not be a proper premise for $m$ in the formal context $\mathbb{K}_2$ which originates from $\mathbb{K}_1$ by adding new objects. Since $P \longrightarrow \{m\}$ is still valid in $\mathbb{K}_2$, there must be a proper subset $S \subsetneq P$ such that $m \in S''_{\mathbb{K}_2}$. But then $m \in S''_{\mathbb{K}_1}$, and $P$ is not a proper premise for $m$ in $\mathbb{K}_1$, a contradiction. □

Obviously, instead of asking implications to the expert which are of the form $P \longrightarrow \{m\}$, one can equivalently ask implications like

$$P \longrightarrow P^{\bullet} \cap C,$$

since then $P$ is a proper premise for all attributes in $P^{\bullet} \cap C$. It may be easier to judge the dependency of all attributes from $P$ in one step. To this purpose, we can replace the check $\mathcal{L} \not\models (P \longrightarrow \{m\})$ by $\mathcal{L} \not\models (P \longrightarrow P^{\bullet} \cap C)$ in line 4 of `algorithm-1`.

## 3.2    Incomplete Counterexamples

For practical applications it may be too much to ask the expert to provide completely specified counterexamples. Indeed, it should be sufficient to just provide as much from the counterexamples as is necessary to refute a given implication.

There have been several attempts to include the possibility of allowing *incompletely specified counterexamples* into the classical attribute exploration with pseudo-intents. The first approach is made in [5] and implemented in ConImp [6] using three-valued Kleene logic. The newer works of [2, 9] introduce the notion of a *partial context*. However, while in [2] a partial context is defined as a suitable generalization of formal contexts, [9] uses two formal contexts, one of which constitutes all *certain* incidences and the other one all *possible* ones. We shall follow the approach of [9] for developing an exploration algorithm with proper premises that in addition to `algorithm-1` from Listing 1.1 allows for incompletely specified counterexamples. They are added to a context as in Table 2, i.e. only the premise attributes of the refuted implications have to be indicated as certain. Possible attributes must not include the conclusion $m \in M$.

Denote with $\mathbb{K}_{back} = (G_{back}, M, I_{back})$ the background context of our exploration. Then, we use two working contexts. The certain context $\mathbb{K}_+ = (G, M, I_+)$ has to satisfy $I_+ \subseteq I_{back} \cap (G \times M)$, i. e. every incidence present in $\mathbb{K}_+$ is correct. On the other hand, the possible context $\mathbb{K}_? = (G, M, I_?)$ satisfies $I_{back} \cap (G \times M) \subseteq I_?$, i. e. every possible incidence (and maybe more) are contained in $\mathbb{K}_?$. According to [9] then, we define a *partial context* as a pair $((G, M, I_+), (G, M, I_?))$ with $I_+ \subseteq I_? \subseteq G \times M$. Finally, we shall call a formal context $(G, M, I)$ a *realizer* of the partial context $((G, M, I_+), (G, M, I_?))$ if and only if $I_+ \subseteq I \subseteq I_?$.

Partial contexts can be thought of as a compact representation of all of its realizers, i. e. as a representation of a set of *possible* contexts, but where it is not known which context is the correct one. An implication $(A \longrightarrow B) \in \mathrm{Imp}(M)$ is thus *refuted* by a partial context $(\mathbb{K}_+, \mathbb{K}_?)$ if and only if it does not hold in *all* its realizers. A counterexample to an implication $A \longrightarrow B$ provided by the expert during the exploration process can then be thought of as consisting of two disjoint sets $N_+, N_- \subseteq M$ such that

$$A \subseteq N_+ \quad \text{and} \quad B \cap N_- \neq \emptyset.$$

The counterexample is then added to $\mathbb{K}_+$ and $\mathbb{K}_?$ as follows: in $\mathbb{K}_+$, a new object is added that has exactly the attributes from $N_+$; in $\mathbb{K}_?$, a new object is added that has exactly the attributes from $M \setminus N_-$. After this modification, $(\mathbb{K}_+, \mathbb{K}_?)$ will refute $A \longrightarrow B$.

Recall that the main aspect of proper premises is that these are $\supseteq$-minimal sets entailing some attribute $m$. To transfer this notion to incomplete counterexamples we can reason as follows: the partial $(\mathbb{K}_+ = (G, M, I_+), \mathbb{K}_? = (G, M, I_?))$ actually represents the set of all its realizers, and each realizer of $(\mathbb{K}_+, \mathbb{K}_?)$ is considered to be equally possible. If now $P \subseteq M$ and $m \in M$, we can say that $P$ *possibly entails* $m$ in $(\mathbb{K}_+, \mathbb{K}_?)$ if and only if there exists a realizer $\mathbb{K} = (G, M, I)$ of $(\mathbb{K}_+, \mathbb{K}_?)$ such that $m \in P''$. If we denote the derivation operator in $\mathbb{K}_+$ by $\cdot^+$ and in $\mathbb{K}_?$ by $\cdot^?$, then we can infer from $I_+ \subseteq I$ that $P'_{\mathbb{K}} \supseteq P^+$, and further from $I \subseteq I_?$ that $P''_{\mathbb{K}} \subseteq P^{+?}$. Therefore, $m \in P^{+?}$.

On the other hand, if $m \in P^{+?}$, then we can find a realizer $\mathbb{K}$ of $(\mathbb{K}_+, \mathbb{K}_?)$ such that $m \in P''_{\mathbb{K}}$ as follows: for all objects $g \in P'_{\mathbb{K}}$ we set $\{g\}'_{\mathbb{K}} := \{g\}^?$, and for $g \in G \setminus P'_{\mathbb{K}}$ we set $\{g\}'_{\mathbb{K}} := \{g\}^+$. Then $m \in P''_{\mathbb{K}}$. Therefore we have shown the following proposition, which is also contained in [9].

**Proposition 4 (Proposition 30 from [9]).** *A set $P \subseteq M$ possibly entails $m \in M$ if and only if $m \in P^{+?}$.*

An immediate consequence of this proposition is

$$P^{+?} = \bigcup_{\mathbb{K} \text{ realizer of } (\mathbb{K}_+, \mathbb{K}_?)} P''_{\mathbb{K}}.$$

With these considerations in mind it is now easy to generalize the definition of proper premises to the setting of incomplete counterexamples.

**Definition 5.** *Let* $\mathbb{K}_+ = (G, M, I_+)$ *and* $\mathbb{K}_? = (G, M, I_?)$ *be two formal contexts with* $I_+ \subseteq I_?$. *Let* $m \in M$. *A set* $P \subseteq M$ *is called a* possible proper premise *for* $m$ *in* $(\mathbb{K}_+, \mathbb{K}_?)$ *if and only if* $P$ *is* $\subseteq$*-minimal with possibly entailing* $m$. $P$ *is called a* possible proper premise *if and only if* $P$ *is a possible proper premise for some* $m \in M$.

Of course, together with Proposition 4 above, we immediately obtain that $P$ is a possible proper premise for $m$ in $(\mathbb{K}_+, \mathbb{K}_?)$ if and only if $P$ is $\subseteq$-minimal with $m \in P^{+?}$.

We are now extending `algorithm-1` from Listing 1.1 to also allow the expert to provide incomplete counterexamples during the exploration. For this we effectively only have to adapt `algorithm-1` to keep two working contexts $\mathbb{K}_+$ and $\mathbb{K}_?$ instead of only one. However, we shall also include some optimizations introduced in [9] to adjust these contexts as soon as the expert confirms a new implication $P \longrightarrow \{m\}$: we then try to extend $\mathbb{K}_+$ by incidences that are entailed by this new implication and the ones already known. Furthermore, for each $g \in G$, we delete all attributes in $m \in g^?$ that would result in impossible incidences, i. e. we remove $m$ from $g^?$ if

$$\mathcal{L}(g^+ \cup \{m\}) \not\subseteq g^?,$$

where $\mathcal{L}$ denotes the set of all implications accepted so far (including $P \longrightarrow \{m\}$).

A formulation in pseudo-code of this generalization of `algorithm-1` is shown as `algorithm-2` in Listing 1.2. The proof of the following theorem is similar to Theorem 3.

**Theorem 6.** *Using the same notation as in Listing 1.2, upon termination of* `algorithm-2` *it is true that the set* $\mathcal{L} \setminus \mathcal{B}$ *is a base of* $\mathrm{Th}_C(\mathbb{K}_{\mathrm{back}})$ *with background knowledge* $\mathcal{B}$ *for the background context* $\mathbb{K}_{\mathrm{back}} = (G_{\mathrm{back}}, M, I_{\mathrm{back}})$ *of the exploration. Furthermore, all premises of implications in* $\mathcal{L} \setminus \mathcal{B}$ *are proper premises in* $\mathbb{K}_{\mathrm{back}}$.

For the proof it is helpful to keep in mind that the background context at any point in a run of `algorithm-2` contains a subcontext which is a realizer of the current partial working context. More precisely, if $G$ denotes the current set of objects, then the formal context

$$\mathbb{K}_{\mathrm{back}} = (G_{\mathrm{back}} \cap G, M, I_{\mathrm{back}} \cap G \times M)$$

is a realizer of the current partial working context.

*Proof.* Seeing that $\mathcal{L} \setminus \mathcal{B}$ is sound for $\mathbb{K}_{\mathrm{back}}$ can be seen as in Theorem 3: all implications in $\mathcal{L} \setminus \mathcal{B}$ have been confirmed by the expert, and thus must be valid in $\mathbb{K}_{\mathrm{back}}$.

Let us consider the completeness of $\mathcal{L} \setminus \mathcal{B}$ for $\mathbb{K}_{\mathrm{back}}$ with background knowledge $\mathcal{B}$. To this end, suppose by contradiction that completeness does not hold. Then there exists $A \subseteq M \setminus C, m \in M \setminus C$ such that $\mathcal{L} \not\models (A \longrightarrow \{m\})$ but $A \longrightarrow \{m\}$ holds in $\mathbb{K}_{\mathrm{back}}$. Then in the iteration for the attribute $m$ we consider the point where we reach line 18. If $\bar{\mathcal{L}}$ denotes the current value of $\mathcal{L}$ at this point, then we still have $\bar{\mathcal{L}} \not\models (A \longrightarrow \{m\})$. On the other hand, for the current values $\bar{\mathbb{K}}_+$ of the certain

**Listing 1.2.** Attribute exploration using proper premises with incomplete couterexamples

```
0   define algorithm-2 (𝕂₊ = (G,M,I₊), 𝕂₍ = (G,M,I₍), C ⊊ M,
1       ℬ ⊆ Th_C(𝕂₊))
2       ℒ := ℬ
3       forall m ∈ C do
4           𝒫 := {P ⊆ M\C | P possible proper premise for m in (𝕂₊,𝕂₍)}
5           while there exists P ∈ 𝒫 with ℒ ⊭ (P ⟶ {m}) do
6               if expert confirms P ⟶ {m} then
7                   ℒ := ℒ∪{P ⟶ {m}}
8                   forall g ∈ G do
9                       g⁺ := ℒ(g⁺)
10                      forall m ∈ g^? \g⁺ where ℒ(g⁺∪{m}) ⊈ g^? do
11                          remove m from g^?
12                      end
13                  end
14              else
15                  ask expert for valid counterexample and augment 𝕂₊ and 𝕂₍
16                  𝒫 := {P ⊆ M\C | P possible proper premise for m in (𝕂₊,𝕂₍)}
17              end
18          end
19      end
20      return ℒ\ℬ
21  end
```

context and $\bar{\mathbb{K}}_?$ of the possible context at the same point in the iteration there exists a subcontext $\bar{\mathbb{K}}$ of $\mathbb{K}_{\text{back}}$ which is a realizer of $(\bar{\mathbb{K}}_+, \bar{\mathbb{K}}_?)$, and $A \longrightarrow \{m\}$ also holds in $\bar{\mathbb{K}}$. Therefore, there exists a possible proper premise $P \subseteq A$ for $m$ in $(\bar{\mathbb{K}}_+, \bar{\mathbb{K}}_?)$. But $\bar{\mathcal{L}} \not\models (P \longrightarrow \{m\})$, so we could not have reached line 18 yet, a contradiction.

It remains to be shown that all premises in $\mathcal{L} \setminus \mathcal{B}$ are proper premises in $\mathbb{K}_{\text{back}}$. To this end, let $(P \longrightarrow \{m\}) \in \mathcal{L} \setminus \mathcal{B}$, and suppose by contradiction that $P$ is not a proper premise for $m$ in $\mathbb{K}_{\text{back}}$. Then there exists a subset $\bar{P} \subsetneq P$ such that $\bar{P} \longrightarrow \{m\}$ is valid in $\mathbb{K}_{\text{back}}$ as well. But then $P$ is never a possible proper premise for $m$ in any partial working context of the exploration, and is never asked to the expert and thus cannot be an element of $\mathcal{L} \setminus \mathcal{B}$, a contradiction. □

Note that a slight variation of the argument of the proof shows that $\mathcal{L}$ is complete for *every* realizer of the final partial working context, and in particular for the final certain context $\mathbb{K}_+$. On the other hand, by line 9 of `algorithm-2` it is true that $\mathcal{L}$ is sound for $\mathbb{K}_+$, and we obtain that $\mathcal{L} \setminus \mathcal{B}$ is a base of $\mathbb{K}_+$ with background knowledge $\mathcal{B}$. The final possible context $\mathbb{K}_?$, however, may still contain incidences invalidating implications in $\mathcal{L}$, contradicting the soundness of $\mathcal{L}$ for $\mathbb{K}_?$. Thus, not all implications valid in $\mathbb{K}_{\text{back}}$ and $\mathbb{K}_+$ necessarily hold in $\mathbb{K}_?$.

Recall that we have required our expert to not make errors, i.e. to neither confirm invalid implications nor to provide false counterexamples. However, as we have already noted, this assumption may not be practical, and it may be necessary for applications to help the expert to detect errors. In our particular setting of `algorithm-2` we can use the same idea as in [9] and check in every iteration whether

$$\forall g \in G: g^+ \subseteq g^?$$

is still valid. If at a certain point in the exploration this constraint is not satisfied, then we have reached an inconsistent state (i.e. $(\mathbb{K}_+, \mathbb{K}_?)$ is not a partial context anymore). We would then have to abort the exploration and have the expert to find the error in the counterexamples and implications given so far. As soon as this is done, te exploration can be started anew, using the data from the previous run as starting partial context and background knowledge, respectively.

We have not yet talked about how to compute possible proper premises, and indeed we shall not do so in detail. The applications we have in mind for this work are sufficiently small such that a naive exhaustive search is feasible. On the other hand, it would be interesting to consider the question whether methods to compute proper premises in formal contexts can be carried over to yield methods for the computation of possible proper premises in partial contexts.

## 4 Main Hypotheses and Exceptions of the Free Radical Theory of Ageing as RDR

We formalized data reported in the literature or collected by our experimental collaborators as a set of RDR. Here, we demonstrate the application of the developed method of knowledge base construction and validation for a small example (compare Section 2.1), the branch of a larger RDR tree that starts with the general rule [11, p. 1]:

$$\text{NoStimulation} \longrightarrow \text{ROS.old.+, Lifespan.−} \qquad (1.)$$

In this branch cases are evaluated with no specific experimental condition like stimulation of cells with ROS (oxidative stess) or reduced ATP production. The general rule represents the basic hypothesis of the FRTA: During ageing, ROS accumulate, and the provoked cell damage reduces the expected lifespan of an individuum. Increasing ROS levels are also observed in several experiments of our biomedical partners. But, of course, there are exceptions. We defined the following cases. Note that classes are defined as sets of attributes, hence a classification by {ROS.old.+} is different to {ROS.old.+, Lifespan.−}.

$$\text{NoStimulation, AntiOx1.+} \longrightarrow \text{ROS.old.−, Lifespan.+} \qquad (1.1)$$

$$\text{NoStimulation, AntiOx1.+, AntiOx2.−} \longrightarrow \text{ROS.old.+} \qquad (1.1.1)$$

$$\text{NoStimulation, Mouse, AntiOx2.−} \longrightarrow \text{ROS.old.+} \qquad (1.2)$$

$$\text{NoStimulation, AntiOx2.−, CElegans} \longrightarrow \text{ROS.old.+} \qquad (1.3)$$

$$\text{NoStimulation, Mutation-ETC, Mouse} \longrightarrow \emptyset \qquad (1.4)$$

**Table 2. Examples (cornerstone cases) related to the RDR knowledge base.**
Observed cases are described by the certain context $\mathbb{K}_+$. Supplementary possible attributes of $\mathbb{K}_?$ are indicated by '?'. 1.S1 denotes a specific case confirming rule (1.): In old rats, enhanced expression of genes associated to the antioxidant glutathione together with increased oxidation of lipids hints at an increased ROS concentration [19]. The cornerstone cases corresponding to RDR (1.2) and (1.3) were merged, since the same case was observed for both species. Rule (1.4) was observed for bone marrow from two mouse strains with mutations in a gene coding for a protein of the ETC [12, 13]. In the hippocampus, however, for the same strains a significant increase of ROS over age was observed [15]. Therefore, RDR (1.4) was not accepted as background knowledge for the subsequent attribute exploration, and the conflicting cases 1.4a and 1.4b were listed in the initial context.

| | AntiOx1.+ | AntiOx1.− | AntiOx2.+ | AntiOx2.− | CElegans | Mouse | Mut-ETC | ROS.old.+ | ROS.old.− | Lifespan.+ | Lifespan.− |
|---|---|---|---|---|---|---|---|---|---|---|---|
| **1.** | ? | ? | ? | ? | × | × | ? | × | | | × |
| **1.S1** | × | ? | ? | ? | × | | × | | | ? | ? |
| **1.1** | × | ? | ? | × | × | | | × | × | | |
| **1.1.1** | × | | × | × | × | | × | | | | ? |
| **1.2–1.3** | ? | ? | | × | × | × | | × | | ? | ? |
| **1.4a** | ? | ? | ? | ? | ? | × | × | | | ? | ? |
| **1.4b** | ? | ? | ? | ? | ? | × | × | × | | ? | ? |

These rules are taken from [11, p. 2] and [12, 13], respectively. We only considered experiments for mice and the worm *C. elegans* often investigated as a model organism for general ageing. + and − denote significantly enhanced / reduced lifespan and concentrations of ROS or antioxidants, respectively. For normal conditions, neither + nor − is attributed. Therefore, the interesting information expressed by rules (1.2) and (1.3) is that ROS increase, but no reduction of lifespan is observed. AntiOx.+/− means a permanently high/low antioxidant level, for instance by a mutation, or a high/low antioxidant concentration during an experiment.

## 4.1    Attribute Exploration of the Developed RDR Knowledge Base

Examples corresponding to the RDR were collected in the certain and possible contexts of Table 2 as initial contexts of an attribute exploration with algorithm-2. The common attribute NoStimulation is not mentioned explicitly. Cases with conflicting values for one variable were excluded by the background implications AntiOx1.+, AntiOx1.− $\longrightarrow \bot$ and AntiOx2.+, AntiOx2.− $\longrightarrow \bot$. Since AntiOx1 and AntiOx2 represent different, but not further specified enzymes, only the attribute set {AntiOx1.+, AntiOx2.−} occurs in observed cases. The symmetric case was excluded via the background implication AntiOx1.−, AntiOx2.+ $\longrightarrow \bot$.

Rules (1.1.1), (1.2) and (1.3) were combined to AntiOx2.$-$ $\longrightarrow$ ROS.old.$+$, the single accepted background implication expressing biological knowledge.

Only four implications with proper premises were accepted during the exploration:

$$AntiOx1.- \longrightarrow ROS.old.+ \qquad (3)$$

$$AntiOx2.+ \longrightarrow ROS.old.- \qquad (4)$$

$$AntiOx1.+, AntiOx2.+, CElegans \longrightarrow Lifespan.+ \qquad (5)$$

$$AntiOx1.-, Mouse, Mut\text{-}ETC \longrightarrow Lifespan.- \qquad (6)$$

Because of the background implication AntiOx1.$-$, AntiOx2.$+$ $\longrightarrow \perp$, rule (3) implies that AntiOx2.$+$ does not hold, rule (4) not AntiOx1.$-$. Thus, the general ROS defense is disturbed / intact, and we accepted the conclusions ROS.old.$+$/-.

Together with (4), (5) specifies RDR (1.1) in the sense that exception (1.1.1) does not apply, since the concentration of both antioxidants is high. The strong conclusion Lifespan.$+$ can be assumed for the short living worm C. elegans: "Although experimental augmentations of antioxidant defenses tend to enhance resistance to induced oxidative stress, such manipulations are generally ineffective in the extension of lifespan of long-lived strains of animals." [17, Abstract] Implication (6) is symmetric: Mutations (deletions) of the mitochondrial DNA can cause lifespan reducing damage for long-lived animals like mice. [11, p. 6]

Finally, from overall 30 rejected implications we mention AntiOx2.$+$ $\longrightarrow$ Lifespan.$+$: Following rule (4), ROS are reduced, but this is not sufficient to extend lifespan, and of course there are other factors like mutations restraining it. The following proposed implication refers to parallel, comparable observations for mouse and C. elegans:

$$AntiOx1.+, Mouse, CElegans, Mut\text{-}ETC \longrightarrow ROS.old.- \qquad (7)$$

The implication was rejected by a counterexample with the certain attributes of the premise, additionally AntiOx2.$-$; ROS.old.$-$ was excluded.

## 5  Discussion and Conclusions

We developed and proved algorithms for the exploration of implications with proper premises, requiring complete or incomplete counterexamples. Such implications have minimal premises and thus highlight necessary conditions. As discussed in [16], the computation of a base of this kind in several cases is faster than computing the classical stem base with pseudo-intents as premises. Finally, the base is iteration-free, which allowed a straightforward restriction of the exploration to implications with conclusions from a basic set $C \subsetneq M$.

A further step in algorithmic development could be an exploration aiming at the generation of a knowledge base consisting only of RDR. For smaller and medium applications, the complexity of the inference problem is not critical, and the rules of the base itself provide insight to the explored domain. Possibly, ideas from attribute exploration with incomplete counterexamples could be used, for instance two contexts of positive and related negative examples. Or implications are explored first, generating the most specific RDR without negated attributes. Then the attributes of their premises are negated in several iterations. In any case, the challenge will be to find an algorithm asking a minimal number of questions and thus alleviating the effects of the inherent combinatorial explosion.

For our small RDR knowledge base, the accepted and the rejected implications provide a structured overview on existing knowledge and helped to filter out interesting experiments. The implications (5) and (6) motivated a more detailed literature search and revealed unexpected differences between long- and short-lived animals. Only five implications were accepted, together with the RDR AntiOx2.− ⟶ ROS.old.+. They are a compact representation of generally accepted facts related to the influence of antioxidants and mutations of the ETC on ROS production and lifespan.

In nature, there is almost no rule without exception, hence even these implications can be challenged by rare counterexamples. Following [18], these could be treated separately by a lattice of exceptions. This could complement our approach to specify conditions of exceptions starting from RDR. In subsequent work, we investigate a larger data and literature set and define observed cases by more specific attributes, in order to obtain more expressive and more reliable rules.

**Acknowledgments.** Beyond the cited experimental works, we thank many members of the ROSAge project groups within the Rostock Departments of Medical Biochemistry and Molecular Biology, of Gastroenterology and of Biostatistics and Informatics in Medicine and Ageing, as well as within the Departments of Dermatology, Lübeck and Leipzig. Presentations and discussions of their data contributed to the definition of the present pilot study and to the selection of interesting biomedical cases.

# Bibliography

[1] Andreyev, A., Kushnareva, Y., Starkov, A.: Mitochondrial metabolism of reactive oxygen species. Biochemistry (Moscow) 70(2), 200–214 (2005)

[2] Baader, F., Sertkaya, B.: Usability Issues in Description Logic Knowledge Base Completion. In: Ferré, S., Rudolph, S. (eds.) ICFCA 2009. LNCS, vol. 5548, pp. 1–21. Springer, Heidelberg (2009)

[3] Bertet, K., Monjardet, B.: The multiple facets of the canonical direct unit implicational basis. Theoretical Computer Science 411(22-24), 2155–2166 (2010)

[4] Borchmann, D.: conexp-clj – A General-Purpose Tool for Formal Concept Analysis, http://github.com/exot/conexp-clj

[5] Burmeister, P.: Merkmalimplikationen bei unvollständigem Wissen. In: Proceedings: Arbeitstagung Begriffsanalyse und Künstliche Intelligenz 1988, pp. 15–46. Technische Universität Clausthal, Clausthal-Zellerfeld (1991)

[6] Burmeister, P.: ConImp, http://www.mathematik.tu-darmstadt.de/~burmeister/

[7] Compton, P., et al.: Ripple down rules: Turning knowledge acquisition into knowledge maintenance. Artificial Intelligence in Medicine 4(6), 463–475 (1992)

[8]  Ganter, B.: Pseudo-models and propositional Horn inference. Preprint (2003)
[9]  Ganter, B., Obiedkov, S.: Conceptual Exploration. Preprint, Dresden (2013)
[10] Ganter, B., Wille, R.: Formal Concept Analysis – Mathematical Foundations. Springer, Heidelberg (1999)
[11] Kirkwood, T., Kowald, A.: The free-radical theory of ageing – older, wiser and still alive. Bioessays (2012)
[12] Kretzschmar, C., et al.: Influence of oxidative stress on hematopoietic cell aging in a mouse model with mitochondrial DNA mutations during aging. Ann. Hematol. 92(suppl. 1), S22 (2013)
[13] Kretzschmar, C., et al.: Polymorphism nt7778g/t in Mitochondrial Atp8 Gene Promotes Protective Effect on Reactive Oxygen Species Level in Murine Hematopoietic Cells During Aging. Blood 122(21), 1196–1197 (2013)
[14] Kwok, R.B.H.: Translations of Ripple Down Rules into logic formalisms. In: Dieng, R., Corby, O. (eds.) EKAW 2000. LNCS (LNAI), vol. 1937, pp. 366–379. Springer, Heidelberg (2000)
[15] Reichart, G., Mayer, J., Köhling, R.: Measurements of mitochondrial superoxide levels in the mouse brain hippocampus, Unpublished data, Oscar Langendorff Institute of Physiology, Rostock (2013)
[16] Ryssel, U., Distel, F., Borchmann, D.: Fast algorithms for implication bases and attribute exploration using proper premises. Annals of Mathematics and Artificial Intelligence, 1–29 (2013)
[17] Sohal, R., Orr, W.: The redox stress hypothesis of aging. Free Radical Biology and Medicine 52(3), 539–555 (2012)
[18] Stumme, G.: Attribute Exploration with Background Implications and Exceptions. In: Bock, H.-H., Polasek, W. (eds.) Data Analysis and Information Systems, pp. 457–469. Springer (1996)
[19] Yang, W., et al.: Age-dependent changes of the antioxidant system in rat liver are accompanied by altered MAPK activation and a decline in mTOR signaling, Tübingen (to be published, 2014)

# Subdirect Decomposition of Concept Lattices[*],[**]

Rudolf Wille

Algebra Universalis
Technische Hochschule Darmstadt, Darmstadt
West Germany

*Dedicated to Garrett Birkhoff on the occasion of his seventieth birthday*

## 1. Introduction

In [1], G. Birkhoff exhibited the subdirect product of algebraic structures as a universal tool, which since has been extensively used in the study of algebraic theories. Although a subdirect product is not uniquely determined by its factors, there are useful construction methods based on subdirect products (cf. Wille [8], [9], [10]). The aim of this paper is to make these methods available for handling the "Determination Problem" of concept lattices as it is exposed in Wille [11]. In particular, a useful method for determining concept lattices via its scaffoldings will be developed under some finiteness condition.

## 2. Concept lattices

First we recall some notions from Wille [11]. A *context* is defined as a triple $(G, M, I)$ where $G$ and $M$ are sets, and $I$ is a binary relation between $G$ and $M$; the elements of $G$ and $M$ are called *objects* and *attributes*, respectively. If $gIm$ for $g \in G$ and $m \in M$ we say: the object $g$ has the attribute $m$. The relation $I$ establishes a *Galois connection* between the power sets of $G$ and $M$ (cf. Birkhoff [2]) which is expressed by the definition

$$A' := \{m \in M \mid gIm \quad \text{for all} \quad g \in A\} \quad \text{for} \quad A \subseteq G,$$
$$B' := \{g \in G \mid gIm \quad \text{for all} \quad m \in B\} \quad \text{for} \quad B \subseteq M.$$

---

[*] Presented by R. P. Dilworth. Received March 4, 1982. Accepted for publication in final form June 8, 1982.

[**] This is a reprint of a paper originally published in Algebra Universalis 17, 275-287 (1983).

C.V. Glodeanu, M. Kaytoue, and C. Sacarea (Eds.): ICFCA 2014, LNAI 8478, pp. 284–296, 2014.

Following traditional philosophy, a *concept* of a context $(G, M, I)$ is defined as a pair $(A, B)$ where $A \subseteq G$, $B \subseteq M$, $A' = B$, and $B' = A$; $A$ and $B$ are called the *extent* and the *intent* of the concept $(A, B)$, respectively. The relation "subconcept–superconcept" is captured by the definition

$$(A_1, B_1) \leq (A_2, B_2) : \Leftrightarrow A_1 \subseteq A_2 (\Leftrightarrow B_1 \supseteq B_2)$$

for concepts $(A_1, B_1)$ and $(A_2, B_2)$ of $(G, M, I)$. $\mathfrak{B}(G, M, I)$ denotes the set of all concepts of the context $(G, M, I)$ and $\underline{\mathfrak{B}}(G, M, I) := (\mathfrak{B}(G, M, I), \leq)$. A subset $D$ of a complete lattice $L$ is called *infimum-dense* (*supremum-dense*) if $L = \{\bigwedge X \mid X \subseteq D\}$ $(L = \{\bigvee X \mid X \subseteq D\})$. Now, we are able to formulate the basic theorem of "concept lattices" (cf. Wille [11]).

THEOREM 1. *Let* $(G, M, I)$ *be a context. Then* $\underline{\mathfrak{B}}(G, M, I)$ *is a complete lattice, called the* concept lattice *of* $(G, M, I)$, *in which infima and suprema can be described as follows:*

$$\bigwedge_{j \in J} (A_j, B_j) = \left( \bigcap_{j \in J} A_j, \left( \bigcap_{j \in J} A_j \right)' \right),$$

$$\bigvee_{j \in J} (A_j, B_j) = \left( \left( \bigcap_{j \in J} B_j \right)', \bigcap_{j \in J} B_j \right).$$

*Conversely, if* $L$ *is a complete lattice then* $L \cong \underline{\mathfrak{B}}(G, M, I)$ *if and only if there are mappings* $\gamma : G \to L$ *and* $\mu : M \to L$ *such that* $\gamma G$ *is supremum-dense in* $L$, $\mu M$ *is infimum-dense in* $L$, *and* $gIm$ *is equivalent to* $\gamma g \leq \mu m$ *for all* $g \in G$ *and* $m \in M$; *in particular,* $L \cong \underline{\mathfrak{B}}(L, L, \leq)$.

For the complete lattice $L := \underline{\mathfrak{B}}(G, M, I)$ the mappings $\gamma : G \to L$ and $\mu : M \to L$ in Theorem 1 are naturally defined by

$$\gamma g := (\{g\}'', \{g\}') \quad \text{for} \quad g \in G,$$
$$\mu m := (\{m\}', \{m\}'') \quad \text{for} \quad m \in M.$$

An important problem is: How can one determine the concept lattice of a given context? One way to approach this problem is based on the idea to construct the concept lattice of a context by the concept lattices of some suitable subcontexts. Here a *subcontext* of a context $(G, M, I)$ is understood as a triple $(H, N, I \cap (H \times N))$ with $H \subseteq G$ and $N \subseteq M$; we often write $(H, N)$ instead of $(H, N, I \cap (H \times N))$. In Wille [11], it has been shown that for a partition $\{N_j \mid j \in J\}$ of $M$ an isomorphism of the $\bigvee$-semilattice $\mathfrak{B}(G, M, I)$ onto a subdirect product of the $\bigvee$-semilattices $\mathfrak{B}(H, N_j, I \cap (G \times N_j))(j \in J)$ is given by $(A, B) \mapsto ((B \cap N_j)', B \cap N_j)_{j \in J}$. Since the construction methods are more powerful if they are based on subdirect products of complete lattices instead of subdirect products of $\vee$-semilattices, we analyse in the following subdirect decompositions of $\mathfrak{B}(G, M, I)$ as a complete lattice. A main purpose is to obtain for contexts satisfying the "chain condition" a method for determining the scaffolding of $\mathfrak{B}(G, M, I)$ directly from the context $(G, M, I)$. Then the concept lattice can be constructed as an isomorphic copy of the ideal lattice of its scaffolding (see Wille [9]).

## 3. Complete congruence relations

Throughout this section $(G, M, I)$ will be a context and $(H, N)$ a subcontext of $(G, M, I)$. $(H, N)$ is said to be *compatible* if $(A' \cap N)' \cap H \subseteq A''$ for all $A \subseteq G$ and $(B' \cap H)' \cap N \subseteq B''$ for all $B \subseteq M$. By $\pi(H, N)(A, B) := (A \cap H, B \cap N)$ we define a map $\pi(H, N)$ from $\mathfrak{B}(G, M, I)$ into $\mathcal{P}(H) \times \mathcal{P}(N)$ where, in general, $\mathcal{P}(S)$ is the complete lattice of all subsets of a set $S$.

PROPOSITION 2. *$(H, N)$ is compatible if and only if $\pi(H, N)$ is a complete lattice homomorphism from $\mathfrak{B}(G, M, I)$ onto $\mathfrak{B}(H, N, I \cap (H \times N))$.*

*Proof.* Let $(H, N)$ be compatible. If $(A, B) \in \mathfrak{B}(G, M, I)$ then $(A \cap H)' \cap N = (B' \cap H)' \cap N = B'' \cap N = B \cap N$ and $(B \cap N)' \cap H = (A' \cap N)' \cap H = A'' \cap H = A \cap H$ wherefore $(A \cap H, B \cap H) \in \mathfrak{B}(H, N, I \cap (H \times N))$. If $(C, D) \in \mathfrak{B}(H, N, I \cap (H \times N))$ then $C'' \cap H = (C' \cap N)' \cap H = D' \cap H = C$ and $C' \cap N = D$. Hence $\pi(H, N)$ is a surjective map from $\mathfrak{B}(G, M, I)$ onto $\mathfrak{B}(H, N, I \cap (H \times N))$. By Theorem 1, $\pi(H, N)$ preserves arbitrary infima and suprema. Conversely, let $\pi(H, N)$ be a complete lattice homomorphism from $\mathfrak{B}(G, M, I)$ onto $\mathfrak{B}(H, N, I \cap (H \times N))$. Then, in particular, $(A'' \cap H, A' \cap N) \in \mathfrak{B}(H, N, I \cap (H \times N))$ for $A \subseteq G$; hence $(A' \cap N)' \cap H = A'' \cap H \subseteq A''$. This shows together with the dual argument that $(H, N)$ is compatible.

The kernel of $\pi(H, N)$ is denoted by $\underline{\Theta}(H, N)$, if $(H, N)$ is compatible Proposition 2 yields that $\Theta(H, N)$ is complete congruence relation of $\mathfrak{B}(G, M, I)$ and that $\mathfrak{B}(H, N, I \cap (H \times N)) \cong \mathfrak{B}(G, M, I)/\underline{\Theta}(H, N)$ (we recall that an equivalence relation $\Theta$ on a complete lattice $L$ is a *complete congruence relation* if $x_j \Theta y_j (j \in J)$ always imply

$$\left(\bigwedge_{j \in J} x_j\right) \Theta \left(\bigwedge_{j \in J} y_j\right) \quad \text{and} \quad \left(\bigvee_{j \in J} x_j\right) \Theta \left(\bigvee_{j \in J} y_j\right).$$

For the rest of this section we assume that the context $(G, M, I)$ satisfies the following *chain condition*: there is no infinite descending chain $\{g_1\}'' \supset \{g_2\}'' \supset \{g_3\}'' \supset \cdots$ with $g_1, g_2, g_3, \ldots$ in $G$ and there is no infinite ascending chain $\{m_1\}' \subset \{m_2\}' \subset \{m_3\}' \subset \cdots$ with $m_1, m_2, m_3, \ldots$ in $M$.

For a complete congruence relation $\Theta$ of $\mathfrak{B}(G, M, I)$ we define

$G(\Theta) := \{g \in G \mid \gamma g$ is the smallest element of a $\Theta$-class$\}$

$M(\Theta) := \{m \in M \mid \mu m$ is the greatest element of a $\Theta$-class$\}$.

PROPOSITION 3. *If $\Theta$ is a complete congruence relation of $\mathfrak{B}(G, M, I)$ then $(G(\Theta), M(\Theta))$ is a compatible subcontext of $(G, M, I)$ and $\Theta = \underline{\Theta}(G(\Theta), M(\Theta))$.*

*Proof.* Suppose $g \in (A' \cap M(\Theta))' \cap G(\Theta)$ but $g \notin A''$ for some $A \subseteq G$. Then there is an $m \in A'$ such that $\{m\}'$ is maximal in $\{\{n\}' \mid n \in M$ and $(g, n) \notin I\}$. Since $\gamma g \leq \bigwedge \{\mu n \mid n \in M$ and $\mu m < \mu n\}$, $\mu m$ is $\wedge$-irreducible and $\gamma g \vee \mu m$ covers $\mu m$. Hence $g \in G(\Theta)$ implies that $\mu m$ is a greatest element of a $\Theta$-class wherefore $m \in A' \cap M(\Theta)$. This contradicts $g \in (A' \cap M(\Theta))'$ and $(g, m) \notin I$. Hence $(A' \cap M(\Theta))' \cap G(\Theta) \subseteq A''$ for all $A \subseteq G$ and dually $(B' \cap G(\Theta))' \cap M(\Theta) \subseteq B''$ for all $B \subseteq M'$, i.e. $(G(\Theta), M(\Theta))$ is compatible. Let $(A, B) \in \mathfrak{B}(G, M, I)$, and let $(\underline{A}, \underline{B})$ and $(\bar{A}, \bar{B})$ be the smallest and the greatest concept in the $\Theta$-class containing $(A, B)$. Then $\gamma g \leq (A, B)$ for $g \in G(\Theta)$ implies $\gamma g \leq (A, B)$. Therefore $A \cap G(\Theta) = \underline{A} \cap G(\Theta)$ and dually $B \cap M(\Theta) = \bar{B} \cap M(\Theta)$. Hence $(A, B)\Theta(C, D)$ implies $A \cap G(\Theta) = C \cap G(\Theta)$ and $B \cap M(\Theta) = D \cap M(\Theta)$, i.e. $(A, B)\underline{\Theta}(G(\Theta), M(\Theta))(C, D)$. The converse implication follows from $(\underline{A}, \underline{B}) = \bigvee \gamma(A \cap G(\Theta))$ and $(\bar{A}, \bar{B}) = \bigwedge \mu(B \cap M(\Theta))$. This proves $\Theta = \underline{\Theta}(G(\Theta), M(\Theta))$.

For a characterization of the subcontexts $(G(\Theta), M(\Theta))$ we introduce the following notion: $(H, N)$ is said to be *saturated* if $\{g\}' = (\{g\}'' \cap H)'$ implies $g \in H$ for all $g \in G$ and if $\{m\}' = (\{m\}'' \cap N)'$ implies $m \in N$ for all $m \in M$. Since $\bigvee \gamma A$ is the smallest element of a $\Theta$-class for each $A \subseteq G(\Theta)$ and since $\bigwedge \mu B$ is the greatest element of a $\Theta$-class for each $A \subseteq G(\Theta)$ and since $\bigwedge \mu B$ is the greatest element of a $\Theta$-class for each $B \subseteq M(\Theta)$, the subcontext $(G(\Theta), M(\Theta))$ is saturated for all complete congruence relations $\Theta$ of $\mathfrak{B}(G, M, I)$.

PROPOSITION 4. *If $(H, N)$ is saturated and compatible then $H = G(\underline{\Theta}(H, N))$ and $N = M(\underline{\Theta}(H, N))$.*

*Proof.* For $h \in H$ and $(A, B) \in \mathfrak{B}(G, M, I)$, $\{h\}'' \cap H = A \cap H$ implies $\{h\}'' \subseteq A$; hence $\gamma h \in G(\underline{\Theta}(H, N))$. If $\gamma g$ is the smallest element of a $\underline{\Theta}(H, N)$-class for $g \in G$ then $\{g\}'' = (\{g\}'' \cap H)''$ wherefore $g \in H$ as $(H, N)$ is saturated. This and the dual argument proves the assertion.

Proposition 3 and 4 yield a one-to-one correspondence between the complete congruence relations of $\underline{\mathfrak{B}}(G, M, I)$ and the saturated compatible subcontexts of $(G, M, I)$. This correspondence is closely related to the duality elaborated in Urquhart [7].

## 4. Weak perspectivity

For the determination of subdirect decompositions of a concrete concept lattice it is commendable to reduce (if possible) the given context $(G, M, I)$ to a minimal compatible subcontext $(H, N)$ for which $\pi(H, N)$ is an isomorphism. Let us call a context *reduced* if $\pi(H, N)$ is not injective for each of its proper compatible subcontexts $(H, N)$. Throughout this section we assume that $(G, M, I)$ is a reduced context satisfying the chain condition (in Section 3). With Theorem 1 we conclude that $\gamma$ is a bijective map from $G$ onto the set of all $\vee$-irreducible elements of $\underline{\mathfrak{B}}(G, M, I)$ and $\mu$ is a bijective map from $M$ onto the set of all $\wedge$-irreducible elements of $\underline{\mathfrak{B}}(G, M, I)$.

Congruence relations of lattices are successfully studied via the notions of weak perspectivity and weak projectivity (cf. Crawley and Dilworth [4], Grätzer [5]). These notions can be carried over to contexts. For $g \in G$ and $m \in M$, $g$ is *weakly perspective* to $m$, in symbols $g \nearrow m$, if $\{m\}'$ is maximal in $\{\{n\}' \mid n \in M$ and $(g, n) \notin I\}$; dually, $m$ is *weakly perspective* to $g$, in symbols $m \searrow g$, if $\{g\}''$ is minimal in $\{\{h\}'' \mid h \in G$ and $(h, m) \notin I\}$. If $g \nearrow m$ and $m \searrow g$, we call $g$ and $m$ *perspective* and write $g \sim m$ or $m \sim g$. In $G \cup M$ an element $x$ is *weakly projective* to an element $y$, in symbols $x \approx_w y$, if $x = y$ or if there are elements $x = x_0, x_1, \ldots, x_k = y$ in $G \cup M$ such that $x_{i-1}$ is weakly perspective to $x_i$ for $i = 1, \ldots, k$. If $x \approx_w y$ and $y \approx_w x$, we call $x$ and $y$ *projective* and write $x \approx y$. For $X \subseteq G \cup M$ we define the *weakly projective closure* by $\langle X \rangle := \{y \in G \cup M \mid x \approx_w y \text{ for some } x \in X\}$.

PROPOSITION 5. *A subcontext $(H, N)$ of $(G, M, I)$ is compatible if and only if $\langle H \cup N \rangle = H \cup N$.*

*Proof.* Let $(H, N)$ be compatible. Suppose $g \in H$, $g \nearrow m$, but $m \notin N$. Then $g \in (\{m\}'' \cap N)' \cap H = \{m\}''' \cap H = \{m\}' \cap H$ what contradicts $(g, m) \notin I$. Therefore $g \in H$ and $g \nearrow m$ imply $m \in N$ and dually $m \in N$ and $m \searrow g$ imply $g \in H$. Hence $\langle H \cup N \rangle = H \cup N$. Conversely, let us assume $\langle H \cup N \rangle = H \cup N$. Suppose $g \in (A' \cap N)' \cap H$ but $g \notin A''$ for some $A \subseteq G$. Then there is an $m \in A'$ such that $\{m\}'$ is maximal in $\{\{n\}' \mid n \in M$ and $(g, n) \notin I\}$. It follows that $g \nearrow m$ and hence $m \in A \cap N$ what contradicts $g \in (A' \cap N)'$ and $(g, m) \notin I$. Therefore $(A' \cap N)' \cap H \subseteq A''$ for all $A \subseteq G$ and dually $(B' \cap H)' \cap N \subseteq B''$ for all $B \subseteq M$, i.e. $(H, N)$ is compatible.

COROLLARY. *The compatible subcontexts of* $(G, M, I)$ *form a complete sublattice* $\mathfrak{D}(G, M, I)$ *of the complete lattice* $\mathcal{P}(G) \times \mathcal{P}(M)$.

THEOREM 6. Let $(G, M, I)$ be a reduced context satisfying the chain condition. Then $\Theta \mapsto (G(\Theta), M(\Theta))$ describes an antiisomorphism from the complete lattice of all complete congruence relations of $\mathfrak{B}(G, M, I)$ onto the complete sublattice of $\mathcal{P}(G) \times \mathcal{P}(M)$ consisting of all compatible subcontexts of $(G, M, I)$.

*Proof.* For $g \in G$ and $H \subseteq G$, $\{g\}' = (\{g\}'' \cap H)'$ implies $\gamma g = \gamma(\{g\}'' \cap H)$ and hence $g \in H$ as $\gamma$ is a bijection from $G$ onto the set of all $\bigvee$-irreducible elements of $\mathfrak{B}(G, M, I)$. This shows together with the dual argument that every subcontext of $(G, M, I)$ is saturated. Therefore, the described mapping is a bijection from the set of all complete congruence relation of $\mathfrak{B}(G, M, I)$ onto $\mathfrak{D}(G, M, I)$ by Proposition 3 and 4. The preceding corollary states that $\mathfrak{D}(G, M, I)$ is a complete sublattice of $\mathcal{P}(G) \times \mathcal{P}(M)$. Obviously, $\Theta_1 \subseteq \Theta_2$ is equivalent to $G(\Theta_1) \supseteq G(\Theta_2)$ and $M(\Theta_1) \supseteq M(\Theta_2)$ for complete congruence relations of $\mathfrak{B}(G, M, I)$. Hence the described mapping is an antiisomorphism.

COROLLARY. *Let* $\leq_w$ *be the (partial) order induced by* $\approx_w$ *on* $G \cup M/\approx$. *Then* $(G \cup M/\approx, \leq_w)$ *is isomorphic to the ordered set of all* $\bigwedge$-*irreducible complete congruence relations of* $\mathfrak{B}(G, M, I)$, *and* $\mathfrak{D}(G, M, I)$ *is isomorphic to the complete lattice of all order filters of* $(G \cup M/\approx, \leq_w)$.

The results of this section show how we may study complete congruence relations of $\mathfrak{B}(G, M, I)$ via the digraph $(G \cup M, \nearrow \cup \searrow)$ which can be easily derived from the context $(G, M, I)$. The digraph $(G \cup M, \nearrow \cup \searrow)$ and the ordered set $(G \cup M/\approx, \leq_w)$ are closely related to the double digraph considered in Urquhart [7]. The connection to the digraph $(J(L), C)$ in Jónsson, Nation [6] should also be mentioned.

## 5. Subdirect product constructions

In this section we elaborate for concept lattices the construction methods developed for subdirect products of complete lattices in Wille [9]. By Theorem 6, the subdirect decompositions of a reduced context satisfying the chain condition are in one-to-one correspondence to the families of compatible subcontexts of which the join is the whole context. In general, for compatible subcontexts $(H_j, N_j)(j \in J)$ of a context $(G, M, I)$, $(A, B) \mapsto (A \cap H_j, B \cap N_j)_{j \in J}$ is an isomorphism from $\mathfrak{B}(G, M, I)$ onto a subdirect product of the $\mathfrak{B}(H_j, N_j, I \cap (H_j \times N_j))$ if and only if $\Theta(\bigcup_{j\in J} H_j, \bigcup_{j\in J} N_j)$ is the identity on $\mathfrak{B}(G, M, I)$; the subcontext $(\bigcup_{j\in J} H_j, \bigcup_{j\in J} N_j)$ is again compatible as $g \in (A' \cap \bigcup_{j\in J} N_j)' \cap \bigcup_{j\in J} H_j$ for $A \subseteq G$ implies $g \in (A' \cap N_k)' \cap H_k \subseteq A''$ for some $k \in J$ and hence $(A' \cap \bigcup_{j\in J} N_j)' \cap \bigcup_{j\in J} H_j \subseteq A''$.

Now, let $(G, M, I)$ be an arbitrary context and let $(H_j, N_j)(j \in J)$ be compatible subcontexts of $(G, M, I)$ such that $\Theta(\bigcup_{j\in J} H_j, \bigcup_{j\in J} N_j)$ is the identity on $\mathfrak{B}(G, M, I)$. We define a map

$$\alpha_{jk} : \mathfrak{B}(H_k, N_k, I \cap (H_k \times N_k)) \to \mathfrak{B}(H_j, N_j, I \cap (H_j \times N_j)) \quad (j, k \in J)$$

by

$$\alpha_{jk}(A_k, B_k) := (A_k'' \cap H_j, A_k' \cap N_j) \quad \text{for all} \quad (A_k, B_k) \in \mathfrak{B}(H_k, N_k, I \cap (H_k \times N_k)).$$

**PROPOSITION 7.** $\alpha_{jk}$ *is the greatest of the* $\bigvee$*-preserving maps*

$$\alpha : \mathfrak{B}(H_k, N_k, I \cap (H_k \times N_k)) \to \mathfrak{B}(H_j, N_j, I \cap (H_j \times N_j))$$

*satisfying*

$$\alpha(\{g\}'' \cap H_k, \{g\}' \cap N_k) \le (\{g\}'' \cap H_j, \{g\}' \cap N_j) \text{ for all } g \in G.$$

*Proof.* By Proposition 2, $\alpha_{jk}$ is a map into $\mathfrak{B}(H_j, N_j, I \cap (H_j \times N_j))$. Using the general formulas $\bigcap_{t\in T} X_t' = (\bigcap_{t\in T} X_t')'' = (\bigcup_{t\in T} X_t)'$ and $(X'' \cap (X \cup Y))' = X'$ and the assumption that $(H_k, N_k)$ is compatible, for $(A_t, B_t) \in \mathfrak{B}(H_k, N_k, I \cap (H_k \times N_k))$ $(t \in T)$ we obtain

$$\left(\left(\bigcap_{t\in T} B_t\right)' \cap H\right)' = \left(\left(\bigcap_{t\in T} (A_t' \cap N)\right)' \cap H\right)' = \left(\left(\bigcap_{t\in T} A_t'\right)'' \cap N\right)' \cap H'$$

$$= \left(\left(\bigcap_{t\in T} A_t'\right)' \cap H\right)' = \left(\left(\bigcup_{t\in T} A_t\right)'' \cap H\right)' = \left(\bigcup_{t\in T} A_t\right)' = \bigcap_{t\in T} A_t'.$$

Therefore

$$\alpha_{jk} \bigvee_{t\in T} (A_t, B_t) = \alpha_{jk}\left(\left(\bigcap_{t\in T} B_t\right)' \cap H, \bigcap_{t\in T} B_t\right)$$

$$= \left(\left(\left(\bigcap_{t\in T} B_t\right)' \cap H\right)'' \cap H_j, \left(\left(\bigcap_{t\in T} B_t\right)' \cap H\right)' \cap N_j\right)$$

$$= \left(\left(\bigcap_{t\in T} A_t'\right)' \cap H_j, \bigcap_{t\in T} A_t' \cap N_j\right) = \bigvee_{t\in T} \alpha_{jk}(A_t, B_t);$$

hence $\alpha_{jk}$ is $\bigvee$-preserving. For an arbitrary $\bigvee$-preserving map $\alpha$ specified in Proposition 7 we have

$$\alpha(A_k, B_k) = \alpha \bigvee_{g \in A_k} (\{g\}'' \cap H_k, \{g\}' \cap N_k)$$

$$= \bigvee_{g \in A_k} \alpha(\{g\}'' \cap H_k, \{g\}' \cap N_k) \leq \bigvee_{g \in A_k} (\{g\}'' \cap H_j, \{g\}' \cap N_j)$$

$$= ((A_k' \cap N_j)', A_k' \cap N_j) = \alpha_{jk}(A_k, B_k);$$

hence $\alpha \leq \alpha_{jk}$.

Now, we are ready to apply Konstruktion I in Wille [9]. This leads to the following theorem.

**THEOREM 8.** *Let $(G, M, I)$ be a context and let $(H_j, N_j)(j \in J)$ be compatible subcontexts of $(G, M, I)$ such that $\Theta(\bigcup_{j \in J} H_j, \bigcup_{j \in J} N_j)$ is the identity on $\underline{\mathfrak{B}}(G, M, I)$. Then $(A, B) \mapsto (A \cap H_j, B \cap N_j)_{j \in J}$ describes an isomorphism from $\underline{\mathfrak{B}}(G, M, I)$ onto a subdirect product of the $\underline{\mathfrak{B}}(H_j, N_j, I \cap (H_j \times N_j))(j \in J)$ which has $G(\alpha_{jk} \mid j, k \in J) := \{\alpha_{jk}(A_k, B_k)_{j \in J} \mid k \in J$ and $(A_k, B_k) \in \mathfrak{B}(H_k, N_k, I \cap (H_k \times N_k))\} \backslash \{(N_k' \cap H_k, N_k)\}\}$ as a supremum-dense subset.*

If $G(\alpha_{jk} \mid j, k \in J)$ is considered as a partial $\bigvee$-semilattice (induced by the join in the direct product) then $\underline{\mathfrak{B}}(G, M, I)$ is isomorphic to the complete lattice of all complete ideals of $G(\alpha_{jk} \mid j, k \in J)$ (we recall that a subset $A$ of a partial $\bigvee$-semilattice is a *complete ideal* if $x \leq a$ and $a \in A$ imply $x \in A$ and if $X \subseteq A$ implies $\bigvee X \in A$ whenever $\bigvee X$ exists). An isomorphic copy of the partial $\bigvee$-semilattice $G(\alpha_{jk} \mid j, k \in J)$ is its inverse image in $\underline{\mathfrak{B}}(G, M, I)$ under the isomorphism of Theorem 8 which can be described by

$$G((H_j, N_j) \mid j \in J) :=$$

$$\{\gamma A_j \mid j \in J \text{ and } (A_j, B_j) \in \mathfrak{B}(H_j, N_j, I \cap (H_j \times N_j)) \backslash \{(N_j' \cap H_j', N_j)\}\}$$

**COROLLARY.** $\underline{\mathfrak{B}}(G, M, I)$ *is isomorphic to the complete lattice of all complete ideals of $G((H_j, N_j) \mid j \in J)$.*

## 6. Scaffoldings

The construction of the concept lattice $\underline{\mathfrak{B}}(G, M, I)$ via $G((H_j, N_j) \mid j \in J)$ or $G(\alpha_{jk} \mid j, k \in J)$ is usually more economical if the compatible subcontexts $(H_j, N_j)(j \in J)$ are smaller. The extreme case is present if all $\Theta(H_j, N_j)(j \in J)$ are $\bigwedge$-irreducible. If $(G, M, I)$ is a reduced context satisfying the chain condition then, by Theorem 6, $\underline{\Theta}(H, N)$ is $\bigwedge$-irreducible for a compatible subcontext $(H, N)$ of $(G, M, I)$ if and only if $H \cup N = \langle\langle g\rangle\rangle$ for some $g \in G$ (notice that for every $m \in M$ there is a $g \in G$ with $g \sim m$ and dually). This together with Theorem 8 yields the following theorem.

THEOREM 9. *Let $(G, M, I)$ be a reduced context satisfying the chain condition. Then $(A, B) \mapsto (A \cap \langle \bar{g}\rangle, B \cap \langle \bar{g}\rangle)_{\bar{g} \in G/\approx}$ describes an isomorphism from $\underline{\mathfrak{B}}$ $(G, M, I)$ onto a subdirect product of the completely subdirect irreducible concept lattices $\mathfrak{B}(\langle\bar{g}\rangle \cap G, \langle\bar{g}\rangle \cap M, I \cap \langle\bar{g}\rangle^2)$ $(\bar{g} \in G/\approx)$ which has $G(\alpha_{\bar{g}\bar{h}} \mid \bar{g}, \bar{h} \in G/\approx)$ as a supremum-dense subset.*

The partial $\bigvee$-semilattice $G(((\langle\bar{g}\rangle \cap G, \langle\bar{g}\rangle \cap M \mid \bar{g} \in G/\approx)$ isomorphic to $G(\alpha_{\bar{g}\bar{h}} \mid \bar{g}, \bar{h} \in G/\approx)$ is called the *scaffolding* of $\underline{\mathfrak{B}}(G, M, I)$ (see Wille [9], [10]) and denoted by $\mathscr{G}(G, M, I)$. We recall that $\underline{\mathfrak{B}}(G, M, I)$ is isomorphic to the complete lattice of all complete ideals of its scaffolding $\mathscr{G}(G, M, I)$. How a concept lattice may be constructed via its scaffolding shall be demonstrated by an example. We choose the following reduced finite context which occurs in the analysis of homomorphisms of partial algebras (see Burmeister and Wojdyło [3]).

|    | $i$ | $a$ | $o$ | $f$ | $s$ | $f_i$ | $c_r$ | $i_n$ | $P_r$ | $P$ |
|----|-----|-----|-----|-----|-----|-------|-------|-------|-------|-----|
| 1  | ×   | ×   | ×   |     |     |       |       |       |       |     |
| 2  | ×   | ×   | ×   |     |     |       |       |       | ×     | ×   |
| 3  | ×   | ×   |     | ×   |     |       | ×     | ×     | ×     |     |
| 4  | ×   | ×   |     | ×   |     |       | ×     | ×     | ×     | ×   |
| 5  | ×   |     |     | ×   | ×   |       | ×     | ×     | ×     | ×   |
| 6  | ×   |     |     | ×   | ×   | ×     | ×     | ×     | ×     | ×   |
| 7  |     | ×   | ×   | ×   | ×   |       |       |       |       |     |
| 8  |     | ×   | ×   | ×   | ×   |       |       |       | ×     | ×   |
| 9  |     | ×   | ×   | ×   | ×   | ×     |       |       | ×     | ×   |
| 10 |     | ×   | ×   | ×   | ×   | ×     | ×     | ×     | ×     | ×   |

The set $G := \{1, 2, \ldots 10\}$ of objects consists of names for concrete homomorphisms. The attributes in the set $M := \{i, a, o, f, s, f_i, c_r, i_n, P_r, P\}$ are explained by $i$: injective, $a$: almost onto, $o$: onto, $f$: full, $s$: strong, $f_i$: final, $c_r$: relatively closed, $i_n$: initial, $P_r$: relatively $P$-closed, $P$: $P$-closed. The crosses of the table indicate the relation $I$. First we determine the digraph $(G \cup M, \nearrow \cup \searrow)$.

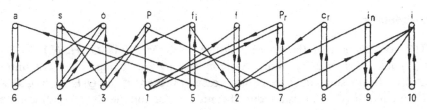

The scaffolding of $\underline{\mathfrak{B}}(G, M, I)$ is already determined by the maximal subcontexts of the form $(\langle\{g\}\rangle \cap G, \langle\{g\}\rangle \cap M)$ with $g \in G$; these subcontexts are given by the following tables (the generator is encircled).

|    | $i$ | $a$ | $o$ | $f$ | $s$ | $P_r$ | $P$ |
|----|-----|-----|-----|-----|-----|-----|-----|
| 1  | ×   | ×   | ×   |     |     |     |     |
| 2  | ×   | ×   | ×   |     |     | ×   | ×   |
| ③  | ×   | ×   |     | ×   |     | ×   |     |
| 4  | ×   | ×   |     | ×   |     | ×   | ×   |
| 6  | ×   |     |     | ×   | ×   | ×   | ×   |
| 7  |     | ×   | ×   | ×   | ×   |     |     |
| 10 |     | ×   | ×   | ×   | ×   | ×   | ×   |

|   | $a$ | $o$ | $f$ | $s$ | $f_i$ |
|---|-----|-----|-----|-----|-----|
| 2 | ×   | ×   |     |     |     |
| 4 | ×   |     | ×   |     |     |
| ⑤ |     |     | ×   | ×   |     |
| 6 |     |     | ×   | ×   | ×   |

|    | $i$ | $f$ | $c_r$ |
|----|-----|-----|-----|
| 2  | ×   |     |     |
| ⑧  |     | ×   |     |
| 10 |     | ×   | ×   |

|    | $i$ | $f$ | $i_n$ |
|----|-----|-----|-----|
| 2  | ×   |     |     |
| ⑨  |     | ×   |     |
| 10 |     | ×   | ×   |

The concept lattices of the four subcontexts are described by the following Hasse diagrams. For the objects $g$ (attributes $m$) labels indicate the smallest (greatest) concept of which the extent (intent) contains $g$ ($m$).

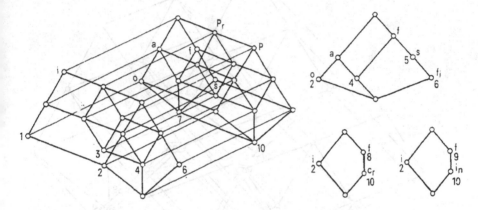

The $\bigvee$-preserving maps $\alpha_{\bar{g}\bar{h}}$ ($g, h \in \{3, 5, 8, 9\}$) are fixed by their images of the $\bigvee$-irreducible elements. Therefore, because of $\alpha_{\bar{g}\bar{h}}(\{k\}'' \cap \langle\bar{h}\rangle, \{k\}' \cap \langle\bar{h}\rangle) = \pi(\langle\bar{g}\rangle \cap G, \langle\bar{g}\rangle \cap M)(\{k\}'', \{k\}')$ for all $k \in \langle\bar{h}\rangle \cap G$, we can read the $\alpha_{\bar{g}\bar{h}}$ from the following table which describes the maps $\pi(\langle\bar{g}\rangle \cap G, \langle\bar{g}\rangle \cap M)$ restricted to $\gamma G$.

| $G$ | 1 | 2 | 3 | 4 | 5 | 6 | 7 | 8 | 9 | 10 |
|-----|---|---|---|---|---|---|---|---|---|----|
| $\bar{3}$ | 1 | 2 | 3 | 4 | 6 | 6 | 7 | 10 | 10 | 10 |
| $\bar{5}$ | 2 | 2 | 4 | 4 | 5 | 6 | 0 | 0 | 0 | 0 |
| $\bar{8}$ | 2 | 2 | 0 | 0 | 0 | 0 | 8 | 8 | 10 | 10 |
| $\bar{9}$ | 2 | 2 | 0 | 0 | 0 | 0 | 9 | 9 | 10 | 10 |

By Theorem 9, an isomorphic copy of $\mathfrak{B}(G, M, I)$ can be obtained by forming all joins in $\Pi(\mathscr{L}(\langle \bar{g} \rangle \cap G, \langle \bar{g} \rangle \cap M, I \cap \langle \bar{g} \rangle^2) \mid g \in \{3, 5, 8, 9\})$ of the ten elements described by the columns of the table. For another construction of $\mathfrak{B}(G, M, I)$ we first determine the scaffolding $\mathscr{G}(G, M, I)$. For this we draw the disjoint union of the Hasse Diagrams above without the least elements. The drawing has to be completed to a diagram of the quasi-order $Q$ defined by $(A_{\bar{g}}, B_{\bar{g}}) Q (A_{\bar{h}}, B_{\bar{h}}) : \Leftrightarrow$ $\alpha_{\bar{g}\bar{h}}(A_{\bar{h}}, B_{\bar{h}}) \geq (A_{\bar{g}}, B_{\bar{g}})$; the equivalence relation $Q \cap Q^{-1}$ may be indicated by encircling. The resulting diagram describes the scaffolding $\mathscr{G}(G, M, I)$ because $(\bigvee \gamma A_1) \vee (\bigvee \gamma A_2)$ is in $\mathscr{G}(G, M, I)$ for $(A_i, B_i) \in \mathfrak{B}(\langle \bar{h}_i \rangle \cap G, \langle \bar{h}_i \rangle \cap M, I \cap \langle \bar{h}_i \rangle^2)$ $(i = 1; 2)$ if and only if there is a $g \in \{3, 5, 8, 9\}$ such that $(A_i, B_i) Q \alpha_{\bar{g}\bar{h}_1}(A_1, B_1) \vee \alpha_{\bar{g}\bar{h}2}(A_2, B_2)$ for $i = 1, 2$ (cf. Wille [10]. Construction II).

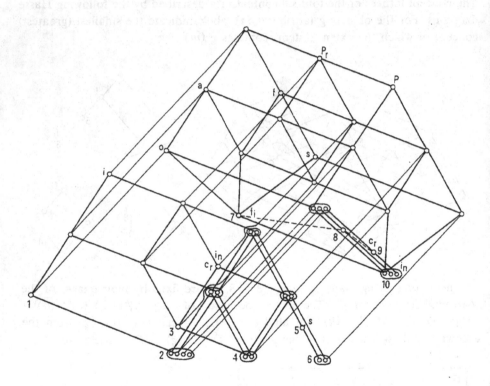

Forming the lattice of all (complete) ideals of $\mathscr{G}(G, M, I)$ leads to the following Hasse diagram of $\mathfrak{B}(G, M, I)$.

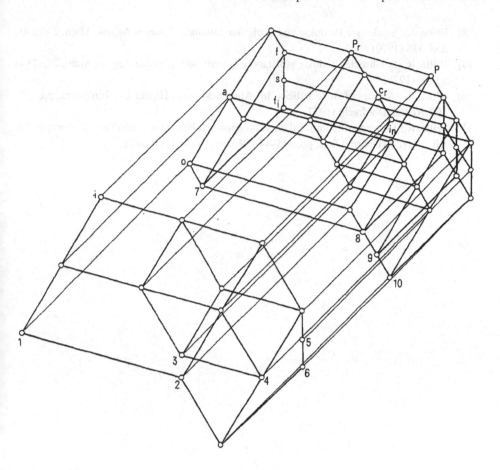

# References

[1]  Birkhoff, G.: Subdirect unions in universal algebras. Bull. Amer. Math. Soc. 50, 764–768 (1944)

[2]  Birkhoff, G.: Lattice theory, 3rd edn. Amer. Math. Soc., Providence (1967)

[3]  Burmeister, P., Wojdyło, B.: Properties of homomorphisms and quomorphisms between partial algebras. Preprint

[4]  Crawley, P., Dilworth, R.P.: Algebraic theory of lattices. Prentice Hall, Englewood Cliffs (1973)

[5]  Grätzer, G.: General lattice theory. Birkhäuser, Basel (1978)

[6]  Jónsson, B., Nation, J.B.: A report of sublattices of free lattices. In: Csákány, B., Schmidt, J. (eds.) Contributions to Universal Algebra, pp. 223–257. North-Holland, Amsterdam (1977)

[7]  Urouhart, A.: A topological representation theory for lattices. Alg. Universalis 8, 45–58 (1978)

[8]   Wille, R.: Subdirekte Produkte und konjunkte Summen. J. Reine Angew. Math. 239/240, 333–338 (1970)

[9]   Wille, R.: Subdirekte Produkte vollständiger Verbände. J. Reine Angew. Math. 283/284, 53–70 (1976)

[10]  Wille, R.: Aspects of finite lattices. In: Aigner, M. (ed.) Higher Combinatorics, pp. 79–100. Reidel, Dordrecht (1977)

[11]  Wille, R.: Restructuring lattice theory: an approach based on hierarchies of concepts. In: Rival, I. (ed.) Ordered Sets, pp. 445–470. Reidel, Dordrecht (1982)

# Author Index